Lecture Notes in Mathematics 2154

More information about this series at http://www.springer.com/series/304

Michèle Loday-Richaud

Divergent Series, Summability and Resurgence II

Simple and Multiple Summability

Michèle Loday-Richaud
LAREMA, Université d'Angers
Angers, France

ISSN 0075-8434 ISSN 1617-9692 (electronic)
Lecture Notes in Mathematics
ISBN 978-3-319-29074-4 ISBN 978-3-319-29075-1 (eBook)
DOI 10.1007/978-3-319-29075-1

Library of Congress Control Number: 2016940058

Printed on acid-free paper

This Springer imprint is published by Springer Nature
The registered company is Springer International Publishing AG Switzerland

To Hervé and Sylvie

Avant-Propos

Le sujet principal traité dans la série de volumes *Divergent Series, Summability and Resurgence* est la théorie des développements asymptotiques et des séries divergentes appliquée aux équations différentielles ordinaires (EDO) et à certaines équations aux différences dans le champ complexe.

Les équations différentielles dans le champ complexe, et dans le cadre holomorphe, sont un sujet très ancien. La théorie a été très active dans la deuxième moitié du XIX-ème siècle. En ce qui concerne les *équations linéaires*, les mathématiciens de cette époque les ont subdivisées en deux classes. Pour la première, celle des équations *à points singuliers réguliers* (ou *de Fuchs*), généralisant les équations hypergéométriques d'Euler et de Gauss, ils ont enregistré *"des succès aussi décisifs que faciles"* comme l'écrivait René Garnier en 1919. En revanche, pour la seconde, celle des équations dites *à points singuliers irréguliers*, comme l'écrivait aussi Garnier, *"leurs efforts restent impuissants à édifier aucune théorie générale"*. La raison centrale de ce vif contraste est que toute série entière apparaissant dans l'écriture d'une solution d'une équation différentielle de Fuchs est *automatiquement convergente* tandis que pour les équations irrégulières ces séries sont génériquement *divergentes* et que l'on ne savait qu'en faire. La situation a commencé à changer grâce à un travail magistral de Henri Poincaré entrepris juste après sa thèse, dans lequel il "donne un sens" aux solutions divergentes des EDO linéaires irrégulières en introduisant un outil nouveau, et qui était appelé à un grand avenir, la théorie des développements asymptotiques. Il a ensuite utilisé cet outil pour donner un sens aux séries divergentes de la mécanique céleste, et remporté de tels succès que presque tout le monde a oublié l'origine de l'histoire, c'est-à-dire les EDO ! Les travaux de Poincaré ont (un peu...) remis à l'honneur l'étude des séries divergentes, abandonnée par les mathématiciens après Cauchy. L'Académie des Sciences a soumis ce sujet au concours en 1899, ce qui fut à l'origine d'un travail important d'Émile Borel. Celui-ci est la source de nombre des techniques utilisées dans *Divergent Series, Summability and Resurgence*. Pour revenir aux EDO irrégulières, le sujet a fait l'objet de nombreux et importants travaux de G.D. Birkhoff et R. Garnier durant le premier quart du XX-ème siècle. On retrouvera ici de nombreux prolongements des méthodes de Birkhoff. Après 1940, le sujet a étrangement presque disparu, la théorie étant, je ne sais trop pourquoi, considérée comme achevée, tout comme celle des équations de Fuchs. Ces dernières ont réémergé au début des années 1970, avec les travaux de Raymond Gérard, puis un livre de Pierre Deligne. Les équations irrégulières ont suivi avec des travaux de l'école allemande et surtout de l'école française. De nombreuses techniques complètement nouvelles ont été introduites (développements asymptotiques Gevrey, k-sommabilité, multisommabilité, fonctions résurgentes...) permettant en particulier une vaste généralisation du *phénomène de Stokes* et sa mise en relation avec la théorie de Galois différentielle et le problème de Riemann-Hilbert généralisé. Tout ceci a depuis reçu de très nombreuses applications dans des domaines très variés, allant de l'intégrabilité des systèmes hamiltoniens aux problèmes de points tournant pour les EDO singulièrement perturbées ou à divers problèmes de modules. On en trouvera certaines dans *Divergent Series, Summability and Resurgence*, comme l'étude résurgente des germes de difféomorphismes analytiques du plan complexe tangents à l'identité ou celle de l'EDO non-linéaire Painlevé I.

Le sujet restait aujourd'hui difficile d'accès, le lecteur ne disposant pas, mis à part les articles originaux, de présentation accessible couvrant *tous les aspects*. Ainsi *Divergent Series, Summability and Resurgence* comble une lacune. Ces volumes présentent un large panorama des recherches les plus récentes sur un vaste domaine classique et passionnant, en pleine renaissance, on peut même dire en pleine explosion. Ils sont néanmoins accessibles à tout lecteur possédant une bonne familiarité avec les fonctions analytiques d'une variable complexe. Les divers outils sont soigneusement mis en place, progressivement et avec beaucoup d'exemples. C'est une belle réussite.

À Toulouse, le 16 mai 2014,

Jean-Pierre Ramis

Preface to the Three Volumes

This three-volume series arose out of lecture notes for the courses we gave together at a CIMPA[1] school in Lima, Peru, in July 2008. Since then, these notes have been used and developed in graduate courses held at our respective institutions, that is, the universities of Angers, Nantes, Strasbourg (France) and the Scuola Normale Superiore di Pisa (Italy). The original notes have now grown into self-contained introductions to problems raised by analytic continuation and the divergence of power series in one complex variable, especially when related to differential equations.

A classical way of solving an analytic differential equation is the power series method, which substitutes a power series for the unknown function in the equation, then identifies the coefficients. Such a series, if convergent, provides an analytic solution to the equation. This is what happens at an ordinary point, that is, when we have an initial value problem to which the Cauchy-Lipschitz theorem applies. Otherwise, at a singular point, even when the method can be applied the resulting series most often diverges; its connection with "actual" local analytic solutions is not obvious despite its deep link to the equation.

The hidden meaning of divergent formal solutions was already pondered in the nineteenth century, after Cauchy had clarified the notions of convergence and divergence of series. For ordinary *linear* differential equations, it has been known since the beginning of the twentieth century how to determine a full set of linearly independent formal solutions[2] at a singular point in terms of a finite number of complex powers, logarithms, exponentials and power series, either convergent or divergent. These formal solutions completely determine the linear differential equation; hence, they contain all information about the equation itself, especially about its analytic solutions. Extracting this information from the divergent solutions was the underly-

[1] Centre International de Mathématiques Pures et Appliquées, or ICPAM, is a non-profit international organization founded in 1978 in Nice, France. It promotes international cooperation in higher education and research in mathematics and related subjects for the benefit of developing countries. It is supported by UNESCO and IMU, and many national mathematical societies over the world.

[2] One says *a formal fundamental solution*.

ing motivation for the theories of summability and, to some extent, of resurgence. Both theories are concerned with the precise structure of the singularities.

Divergent series may appear in connection with any local analytic object. They either satisfy an equation, or are attached to given objects such as formal first integrals in dynamical systems or formal conjugacy maps in classification problems. Besides linear and non-linear ordinary differential equations, they also arise in partial differential equations, difference equations, q-difference equations, etc. Such series, issued from specific problems, call for suitable theories to extract valuable information from them.

A theory of *summability* is a theory that focuses on a certain class of power series, to which it associates analytic functions. The correspondence should be injective and functorial: one expects for instance a series solution of a given functional equation to be mapped to an analytic solution of the same equation. In general, the relation between the series and the function –the latter is called its sum– is *asymptotic*, and depends on the direction of summation; indeed, with non-convergent series one cannot expect the sums to be analytic in a full neighborhood, but rather in a "sectorial neighborhood" of the point at which the series is considered.

One summation process, commonly known as the Borel-Laplace summation, was already given by Émile Borel in the nineteenth century; it applies to the classical Euler series and, more generally, to solutions of linear differential equations with a single "level", equal to one, although the notion of level was by then not explicitly formulated. It soon appeared that this method does not apply to all formal solutions of differential equations, even linear ones. A first generalization to series solutions of linear differential equations with a single, arbitrary level $k > 0$ was given by Le Roy in 1900 and is called *k-summation*. In the 1980's, new theories were developed, mainly by J.-P. Ramis and Y. Sibuya, to characterize k-summable series, a notion a priori unrelated to equations, but which applies to all solutions of linear differential equations with the single level k. The question of whether any divergent series solution of a linear differential equation is k-summable, known as the *Turrittin problem*, was an open problem until J.-P. Ramis and Y. Sibuya in the early 1980's gave a counterexample. In the late 1980's and in the 1990's *multisummability theories* were developed, in particular by J.-P. Ramis, J. Martinet, Y. Sibuya, B. Malgrange, W. Balser, M. Loday-Richaud and G. Pourcin, which apply to all series solution of linear differential equations with an arbitrary number of levels. They provide a unique sum of a formal fundamental solution on appropriate sectors at a singular point.

It was proved that these theories apply to solutions of non-linear differential equations as well: given a series solution of a non-linear differential equation, the choice of the right theory is determined by the linearized equation along this series. On the other hand, in the case of difference equations, not all solutions are multisummable; new types of summation processes are needed, for instance those introduced by J. Écalle in his theory of resurgence and considered also by G. Immink and B. Braaksma. Solutions of q-difference equations are not all multisummable either; specific processes in this case have been introduced by F. Marotte and C. Zhang in the late 1990's.

Summation sheds new light on the *Stokes phenomenon*. This phenomenon occurs when a divergent series has several sums, with overlapping domains, which correspond to different summability directions and differ from one another by exponentially small quantities. The question then is to describe these quantities. A precise analysis of the Stokes phenomenon is crucial for classification problems. For systems of linear differential equations, the meromorphic classification easily follows from the characterization of the Stokes phenomenon by means of the *Stokes cocycle*. The Stokes cocycle is a 1-cocycle in non-abelian Čech cohomology. It is expressed in terms of finitely many automorphisms of the normal form, the *Stokes automorphisms*, which select and organize the "exponentially small quantities". In practice, the Stokes automorphisms are represented by constant unipotent matrices called the *Stokes matrices*. It turned out that these matrices are precisely the correction factors needed to patch together two contiguous sums, that is, sums taken on the two sides of a singular direction, of a formal fundamental solution.[3]

The theory of *resurgence* was independently developed in the 1980's by J. Écalle, with the goal of providing a theory with a large range of applications, including the summation of divergent solutions of a variety of functional equations, differential, difference, differential-difference, etc. Basically, resurgence theory starts with the Borel-Laplace summation in the case of a single level equal to one, and this is the only situation we consider in these volumes. Let us mention however that there are extensions of the theory based on more general kernels.

The theory focuses on what happens in the Borel plane, that is, after one applies a Borel transform. The results are then pulled back via a Laplace transform to the plane of the initial variable also called the Laplace plane. In the Borel plane one typically gets functions, called *resurgent functions*, which are analytic in a neighborhood of the origin and can be analytically continued along various paths in the Borel plane, yet they are not entire functions: one needs to avoid a certain set Ω of possible singular points and analytic continuation usually gives rise to multiple-valuedness, so that these Borel-transformed functions are best seen as holomorphic functions on a Riemann surface akin to the universal covering of $\mathbb{C}\setminus\{0\}$. Of crucial importance are the singularities[4] which may appear at the points of Ω, and *Écalle's alien operators* are specific tools designed to analyze them.

The development of resurgence theory was aimed at non-linear situations where it reveals its full power, though it can be applied to the formal solutions of linear differential equations (in which case the singular support Ω is finite and the Stokes matrices, hence the local meromorphic classification, determined by the action of finitely many alien operators). The non-linearity is taken into account via the convolution product in the Borel plane. More precisely, we mean here the complex convolution which is turned into pointwise multiplication when returning to the original variable by means of a Laplace transform. Given two resurgent functions, analytic

[3] A less restrictive notion of Stokes matrices exists in the literature, which patch together any two sectorial solutions with same asymptotic expansion, but they are not local meromorphic invariants in general.

[4] The terms *singularity* in Écalle's resurgence theory and *microfunction* in Sato's microlocal analysis have the same meaning.

continuation of their convolution product is possible, but new singularities may appear at the sum of any two singular points of the factors; hence, Ω needs to be stable by addition (in particular, it must be infinite; in practice, one often deals with a lattice in $\mathbb{C}$). All operations in the Laplace plane have an explicit counterpart in the Borel plane: addition and multiplication of two functions of the initial variable, as well as non-linear operations such as multiplicative inversion, substitution into a convergent series, functional composition, functional inversion, which all leave the space of resurgent functions invariant.

To have these tools well defined requires significant work. The reward of setting the foundations of the theory in the Borel plane is greater flexibility, due to the fact that one can work with an *algebra* of resurgent functions, in which the analysis of singularities is performed through *alien derivations*[5].

Écalle's important achievement was to obtain the so-called *bridge equation*[6] in many situations. For a given problem, the bridge equation provides an all-in-one description of the action on the solutions of the alien derivations. It can be viewed as an *infinitesimal version of the Stokes phenomenon*: for instance, for a linear differential system with level one it is possible to prove that the set of Stokes automorphisms in a given formal class naturally has the structure of a unipotent Lie group and the bridge equation gives infinitesimal generators of its Lie algebra.

Summability and resurgence theories have useful interactions with the algebraic and geometrical approaches of linear differential equations such as *differential Galois theory* and the *Riemann-Hilbert problem*. The local differential Galois group of a meromorphic linear differential equation at a singular point is a linear algebraic group, the structure of which reflects many properties of the solutions. At a "regular singular" point [7] for instance, it contains a Zariski-dense subgroup finitely generated by the monodromy. However, at an "irregular singular" point, one needs to introduce further automorphisms, among them the Stokes automorphisms, to generate a Zariski-dense subgroup. For linear differential equations with rational coefficients, when all the singular points are regular, the classical Riemann-Hilbert correspondence associates with each equation a monodromy representation of the fundamental group of the Riemann sphere punctured at the singular points; conversely, from any representation of this fundamental group, one recovers an equation with prescribed regular singular points.[8] In the case of possibly irregular singular points, the monodromy representation alone is insufficient to recover the equation; here too one has to introduce the Stokes automorphisms and to connect them via "analytic continuation" of the divergent solutions, that is, via summation processes.

---

[5] Alien derivations are suitably weighted combinations of alien operators which satisfy the Leibniz rule.

[6] Its original name in French is *équation du pont*.

[7] This means that the formal solutions at that point may contain powers and logarithms but no exponential.

[8] The Riemann-Hilbert problem more specifically requires that the singular points in this restitution be *Fuchsian*, that is, simple poles only, which is not always possible.

These volumes also include an application of resurgence theory to the first Painlevé equation. Painlevé equations are nonlinear second-order differential equations introduced at the turn of the twentieth century to provide new transcendents, that is, functions that can neither be written in terms of the classical functions nor in terms of the special functions of physics. A reasonable request was to ask that all the movable singularities[9] be poles and this constraint led to a classification into six families of equations, now called Painlevé I to VI. Later, these equations appeared as conditions for isomonodromic deformations of Fuchsian equations on the Riemann sphere. They occur in many domains of physics, in chemistry with reaction-diffusion systems and even in biology with the study of competitive species. Painlevé equations are a perfect non-linear example to be explored with the resurgent tools.

We develop here the particular example of Painlevé I and we focus on its now classical truncated solutions. These are characterized by their asymptotics as well as by the fact that they are free of poles within suitable sectors at infinity. We determine them from their asymptotic expansions by means of a Borel-Laplace procedure after some normalization. The non-linearity generates a situation which is more intricate than in the case of linear differential equations. Playing the role of the formal fundamental solution is the so-called *formal integral* given as a series in powers of logarithm-exponentials with power series coefficients. More generally, such expansions are called *transseries* by J. Écalle or *multi-instanton expansions* by physicists. In general, the series are divergent and lead to a Stokes phenomenon. In the case of Painlevé I we prove that they are resurgent. Although the Stokes phenomenon can no longer be described by Stokes matrices, it is still characterized by the alien derivatives at the singular points in the Borel plane (see O. Costin *et al.*). The local meromorphic class of Painlevé I at infinity is the class of all second-order equations locally meromorphically equivalent at infinity to this equation. The characterization of this class requires *all* alien derivatives in all higher sheets of the resurgence surface. These extra invariants are also known as *higher order Stokes coefficients* and they can be given a numerical approximation using the *hyperasymptotic theory* of M. Berry and C. Howls. The complete resurgent structure of Painlevé I is given by its *bridge equation* which we state here, seemingly for the first time.

Recently, in quantum field and string theories, the resurgent structure has been used to describe the instanton effects, in particular for quartic matrix models which yield Painlevé I in specific limits. In the late 1990's, following ideas of A. Voros and J. Écalle, applications of the resurgence theory to problems stemming from quantum mechanics were developed by F. Pham and E. Delabaere. Influenced by M. Sato, this was also the starting point by T. Kawai and Y. Takei of the so-called *exact semi-classical analysis* with applications to Painlevé equations with a large parameter and their hierarchies, based on isomonodromic methods.

---

[9] The fixed singular points are those appearing on the equation itself; they are singular for the solutions generically. The movable singular points are singular points for solutions only; they "move" from one solution to another. They are a consequence of the non-linearity.

Summability and resurgence theories have been successfully applied to problems in analysis, asymptotics of special functions, classification of local analytic dynamical systems, mechanics, and physics. They also generate interesting numerical methods in situations where the classical methods fail.

In these volumes, we carefully introduce the notions of analytic continuation and monodromy, then the theories of resurgence, k-summability and multisummability, which we illustrate with examples. In particular, we study tangent-to-identity germs of diffeomorphisms in the complex plane both via resurgence and summation, and we present a newly developed resurgent analysis of the first Painlevé equation. We give a short introduction to differential Galois theory and a survey of problems related to differential Galois theory and the Riemann-Hilbert problem. We have included exercises with solutions. Whereas many proofs presented here are adapted from existing ones, some are completely new. Although the volumes are closely related, they have been organized to be read independently. All deal with power series and functions of a complex variable; the words *analytic* and *holomorphic* are used interchangeably, with the same meaning.

This book is aimed at graduate students, mathematicians and theoretical physicists who are interested in the theories of monodromy, summability or resurgence and related problems.

Below is a more detailed description of the contents.

- Volume 1: *Monodromy and Resurgence* by C. Mitschi and D. Sauzin.
  An essential notion for the book and especially for this volume is the notion of analytic continuation "à la Cauchy-Weierstrass". It is used both to define the monodromy of solutions of linear ordinary differential equations in the complex domain and to derive a definition of resurgence.
  Once monodromy is defined, we introduce the Riemann-Hilbert problem and the differential Galois group. We show how the latter is related to analytic continuation by defining a set of automorphisms, including the Stokes automorphisms, which together generate a Zariski-dense subgroup of the differential Galois group. We state the inverse problem in differential Galois theory and give its particular solution over $\mathbb{C}(z)$ due to Tretkoff, based on a solution of the Riemann-Hilbert problem. We introduce the language of vector bundles and connections in which the Riemann-Hilbert problem has been extensively studied and give the proof of Plemelj-Bolibrukh's solution when one of the prescribed monodromy matrices is diagonalizable.
  The second part of the volume begins with an introduction to the 1-summability of series by means of Borel and Laplace transforms (also called Borel or Borel-Laplace summability) and provides non-trivial examples to illustrate this notion. The core of the subject follows, with definitions of resurgent series and resurgent functions, their singularities and their algebraic structure. We show how one can analyse the singularities via the so-called *alien calculus* in resurgent algebras; this includes the *bridge equation* which usefully connects alien and ordinary derivations. The case of tangent-to-identity germs of diffeomorphisms in the complex plane is given a thorough treatment.

- Volume 2: *Simple and Multiple Summability* by M. Loday-Richaud.
  The scope of this volume is to thoroughly introduce the various definitions of k-summability and multisummability developed since the 1980's and to illustrate them with examples, mostly but not only, solutions of linear differential equations. For the first time, these theories are brought together in one volume. We begin with the study of basic tools in Gevrey asymptotics, and we introduce examples which are reconsidered throughout the following sections. We provide the necessary background and framework for some theories of summability, namely the general properties of sheaves and of abelian or non-abelian Čech cohomology. With a view to applying the theories of summability to solutions of differential equations we review fundamental properties of linear ordinary differential equations, including the main asymptotic expansion theorem, the formal and the meromorphic classifications (formal fundamental solution and linear Stokes phenomenon) and a chapter on index theorems and the irregularity of linear differential operators. Four equivalent theories of k-summability and six equivalent theories of multisummability are presented, with a proof of their equivalence and applications. Tangent-to-identity germs of diffeomorphisms are revisited from a new point of view.
- Volume 3: *Resurgent Methods and the First Painlevé equation* by E. Delabaere.
  This volume deals with ordinary non-linear differential equations and begins with definitions and phenomena related to the non-linearity. Special attention is paid to the first Painlevé equation, or Painlevé I, and to its tritruncated and truncated solutions. We introduce these solutions by proving the Borel-Laplace summability of transseries solutions of Painlevé I. In this context resonances occur, a case which is scarcely studied. We analyse the effect of these resonances on the formal integral and we provide a normal form. Additional material in resurgence theory is needed to achieve a resurgent analysis of Painlevé I up to its bridge equation.

Acknowledgements. We would like to thank the CIMPA institution for giving us the opportunity of holding a winter school in Lima in July 2008. We warmly thank Michel Waldschmidt and Michel Jambu for their support and advice in preparing the application and solving organizational problems. The school was hosted by IMCA (Instituto de Matemática y Ciencias Afines) in its new building of La Molina, which offered us a perfect physical and human environment, thanks to the colleagues who greeted and supported us there. We thank all institutions that contributed to our financial support: UNI and PUCP (Peru), LAREMA (Angers), IRMA (Strasbourg), IMT (Toulouse), ANR Galois (IMJ Paris), IMPA (Brasil), Universidad de Valladolid (Spain), Ambassade de France au Pérou, the International Mathematical Union, CCCI (France) and CIMPA. Our special thanks go to the students in Lima and in our universities, who attended our classes and helped improve these notes via relevant questions, and to Jorge Mozo Fernandez for his pedagogical assistance.

Angers, Strasbourg, Pisa, November 2015

Éric Delabaere, Michèle Loday-Richaud, Claude Mitschi, David Sauzin

Introduction to this Volume

Divergent series may diverge in many ways. They occur in various situations: attached to analytic objects in dynamical systems or as solutions of differential equations, ordinary or partial, linear or non-linear, or of difference equations, of q-difference equations, etc. In each case, they must satisfy specific conditions which reduce the range of possibilities. It is well known from Maillet's theorem, for instance, that if a series satisfies a differential equation then its coefficients satisfy growth estimates of Gevrey type.

The aim of this volume is to give formal power series an analytic meaning in the same way as the sum of a convergent series turns it into an analytic function. However, in the case of divergent power series, say, at 0, the sums, although analytic, are not analytic in a full neighborhood of 0 but on sectors with vertex 0. Because of the large variety of divergent series, theories of summability have been developed to suit particular classes of series. Indeed, a general theory applying to all divergent series does not exist and certainly never will.

In this volume we focus on the best-known class of divergent series, namely the class of so-called *k-summable series* and, more generally, the class of *multi-summable series*. This class depends on positive parameters called levels, a single parameter k in the case of simple summability (called k-summability) and parameters $k_1 < k_2 < \cdots < k_\nu$ in the case of multiple summability (called multisummability or $(k_1,k_2,\cdots,k_\nu)$-summability). Unsurprisingly, this class contains all power series that are solutions of analytic ordinary differential equations since these theories of simple and multiple summability were originally developed to understand the behavior of solutions of differential equations at their singular points. There are various equivalent approaches of both k-summability and multisummability, which we present in detail in this volume (at least those we know of). Among the numerous techniques being used to provide coherent definitions of these theories of summability let us mention the central role of Gevrey asymptotic expansions and the less classical interpretation of series in terms of 0-cochains.

In the particular case of linear differential equations the singular points are isolated and any analytic solution at an ordinary point can be continued as close as needed to a singular point. However, approaching a singular point via analytic continuation tells little about the behavior of solutions in its neighborhood, whereas the summation of formal solutions provides significantly more information. This is even true for numerical approaches since the usual procedures stop being efficient near a singular point, whereas a theory of summability allows, in many cases, numerical calculation of the solutions and their invariants.

Acknowledgements. I am grateful to Jean-Pierre Ramis for introducing me into this subject with kind attention to my questions. I also wish to thank all those who read and commented all or part of the manuscript: especially Sergio Carillo, Anne Duval, Pascal Remy, Duncan Sands, Jacques Sauloy, Michael Singer, the anonymous referees, as well as my co-authors Eric Delabaere, Claude Mitschi and David Sauzin for our many exchanges and Raymond Séroul for his technical support.

Angers, November 2015.

Michèle Loday-Richaud

Abstract In volume I of the book *Divergent Series, Summability and Resurgence* [MS16] analytic functions, called 1-sums, were attached to certain divergent power series as a substitute for the usual sum of a convergent series. These 1-sums are not analytic in a full neighborhood of the origin, but only in sectors with almost arbitrary bisecting direction and large enough opening (in this case, larger than π), and they satisfy 1-Gevrey asymptotic conditions in these sectors. The method used in volume I is based on the fact that one can apply some Fourier-like Borel and Laplace operators, and it is thought as a first step towards Écalle's resurgence theory.

In this volume we give a detailed construction of theories of summability which extend 1-summability to much larger classes of divergent series, the so-called classes of *k-summable series* for any level $k > 0$ and the classes of *multisummable series* for any ν-tuple of positive levels $k_1 < k_2 < \cdots < k_\nu$. Together, these classes contain all series that are solutions of analytic ordinary differential equations.

The methods we use to give coherent definitions are diverse; some generalize the approach by Borel and Laplace operators, some require appropriate asymptotic conditions whereas some others rely on cohomological arguments. The equivalence of the different methods is proved. Gathered together, these various approaches provide a well stocked toolbox to handle summable divergent series.

The necessary material to establish these theories is thoroughly developed. We begin with Poincaré and Gevrey asymptotics both in terms of analytic functions and in terms of sheaves after we have recalled the definition and main properties of sheaves, describing in particular the sheaves we need.

With a view to applying the theories to solutions of ordinary differential equations we dedicate two chapters to formal and local properties of linear differential equations: we compare equations, systems and $\mathscr{D}$-modules. We explain the formal classification of systems by means of a formal fundamental solution and we pay some attention to the case of equations, explaining the calculation of the formal invariants by means of Newton polygons. We include the main asymptotic existence theorem. The meromorphic classification is performed via the Malgrange-Sibuya theorem and the Stokes cocycle theorem, which both characterize the linear Stokes phenomenon; Sibuya's proof of the Malgrange-Sibuya theorem is included. We define infinitesimal neighborhoods suited to the analysis of solutions at irregular singular points. We prove index theorems and we derive the calculation of the irregularity of linear differential operators. Problems of analytic continuation and global properties of solutions of linear differential equations were studied in volume I of the book [MS16] with links with summability. As for the resurgence of solutions of differential equations, a section of the present volume is devoted to the linear case, and the reader is referred to volume III of the book [Dela16] for a discussion of the nonlinear case.

The volume contains applications such as Maillet's theorem or Tauberian theorems. Numerous examples and exercises are given to illustrate the subject. The example of tangent-to-identity germs of diffeomorphisms studied in volume I of the book [MS16] is revisited from a cohomological point of view.

Contents

Chapter 1
Asymptotic Expansions in the Complex Domain

Abstract We introduce asymptotic expansions on sectors with vertex 0 in the complex plane, both in the Poincaré and the Gevrey sense. Gevrey asymptotics are an essential link between divergent series and their sums. We prove basic properties of asymptotic expansions and pay special attention to the case of flat Gevrey-asymptotic functions, that is, asymptotic to 0 in the Gevrey sense. We look at the action of a finite extension of the variable $x = t^r$ and of rank reduction. We prove the Borel-Ritt theorem which extends to the case of a complex variable the classical Borel theorem for a real variable. We end the chapter with the Cauchy-Heine theorem which links (non zero) asymptotic expansions to flat functions, considering both the case of Poincaré and Gevrey asymptotics.

In the whole volume we consider functions of a complex variable x and their asymptotic expansions at a given point x_0 of the Riemann sphere. Without loss of generality we assume that $x_0 = 0$ although for some examples classically studied at infinity we keep $x_0 = \infty$. Indeed, asymptotic expansions at infinity reduce to asymptotic expansions at 0 after the change of variable $x \mapsto z = 1/x$ and asymptotic expansions at $x_0 \in \mathbb{C}$ after the change of variable $x \mapsto t = x - x_0$. The point 0 must belong to the closure of the domain where the asymptotics are studied. In general,

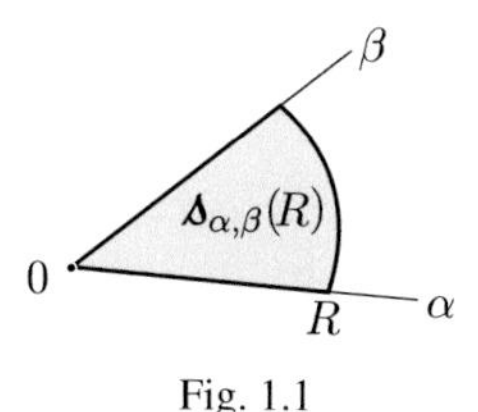

Fig. 1.1

we consider sectors with vertex 0, or germs of such sectors when the radius tends to 0. The sectors are drawn either in the complex plane $\mathbb{C}$, precisely, in $\mathbb{C}^* = \mathbb{C} \setminus \{0\}$ (the functions are then *single-valued* or *univalued*) or on the Riemann surface of the logarithm $\widetilde{\mathbb{C}}$ (the functions are *multivalued* or given in terms of polar coordinates).

M. Loday-Richaud, *Divergent Series, Summability and Resurgence II*,
Lecture Notes in Mathematics, DOI 10.1007/978-3-319-29075-1_1

Notation 1.0.1 We denote by

▷ $\Delta = \Delta_{\alpha,\beta}(R)$ the open sector with vertex 0 consisting of all points $x \in \mathbb{C}$ satisfying the conditions $\alpha < \arg(x) < \beta$ and $0 < |x| < R$;

▷ $\overline{\Delta} = \overline{\Delta}_{\alpha,\beta}(R)$ its closure in $\mathbb{C}^* = \mathbb{C} \setminus \{0\}$ or in $\widetilde{\mathbb{C}}$ (0 is always excluded) and we use the term closed sector;

▷ $\mathscr{O}(\Delta)$ the set of all holomorphic functions on Δ.

Recall that, for functions of one complex variable, we use the words holomorphic and analytic interchangeably, with the same meaning.

Definition 1.0.2 *A sector* $\Delta_{\alpha',\beta'}(R')$ *is said to be a* proper sub-sector *of (or to be* properly included *in) the sector* $\Delta_{\alpha,\beta}(R)$ *and one denotes*

$$\Delta_{\alpha',\beta'}(R') \Subset \Delta_{\alpha,\beta}(R)$$

if its closure $\overline{\Delta}_{\alpha',\beta'}(R')$ *in* $\mathbb{C}^*$ *or* $\widetilde{\mathbb{C}}$ *is included in* $\Delta_{\alpha,\beta}(R)$.

Thus, the notation $\Delta_{\alpha',\beta'}(R') \Subset \Delta_{\alpha,\beta}(R)$ means $\alpha < \alpha' < \beta' < \beta$ and $R' < R$.

1.1 Poincaré asymptotics

Poincaré[1] asymptotic expansions, or for short, asymptotic expansions, are kind of Taylor expansions which provide successive approximations of the functions. Unless otherwise mentioned we consider functions of a complex variable and asymptotic expansions in the complex domain, and this allows us to use the methods of complex analysis. As we will see, the properties of asymptotic expansions in the complex domain may differ significantly of those in the real domain.

In what follows Δ always denotes an open sector with vertex 0 either in $\mathbb{C}^*$ or in $\widetilde{\mathbb{C}}$, the Riemann surface of the logarithm.

1.1.1 Definition

Definition 1.1.1 *A function* $f \in \mathscr{O}(\Delta)$ *is said to* admit a series $\sum_{n\geq 0} a_n x^n$ as asymptotic expansion *(or to be* asymptotic to the series) on a sector Δ *if for all proper sub-sector* $\Delta' \Subset \Delta$ *of* Δ *and all* $N \in \mathbb{N}$, *there exists a constant* $C > 0$ *such that*

$$\boxed{\left| f(x) - \sum_{n=0}^{N-1} a_n x^n \right| \leq C\,|x|^N} \qquad \textit{for all } x \in \Delta'$$

[1] Henri Poincaré, 1854-1912, one of the most eminent mathematicians and cousin of the French President Raymond Poincaré, was also a physicist, a philosopher and an engineer. He had a deep influence in algebraic geometry, dynamical systems, celestial mechanics, special relativity, etc...

The constant $C = C_{N,\mathcal{S}'}$ may depend on N and $\mathcal{S}'$ but no condition is required on the nature of this dependence.

The technical condition *"for all $\mathcal{S}' \Subset \mathcal{S}$"* plays a fundamental role of which we will take benefit soon (cf. Rem. 1.1.10, p. 12).

Observe that the definition includes infinitely many estimates in each of which N is fixed. The conditions have nothing to do with the convergence or the divergence of the series as N goes to infinity. For $N = 1$ the condition tells that f can be continuously continued at 0 on $\mathcal{S}$. For $N = 2$ it says that the function f has a derivative at 0 on $\mathcal{S}$ and for general N, that f has a "*Taylor expansion*" of order N. Like in the case of a real variable, asymptotic expansions of functions of a complex variable, when they exist, are unique and they satisfy the same algebraic rules on sums, products, anti-derivatives and compositions. The proofs are similar and we leave them to the reader. The main difference between the real and the complex case appears when considering the derivation (cf. Prop. 1.1.9, p. 11 and Rem. 1.1.10, p. 12).

Notation 1.1.2 We denote by

▷ $\overline{\mathscr{A}}(\mathcal{S})$ the set of functions of $\mathscr{O}(\mathcal{S})$ having an asymptotic expansion at 0 on $\mathcal{S}$;

▷ $\overline{\mathscr{A}}^{<0}(\mathcal{S})$ the subset of functions of $\overline{\mathscr{A}}(\mathcal{S})$ asymptotic to zero at 0 on $\mathcal{S}$. Such functions are called flat functions at 0 on $\mathcal{S}$;

▷ $T = T_{\mathcal{S}} : \overline{\mathscr{A}}(\mathcal{S}) \to \mathbb{C}[[x]]$ the map assigning to each $f \in \overline{\mathscr{A}}(\mathcal{S})$ its asymptotic expansion at 0 on $\mathcal{S}$.

Due to the uniqueness of the asymptotic expansion, the map $T_{\mathcal{S}}$ is well defined and is called the *Taylor map* on $\mathcal{S}$ (cf. Exa. 1.1.3 below). Due to the algebraic properties of asymptotic expansions the sets $\overline{\mathscr{A}}(\mathcal{S})$ and $\overline{\mathscr{A}}^{<0}(\mathcal{S})$ are endowed with a natural structure of vector spaces and the Taylor map is a linear map with kernel $\overline{\mathscr{A}}^{<0}(\mathcal{S})$. Proposition 1.1.9, p. 11, will improve this result.

It is worth to notice that $\overline{\mathscr{A}}^{<0}(\mathcal{S})$ does not reduce to the null function: exponential functions of various powers of x provide examples of non-zero functions of $\overline{\mathscr{A}}^{<0}(\mathcal{S})$ for any $\mathcal{S}$. For instance, if $\mathcal{S}$ denotes the half-plane $\mathcal{S} = \{x\,;\,|\arg(x)| < \pi/2\}$, the function $\exp(-1/x)$ belongs to $\overline{\mathscr{A}}^{<0}(\mathcal{S})$; if $\mathcal{S} = \{x\,;\,|\arg(x)| < \pi\}$ denotes the plane $\mathbb{C}$ slit along the negative real axis, the function $\exp(-1/\sqrt{x})$ where $\sqrt{x}$ stands for the principal determination of $x^{1/2}$ belongs to $\overline{\mathscr{A}}^{<0}(\mathcal{S})$.

1.1.2 Examples

The examples below will be reconsidered throughout the volume to illustrate our assertions.

Example 1.1.3 (A trivial Example: Convergent Series) Let $\mathcal{S}$ be a punctured disc D^* around 0 (i.e., a sector of opening $> 2\pi$ in $\mathbb{C}$). If f is an analytic function on D then f is asymptotic to its Taylor series at 0 on D^*. Reciprocally, if f is an analytic function on D^* that has an asymptotic expansion at 0 on D^* then, f is bounded near 0. According to the removable singularity theorem, the function f is analytic on all of D [Rud87, Thm. 10.20].

Example 1.1.4 (A Fundamental Example: the Euler Function)

Consider the Euler equation

$$x^2 \frac{\mathrm{d}y}{\mathrm{d}x} + y = x. \tag{1.1}$$

▷ Looking for a power series solution one finds the unique series

$$\widetilde{\mathrm{E}}(x) = \sum_{n\geq 0}(-1)^n n!\, x^{n+1} \tag{1.2}$$

called the *Euler series*. The Euler series is clearly divergent for all $x \neq 0$ and thus, it does not provide an analytic solution near 0 by the usual Cauchy-Weierstrass summation. However, an actual solution can be found by applying the Lagrange method on $\mathbb{R}^+$. We notice that 0 is a singular point of the equation and the Lagrange method must be applied on a domain (i.e., a connected open set) containing no singular point ($\mathbb{R}^+$ is connected, open in $\mathbb{R}$ and does not contain 0). The fact that the series $\widetilde{\mathrm{E}}(x)$ diverges indicates that 0 is an irregular singular point of the Euler equation (Prop. 3.5.3, p. 93). Among the infinitely many solutions given by the method we choose the only one which is bounded as x tends to 0^+; it reads

$$\mathrm{E}(x) = \int_0^x \exp\Big(-\frac{1}{t} + \frac{1}{x}\Big)\frac{\mathrm{d}t}{t} = \int_0^{+\infty} \frac{\mathrm{e}^{-\xi/x}}{1+\xi}\,\mathrm{d}\xi$$

and is not only a solution on $\mathbb{R}^+$ but also a well defined solution on $\Re(x) > 0$.

Actually, the function $\mathrm{E}(x)$ is asymptotic to the Euler series $\widetilde{\mathrm{E}}(x)$ on $\{x \in \mathbb{C}\,;\, \Re(x) > 0\}$. A proof is as follows: writing

$$\frac{1}{1+\xi} = \sum_{n=0}^{N-2} (-1)^n \xi^n + (-1)^{N-1} \frac{\xi^{N-1}}{1+\xi}$$

and using $\int_0^{+\infty} u^n \mathrm{e}^{-u}\,\mathrm{d}u = \Gamma(1+n)$, we get the relation

$$\mathrm{E}(x) = \sum_{n=0}^{N-2} (-1)^n \Gamma(1+n) x^{n+1} + (-1)^{N-1} \int_0^{+\infty} \frac{\xi^{N-1}\mathrm{e}^{-\xi/x}}{1+\xi}\,\mathrm{d}\xi$$

and we are left to bound the integral remainder term.

Choose $0 < \delta < \pi/2$ and consider the (unlimited) proper sub-sector

$$\boldsymbol{\Delta}_\delta = \big\{x\,;\, |\arg(x)| < \pi/2 - \delta\big\}$$

of the half-plane $\boldsymbol{\Delta} = \{x\,;\, \Re(x) > 0\}$.

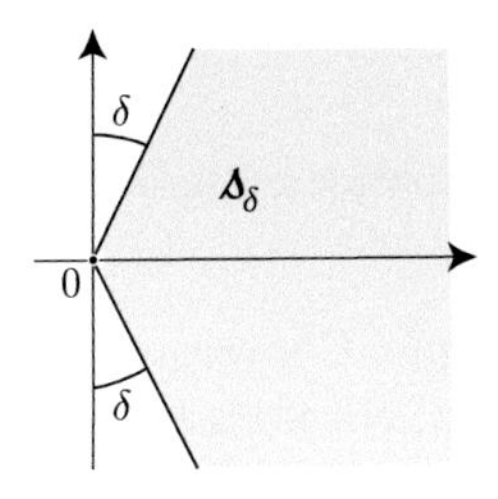

Fig. 1.2

For all $x \in \boldsymbol{\Delta}_\delta$, we can write

$$\Big|\mathrm{E}(x) - \sum_{n=0}^{N-2}(-1)^n n!\, x^{n+1}\Big| \le \int_0^{+\infty} \xi^{N-1}\,\mathrm{e}^{-\Re(\xi/x)}\,\mathrm{d}\xi$$

$$\le \int_0^{+\infty} \xi^{N-1}\,\mathrm{e}^{-\xi\sin(\delta)/|x|}\,\mathrm{d}\xi$$

$$= \frac{|x|^N}{(\sin\delta)^N}\int_0^{+\infty} u^{N-1}\,\mathrm{e}^{-u}\,\mathrm{d}u = C|x|^N$$

with $C = \Gamma(N)/(\sin\delta)^N$. This proves that the function $\mathrm{E}(x)$ is asymptotic to the Euler series $\widetilde{\mathrm{E}}(x)$ at 0 on the half plane $\boldsymbol{\Delta}$. Observe that the constant C does not depend on x but it depends on N and $\boldsymbol{\Delta}_\delta$ and it tends to infinity as δ tends to 0. Thus, the estimate is no longer valid on the whole sector $\boldsymbol{\Delta} = \{x\,;\,\Re(x) > 0\}$.

If we slightly rotate the line of integration to the line d_θ with argument θ (one says *in the θ direction*) then, the same calculation remains valid and provides a function $\mathrm{E}_\theta(x)$ with the same asymptotic expansion on the half plane bisected by d_θ. By Cauchy's theorem, $\mathrm{E}_\theta(x)$ is the analytic continuation of $\mathrm{E}(x)$. Denote by $\mathrm{E}(x)$ the largest analytic continuation of the initial function $\mathrm{E}(x)$ by such a method. Its domain of definition is easily determined: we can rotate the line d_θ as long as it does not meet the pole $\xi = -1$ of the integrand, that is, we can rotate it from $\theta = -\pi$ to $\theta = +\pi$ (the directions $\pm\theta$ being excluded). We get so an analytic continuation of the initial function E on the sector

$$\boldsymbol{\Delta}_{\mathrm{E}} = \{x \in \widetilde{\mathbb{C}}\,;\,-3\pi/2 < \arg(x) < +3\pi/2\}$$

of the universal covering $\widetilde{\mathbb{C}}$ of $\mathbb{C}^*$. On the sector $\boldsymbol{\Delta}_{\mathrm{E}}$ the function $\mathrm{E}(x)$ is asymptotic to the Euler series $\widetilde{\mathrm{E}}(x)$.

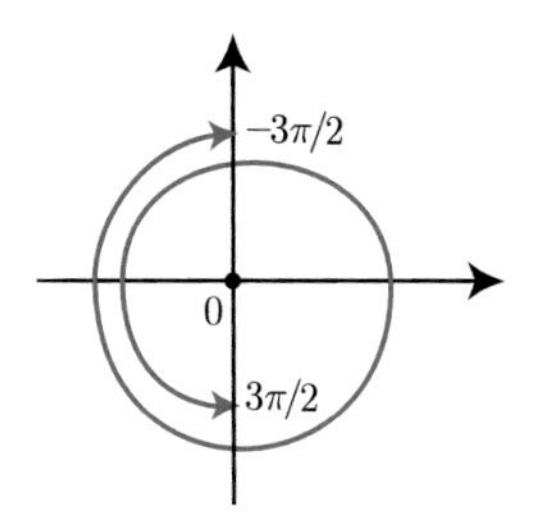

Fig. 1.3. Domain where $\mathrm{E}(x)$ is asymptotic to $\widetilde{\mathrm{E}}(x)$

With this construction we are given on $\{x \in \mathbb{C}^*\,;\,\Re(x) < 0\}$ two determinations $\mathrm{E}^+(x)$ and $\mathrm{E}^-(x)$ of $\mathrm{E}(x)$ when the θ direction approaches $+\pi$ and $-\pi$ respectively.

Let us observe the following two facts:

▷ The determinations $\mathrm{E}^+(x)$ and $\mathrm{E}^-(x)$ are distinct since otherwise the function $\mathrm{E}(x)$ would be analytic at 0 and the Euler series $\widetilde{\mathrm{E}}(x)$ would be convergent.

▷ Although $E(x)$ admits an analytic continuation as a solution of the Euler equation on all of the universal covering $\widetilde{\mathbb{C}}$ of $\mathbb{C}^*$ (by the Cauchy-Lipschitz theorem) its stops having an asymptotic expansion on any sector $\boldsymbol{\Delta}$ larger than $\boldsymbol{\Delta}_{\mathrm{E}}$ (i.e., $\boldsymbol{\Delta}_{\mathrm{E}} \subsetneq \boldsymbol{\Delta}$). Indeed, the two determinations $\mathrm{E}^+(x)$ and $\mathrm{E}^-(x)$ satisfy the relation

$$\mathrm{E}^+(x) - \mathrm{E}^-(x) = 2\pi\mathrm{i}\,\mathrm{e}^{1/x}. \tag{1.3}$$

(see [Lod90] or the calculation of the variation of $\mathrm{E}(x)$ in remark 1.4.3, p. 30) Thus, the determination $\mathrm{E}^+(x)$ can be continued through the negative imaginary axis by setting $\mathrm{E}^+(x) = \mathrm{E}(x) + 2\pi\mathrm{i}\mathrm{e}^{1/x}$ and symmetrically for $\mathrm{E}^-(x)$ through the positive imaginary axis. Any asymptotic condition fails when one reaches the directions in $\Re(x) > 0$ since $\mathrm{e}^{1/x}$ is unbounded at 0 on $\Re(x) > 0$. Such a phenomenon of discontinuity in the asymptotics is called *Stokes phenomenon* (cf. end of Rem. 1.4.3, p. 30 and Sect. 3.5, p. 90).

The function $\mathrm{E}(x)$ is called the *Euler function*.

Unless otherwise specified we consider it as a function defined on $\{x \in \widetilde{\mathbb{C}};|\arg(x)| < 3\pi/2\}$.

Example 1.1.5 (A Classical Example: the Exponential Integral)

The exponential integral $\mathrm{Ei}(x)$ is the function given by

$$\mathrm{Ei}(x) = \int_x^{+\infty} \mathrm{e}^{-t}\frac{\mathrm{d}t}{t}. \tag{1.4}$$

The integral being well defined on horizontal lines avoiding 0 the function $\mathrm{Ei}(x)$ is well defined and analytic on the plane $\mathbb{C}$ slit along the real non-positive axis.

Let us first determine its asymptotic behavior at the origin 0 on the right half plane

$$\boldsymbol{\Delta} = \{x\,;\Re(x) > 0\}.$$

To do so, we start with the asymptotic expansion of its derivative $\mathrm{Ei}'(x) = -\mathrm{e}^{-x}/x$. The Taylor expansion with integral remainder for e^{-x} gives

$$\mathrm{e}^{-x} = \sum_{n=0}^{N-1}(-1)^n\frac{x^n}{n!} + (-1)^N\frac{x^N}{(N-1)!}\int_0^1 (1-u)^{N-1}\,\mathrm{e}^{-ux}\,\mathrm{d}u$$

and then, since $\Re(-ux) < 0$,

$$\Big|\mathrm{Ei}'(x) + \frac{1}{x} + \sum_{n=1}^{N-1}(-1)^n\frac{x^{n-1}}{n!}\Big| \le \frac{|x|^{N-1}}{N!}.$$

We see that a negative power of x occurs which generates a logarithm when one takes its antiderivative. Integrating from $\varepsilon > 0$ to x and letting ε tend to 0 we obtain

$$\Big|\mathrm{Ei}(x) + \ln(x) + \gamma + \sum_{n=1}^{N-1}(-1)^n\frac{x^n}{n\cdot n!}\Big| \le \frac{|x|^N}{N!} \quad \text{with } \gamma = -\lim_{x\to 0^+}\big(\mathrm{Ei}(x) + \ln(x)\big).$$

To fit our definition of an asymptotic expansion we must consider the function $\mathrm{Ei}(x)+\ln(x)$. By extension, one says that $\mathrm{Ei}(x)$ has the asymptotic expansion

$$-\ln(x) - \gamma - \sum_{n=1}^{\infty}(-1)^n\frac{x^n}{n\cdot n!}.$$

We leave the fact that γ is indeed the Euler constant $\lim_{n\to+\infty}\sum_{p=1}^{n} 1/p - \ln(n)$ as an exercise. Notice that, this time, we did not need to shrink the sector $\boldsymbol{\Delta}$.

Now let us look at what happens at infinity.

One could obtain asymptotic expansions of $\mathrm{Ei}(z)$ at infinity by stating estimates like those given in definition 1.1.1. Instead, we notice that the function $y(x) = \mathrm{e}^{1/x}\mathrm{Ei}(1/x)$ is the Euler function $f(x)$. Hence, it has on $\boldsymbol{\Delta}$ at 0 the same asymptotics as $f(x)$ studied in the previous example. Turning back to the variable $z = 1/x \approx \infty$ we can state that $\mathrm{e}^z\,\mathrm{Ei}(z)$ has the series $\sum_{n\ge 0}(-1)^n n!/z^{n+1}$ as asymptotic expansion at infinity on $\boldsymbol{\Delta}$. By extension, one says that $\mathrm{Ei}(z)$ is asymptotic to $\mathrm{e}^{-z}\sum_{n\ge 0}(-1)^n n!/z^{n+1}$ on $\boldsymbol{\Delta}$ at infinity.

Example 1.1.6 (A Generalized Hypergeometric Series ${}_3F_0$)

We consider a generalized hypergeometric equation with given values of the parameters, say,

$$D_{3,1}y \equiv \left\{ z\left(z\frac{\mathrm{d}}{dz}+4\right) - z\frac{\mathrm{d}}{dz}\left(z\frac{\mathrm{d}}{dz}+1\right)\left(z\frac{\mathrm{d}}{dz}-1\right) \right\} y = 0. \tag{1.5}$$

This equation has an irregular singular point at infinity (cf. Def. 3.3.2, p. 75, Prop. 3.5.3, p. 93) and a unique series solution

$$\widetilde{g}(z) = \frac{1}{z^4} \sum_{n\geq 0} \frac{(n+2)!(n+3)!(n+4)!}{2!3!4!n!} \frac{1}{z^n}. \tag{1.6}$$

Using the standard notation for the hypergeometric series, the series $\widetilde{g}$ reads

$$\widetilde{g}(z) = z^{-4}\,{}_3F_0\left(\{3,4,5\} \,\Big|\, \frac{1}{z}\right).$$

By abuse of language, we will also call $\widetilde{g}$ an hypergeometric series.

One can check that, for $-3\pi < \arg(z) < +\pi$, the equation admits the solution

$$g(z) = \frac{1}{2\pi\mathrm{i}\,2!3!4!} \int_\gamma \Gamma(1-s)\Gamma(-s)\Gamma(-1-s)\Gamma(4+s)\,\mathrm{e}^{\mathrm{i}\pi s} z^s\,\mathrm{d}s$$

where γ is a path from $-3-\mathrm{i}\infty$ to $-3+\mathrm{i}\infty$ along the line $\Re(s) = -3$. This follows from the fact that the integrand $G(s,z)$ satisfies the order one difference equation deduced from $D_{3,1}$ by applying a Mellin transform. We leave the proof to the reader. Instead, let us check that the integral is well defined. The integrand $G(s,z)$ being continuous along γ we just have to check the behavior of $G(s,z)$ as s tends to infinity along γ. An asymptotic expansion of $\Gamma(t+\mathrm{i}u)$ for $t \in \mathbb{R}$ fixed and $u \in \mathbb{R}$ large is given by (see [BH86, p. 83]):

$$\Gamma(t+\mathrm{i}u) = |u|^{t-\frac{1}{2}}\,\mathrm{e}^{-\frac{\pi}{2}|u|} \left\{ \sqrt{2\pi}\,\mathrm{e}^{\mathrm{i}\frac{\pi}{2}(t-\frac{1}{2})\,\mathrm{sgn}(u)-\mathrm{i}u}\,|u|^{\mathrm{i}u} \right\} \left\{ 1+O(1/u) \right\}. \tag{1.7}$$

It follows that $G(t+\mathrm{i}u,z)$ satisfies

$$\left|G(t+\mathrm{i}u,z)\right| = (2\pi)^2\,|u|^{-2t+2}\,|z|^t\,\mathrm{e}^{-2\pi|u|-\pi u-u\arg(z)} \left\{ 1+O(1/u) \right\}. \tag{1.8}$$

The exponent of the exponential being negative for $-3\pi < \arg(z) < \pi$ the integral is convergent and it defines an analytic function $g(z)$.

Now, let us prove that the function $g(z)$ is asymptotic to $\widetilde{g}(z)$ at infinity on the sector

$$\boldsymbol{\Delta} = \{z\,;\, -3\pi < \arg(z) < +\pi\}.$$

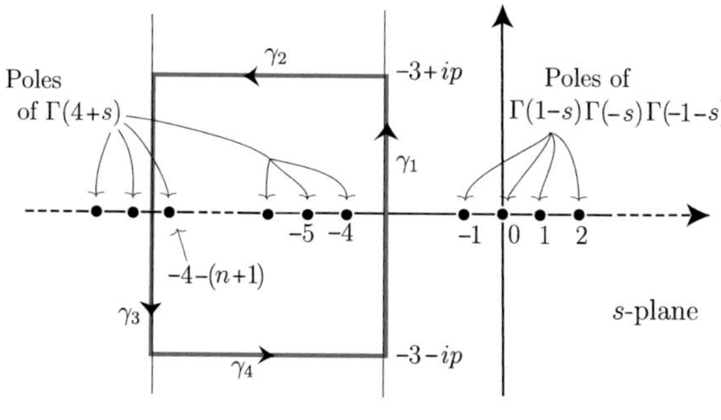

Fig. 1.4. Path $\gamma_{n,p}$

To do so, consider paths $\gamma_{n,p} = \gamma_1 \cup \gamma_2 \cup \gamma_3 \cup \gamma_4$ with positive integers n, p as drawn on Fig. 1.4.

The path $\gamma_{n,p}$ encloses the poles $s_m = -4-m$ for $m = 0,\ldots,n+1$ of $G(s,z)$. The residues are

$$\mathrm{Res}\big(G(s,z); s=-4-m\big) = (2+m)!(3+m)!(4+m)!z^{-4-m}/m! = 2!3!4!a_m.$$

Indeed, $\Gamma\big(4+(-4-m+t)\big) = \Gamma(-m+t) = \frac{(-1)^m}{m!}t^{-1} + O(1)$; in particular, the function $\Gamma(4+s)$ has a simple pole at $s=-4-m$. All other factors of G are non-zero analytic functions.

It follows that

$$\frac{1}{2\pi \mathrm{i}\, 2!3!4!}\int_{\gamma_{n,p}} G(s,z)\,\mathrm{d}s = \frac{1}{z^4}\sum_{m=0}^{n+1}\frac{a_m}{z^m}.$$

Formula (1.8), p. 7 implies the estimate

$$\big|G(t+\mathrm{i}\varepsilon p, z)\big| \le Cp^{2n+5}\mathrm{e}^{-(2\pi+\varepsilon\pi+\varepsilon\arg(z))p}, \quad \varepsilon = \pm 1,$$

valid for $|z|>1$ all along $\gamma_2 \cup \gamma_4$, the constant C depending on n and z but not on p. This shows that the integral along $\gamma_2 \cup \gamma_4$ tends to zero as p tends to infinity and therfore, that

$$g(z) = \frac{1}{z^4}\sum_{m=0}^{n+1}\frac{a_m}{z^m} + g_n(z)$$

where $g_n(z) = \frac{1}{2\pi\mathrm{i}\,2!3!4!}\int_{\gamma^n} G(s,z)\,\mathrm{d}s$ and $\gamma^n = \{s\in\mathbb{C}\,;\, \Re(s) = -4-n-\frac{3}{2}\}$ oriented upwards.

For any (small enough) $\delta > 0$ consider the proper sub-sector $\boldsymbol{\Delta}_\delta$ of $\boldsymbol{\Delta}$ defined by

$$\boldsymbol{\Delta}_\delta = \big\{z\in\mathbb{C}\,;\, |z|>1 \text{ and } -3\pi+\delta < \arg(z) < \pi-\delta\big\}.$$

For $z\in\boldsymbol{\Delta}_\delta$ and $s = -4-n-\frac{3}{2}+\mathrm{i}u \in \gamma^n$, the factor z^s satisfies

$$|z^s| \le \frac{1}{|z|^{4+n+\frac{3}{2}}}\cdot\begin{cases} \mathrm{e}^{-u(\pi-\delta)} & \text{if } u<0,\\ \mathrm{e}^{u(3\pi-\delta)} & \text{if } u>0.\end{cases}$$

and using again formula (1.7), p. 7 we obtain

$$\Big|G\big(-4-n-\frac{3}{2}+\mathrm{i}u, z\big)\Big| \le \frac{\mathrm{Const}_{n,\delta}}{|z|^{(4+n)+1}}|u|^{13+n}\mathrm{e}^{-|u|\delta}.$$

Hence, there exists a constant $C = C(n,\delta)$ depending on n and δ but not on z such that

$$\Big|g(z) - \frac{1}{z^4}\sum_{m=0}^{n+1}\frac{a_m}{z^m}\Big| = |g_n(z)| \le \frac{C}{|z|^{(4+n)+1}} \quad \text{for all } z\in\boldsymbol{\Delta}_\delta.$$

Rewriting this estimate in the form

$$\Big|g(z) - \frac{1}{z^4}\sum_{m=0}^{n}\frac{a_m}{z^m}\Big| = \Big|g_n(z) + \frac{a_{n+1}}{z^{(4+n)+1}}\Big| \le \frac{C+|a_{n+1}|}{|z|^{(4+n)+1}} \quad \text{for all } z\in\boldsymbol{\Delta}_\delta$$

we satisfy definition 1.1.1, p. 2, for g at the order $4+n$.

By this method we do not know how the constant C depends on n but we know that $|a_{n+1}|$ grows like $(n!)^2$ and then $C+|a_{n+1}|$ itself grows at least like $(n!)^2$.

Example 1.1.7 (A Series Solution of a Mild Difference Equation)

Consider the order one difference equation

$$h(z+1) - 2h(z) = \frac{1}{z}\cdot \tag{1.9}$$

A difference equation is said to be *mild* when its companion system, here

$$\begin{bmatrix} y_1(z+1) \\ y_2(z+1) \end{bmatrix} = \begin{bmatrix} 2 & 1/z \\ 0 & 1 \end{bmatrix} \begin{bmatrix} y_1(z) \\ y_2(z) \end{bmatrix}$$

has an invertible leading term; in our case, $\left[\begin{smallmatrix} 2 & 0 \\ 0 & 1 \end{smallmatrix}\right]$ is invertible. The term "mild" and its opposite "wild" were introduced by M. van der Put and M. Singer [vdPS97].

Let us look at what happens at infinity. By identification, we see that equation (1.9), p. 9, has a unique power series solution in the form $\widetilde{h}(z) = \sum_{n\geq 1} h_n/z^n$. The coefficients h_n are defined by the recurrence relation

$$h_n = \sum_{\substack{m+p=n \\ m,p\geq 1}} (-1)^p h_m \frac{(m+p-1)!}{(m-1)!\,p!}$$

starting from the initial value $h_1 = -1$. It follows that the sequence h_n is alternate and satisfies

$$|h_n| \geq n|h_{n-1}|.$$

Consequently, $|h_n| \geq n!$ and the series $\widetilde{h}$ is divergent. The recurrence relation can actually be solved as follows. Consider the Borel transform

$$\widehat{h}(\zeta) = \sum_{n\geq 1} h_n \frac{\zeta^{n-1}}{(n-1)!}$$

of $\widetilde{h}$ (cf. Def. 5.3.1, p. 145). It satisfies the Borel transformed equation

$$\mathrm{e}^{-\zeta}\widehat{h}(\zeta) - 2\widehat{h}(\zeta) = 1$$

and then, $\widehat{h}(\zeta) = 1/(\mathrm{e}^{-\zeta} - 2)$. Its Taylor series at 0 reads

$$T_0\widehat{h}(\zeta) = \sum_{n\geq 0} \frac{(-1)^{n+1}}{n!} \sum_{p\geq 0} \frac{p^n}{2^{p+1}} \zeta^n$$

which implies that $h_n = (-1)^n \sum_{p\geq 0} p^{n-1}/2^{p+1}$. Again, we see that the series $\widetilde{h}$ is divergent since $|h_n| \geq n^{n-1}/2^{n+1}$.

We claim that the function

$$h(z) = \int_0^{+\infty} \widehat{h}(\zeta)\mathrm{e}^{-z\zeta} d\zeta$$

is asymptotic to $\widetilde{h}(z)$ at infinity on the sector $\boldsymbol{\Delta} = \{z\,;\,\Re(z) > 0\}$ (right half-plane). Indeed, choose $N \in \mathbb{N}$ and a proper sub-sector $\boldsymbol{\Delta}_\delta = \{z\,;\,-\frac{\pi}{2}+\delta < \arg(z) < \frac{\pi}{2}-\delta\}$ of $\boldsymbol{\Delta}$.
From the Taylor expansion with integral remainder of $\widehat{h}(\zeta)$ at 0

$$\widehat{h}(\zeta) = \sum_{n=1}^{N} h_n \frac{\zeta^{n-1}}{(n-1)!} + \frac{\zeta^N}{(N-1)!} \int_0^1 (1-t)^{N-1} \widehat{h}^{(N)}(\zeta t)\,\mathrm{d}t$$

we obtain

$$h(z) = \sum_{n=1}^{N} \frac{h_n}{z^n} + \int_0^{+\infty} \frac{\zeta^N}{(N-1)!} \int_0^1 (1-t)^{N-1} \widehat{h}^{(N)}(\zeta t)\, \mathrm{d}t\, \mathrm{e}^{-z\zeta}\, \mathrm{d}\zeta.$$

To bound $\widehat{h}^{(N)}(\zeta t)$ we use the Cauchy Integral Formula

$$\widehat{h}^{(N)}(\zeta t) = \frac{N!}{2\pi \mathrm{i}} \int_{C_{\zeta t}} \frac{\widehat{h}(u)}{(u-\zeta t)^{N+1}}\, \mathrm{d}u$$

where $C_{\zeta t}$ denotes the circle with center ζt, radius $1/2$, oriented counterclockwise. For $t \in [0,1]$ and $\zeta \in [0,+\infty[$ then ζt is non negative and $\Re(u) \geq -1/2$ when u runs over any $C_{\zeta t}$. Hence, we obtain

$$\left|\widehat{h}^{(N)}(\zeta t)\right| \leq \frac{N! 2^N}{2-\mathrm{e}^{1/2}} \quad \text{and} \quad \left|\int_0^1 (1-t)^{N-1} \widehat{h}^{(N)}(\zeta t)\, \mathrm{d}t\right| \leq \frac{(N-1)! 2^N}{2-\mathrm{e}^{1/2}}.$$

Finally, from the identity above we can conclude that, for all $z \in \boldsymbol{\Delta}_\delta$,

$$\left|h(z) - \sum_{n=1}^{N} \frac{h_n}{z^n}\right| \leq \frac{2^N}{2-\mathrm{e}^{1/2}} \int_0^{+\infty} \zeta^N \left|\mathrm{e}^{-\zeta z}\right| \mathrm{d}\zeta = \frac{C}{|z|^{N+1}} \tag{1.10}$$

with $C = \frac{1}{2-\mathrm{e}^{1/2}} N! \frac{2^N}{(\sin\delta)^{N+1}}$. This proves that $h(z)$ is asymptotic to $\widetilde{h}(z)$ at infinity on $\boldsymbol{\Delta}$.

Example 1.1.8 (A Series Solution of a Wild Difference Equation)

Consider the order one inhomogeneous wild difference equation

$$\frac{1}{z}\ell(z+1) + \left(1+\frac{1}{z}\right)\ell(z) = \frac{1}{z}. \tag{1.11}$$

An identification of terms of equal power shows that it admits a unique series solution

$$\widetilde{\ell}(z) = \sum_{n\geq 1} \ell_n z^{-n}$$

whose coefficients ℓ_n are given by the recurrence relation

$$\ell_{n+1} = -2\ell_n - \sum_{\substack{m+p=n \\ m\geq 1, p\geq 1}} (-1)^p \ell_m \frac{(m+p-1)!}{p!\,(m-1)!}$$

from the initial value $\ell_1 = 1$. It follows that the sequence $(\ell_n)_{n\geq 1}$ is alternate and satisfies

$$|\ell_{n+1}| \geq (n-1)|\ell_{n-1}|.$$

Hence, $\ell_{2n} \geq 2^n\,(n-1)!$ for all n and consequently, the series is divergent. The Borel transform $\widehat{\ell}(\zeta)$ of the series $\widetilde{\ell}(z)$ satisfies the equation

$$\int_0^\zeta \mathrm{e}^{-\xi}\widehat{\ell}(\xi)\, \mathrm{d}\xi + \int_0^\zeta \widehat{\ell}(\xi)\, \mathrm{d}\xi + \widehat{\ell}(\xi) = 1$$

equivalent to the two conditions $\widehat{\ell}(0) = 1$ and $\widehat{\ell}'(\zeta) = (-\mathrm{e}^{-\zeta} - 1)\widehat{\ell}(\zeta)$. Hence,

$$\widehat{\ell}(\zeta) = \frac{1}{\mathrm{e}}\mathrm{e}^{-\zeta + \mathrm{e}^{-\zeta}}.$$

We leave as an exercise tthe proof of the fact that the Laplace integral $\int_0^{+\infty} \widehat{\ell}(\zeta)\,\mathrm{e}^{-z\zeta}\,\mathrm{d}\zeta$ is a solution of (1.11) asymptotic to $\widehat{\ell}(z)$ at infinity on the sector $\Re(z) > -1$ (follow the same method as in the previous example and estimate the constant C).

1.1.3 Algebras of Asymptotic Functions

Recall that Δ denotes a given open sector with vertex 0 in $\mathbb{C} \setminus \{0\}$ or in the Riemann surface of the logarithm $\widetilde{\mathbb{C}}$. Unless otherwise mentioned we refer to the usual derivation $\mathrm{d}/\mathrm{d}x$ and to notation 1.1.2, p. 3.

Proposition 1.1.9 (Differential Algebra and Taylor Map)

▷ *The set* $\overline{\mathscr{A}}(\Delta)$ *endowed with the usual algebraic operations and the usual derivation* $\mathrm{d}/\mathrm{d}x$ *is a differential algebra.*

▷ *The Taylor map* $T = T_\Delta : \overline{\mathscr{A}}(\Delta) \to \mathbb{C}[[x]]$ *is a morphism of differential algebras with kernel* $\overline{\mathscr{A}}^{<0}(\Delta)$.

Proof. From the algebraic rules on asymptotic expansions it follows that $\overline{\mathscr{A}}(\Delta)$ is a subalgebra of $\mathscr{O}(\Delta)$. We are left to prove that $\overline{\mathscr{A}}(\Delta)$ is stable under derivation with respect to x and that the Taylor map T_Δ commutes with derivation.

Let $f \in \overline{\mathscr{A}}(\Delta)$ have an asymptotic expansion $T_\Delta f(x) = \sum_{n\geq 0} a_n x^n$. Since f belongs to $\mathscr{O}(\Delta)$ it admits a derivative $f' \in \mathscr{O}(\Delta)$. Moreover, for all $\Delta' \Subset \Delta$ and all $N \geq 0$, there exists $C > 0$ such that, for all $x \in \Delta'$,

$$\Big| f(x) - \sum_{n=0}^{N} a_n x^n \Big| \leq C|x|^{N+1}$$

and we want to prove that for all $\Delta'' \Subset \Delta$, for all $N > 0$, there exists $C' > 0$ such that, for all $x \in \Delta''$,

$$\Big| f'(x) - \sum_{n=0}^{N-1} (n+1)a_{n+1} x^n \Big| \leq C'|x|^N.$$

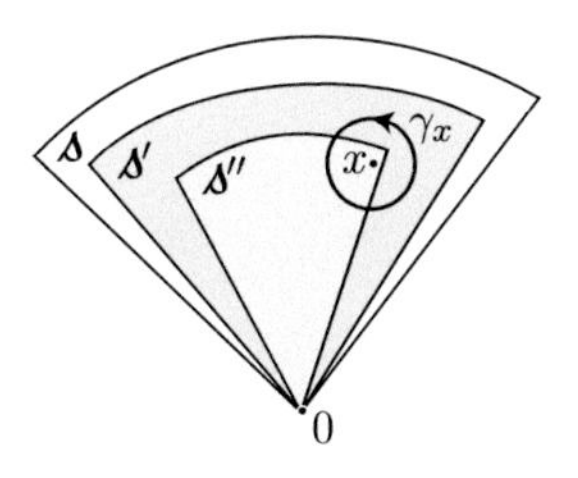

Fig. 1.5

Fix $N > 0$ and consider the function $g(x) = f(x) - \sum_{n=0}^{N} a_n x^n$.

We must prove that the condition

for all $\mathcal{S}' \Subset \mathcal{S}$*, there exists* $C > 0$ *such that* $|g(x)| \leq C|x|^{N+1}$ *for all* $x \in \mathcal{S}'$

implies the condition

for all $\mathcal{S}'' \Subset \mathcal{S}$*, there exists* $C' > 0$ *such that* $|g'(x)| \leq C'|x|^N$ *for all* $x \in \mathcal{S}''$.

Given $\mathcal{S}'' \Subset \mathcal{S}$, choose a sector $\mathcal{S}'$ such that $\mathcal{S}'' \Subset \mathcal{S}' \Subset \mathcal{S}$ (see Fig 1.5) and let δ be so small that, for all $x \in \mathcal{S}''$, the closed disc $\bar{B}(x, |x|\delta)$ centered at x with radius $|x|\delta$ be contained in $\mathcal{S}'$. Denote by γ_x the boundary of $\bar{B}(x, |x|\delta)$.

By assumption, the function g satisfies $|g(t)| \leq C|t|^{N+1}$ for all $t \in \bar{B}(x, |x|\delta)$, in particular for all $t \in \gamma_x$. From Cauchy's integral formula $g'(x) = \frac{1}{2\pi\mathrm{i}}\int_{\gamma_x}\frac{g(t)}{(t-x)^2}\,\mathrm{d}t$ we deduce that, for all $x \in \mathcal{S}''$, the derivative g' satisfies

$$|g'(x)| \leq \frac{1}{2\pi}\max_{t\in\gamma_x}|g(t)|\,\frac{2\pi\,|x|\,\delta}{(|x|\,\delta)^2} \leq \frac{C}{|x|\,\delta}\big(|x|\,(1+\delta)\big)^{N+1} = C'\,|x|^N$$

with $C' = C(1+\delta)^{N+1}/\delta$. Hence, the result. □

Remarks 1.1.10 Let us insist on the role of Cauchy's integral formula.

▷ The proof of proposition 1.1.9, p. 11 does require that the estimates in definition 1.1.1, p. 2 be satisfied for all $\mathcal{S}' \Subset \mathcal{S}$ instead of $\mathcal{S}$ itself. Otherwise, we could not apply Cauchy's integral formula and we could not assert any longer that the algebra $\overline{\mathscr{A}}(\mathcal{S})$ is differential. Algebras of asymptotic functions would not be appropriate to handle solutions of differential equations.

▷ Theorem 1.1.9, p. 11 is no longer valid in real asymptotics, where Cauchy's integral formula does not hold, as is shown by the following counter-example.

The function $f(x) = \mathrm{e}^{-1/x}\sin(\mathrm{e}^{1/x})$ is asymptotic to 0 (the null series) on $\mathbb{R}^+$ at 0. Its derivative $f'(x) = \frac{1}{x^2}\mathrm{e}^{-1/x}\sin(\mathrm{e}^{1/x}) - \frac{1}{x^2}\cos(\mathrm{e}^{1/x})$ has no limit at 0 on $\mathbb{R}^+$ and therefore no asymptotic expansion. This proves that the set of real analytic functions admitting an asymptotic expansion at 0 on $\mathbb{R}^+$ is *not a differential algebra*.

The following proposition provides in particular a proof of the uniqueness of the asymptotic expansion, if any exists.

Proposition 1.1.11 *A function* f *belongs to* $\overline{\mathscr{A}}(\mathcal{S})$ *if and only if* f *belongs to* $\mathscr{O}(\mathcal{S})$ *and a sequence* $(a_n)_{n\in\mathbb{N}}$ *exists such that*

$$\frac{1}{n!}\lim_{\substack{x\to 0\\ x\in\mathcal{S}'}} f^{(n)}(x) = a_n \quad \textit{for all } \mathcal{S}' \Subset \mathcal{S}.$$

Proof. The *only if* part follows from proposition 1.1.9, p. 11. To prove the *if* part consider $\mathcal{S}' \Subset \mathcal{S}$. For all x and $x_0 \in \mathcal{S}'$, the function f admits the Taylor expansion with integral remainder

$$f(x) - \sum_{n=0}^{N-1}\frac{1}{n!}f^{(n)}(x_0)(x-x_0)^n = \int_{x_0}^{x}\frac{1}{(N-1)!}(x-t)^{N-1}f^{(N)}(t)\,\mathrm{d}t.$$

Note that, a priori, we cannot write such a formula for $x_0 = 0$ since 0 does not even belong to the definition set of f. However, by assumption, the limit of the left-hand side as x_0 tends to 0 in Δ' exists; hence, the limit of the right-hand side exists too and we can write

$$f(x) - \sum_{n=0}^{N-1} a_n x^n = \int_0^x \frac{1}{(N-1)!}(x-t)^{N-1} f^{(N)}(t)\,\mathrm{d}t.$$

Then,
$$\begin{aligned}\Big|f(x) - \sum_{n=0}^{N-1} a_n x^n\Big| &\leq \tfrac{1}{(N-1)!}\Big|\int_0^x (x-t)^{N-1} f^{(N)}(t)\,\mathrm{d}t\Big| \\ &\leq \tfrac{|x|^N}{N!}\sup_{t\in\Delta'}\big|f^{(N)}(t)\big| \leq C|x|^N,\end{aligned}$$

the constant $C = \frac{1}{N!}\sup_{t\in\Delta'}|f^{(N)}(t)|$ being finite by assumption. Hence, the conclusion □

1.2 Gevrey Asymptotics

When working with differential equations for instance, it appears soon that the conditions required in Poincaré asymptotics are too weak to fit some natural requests, say for instance, to provide asymptotic functions that are solutions of the equation when the asymptotic series themselves are solution or, better, to set a 1-to-1 correspondence between the series solution and their asymptotic expansion. A precise answer to these questions is found in the theories of summation (cf. Chaps. 5 and 7). A first step towards that aim is given by strengthening Poincaré asymptotics into Gevrey[2] asymptotics.

From now on, we are given $k > 0$ and we denote its inverse by

$$\boxed{s = 1/k} \tag{1.12}$$

If $k > 1/2$ then $\pi/k < 2\pi$ and the sectors of the critical opening π/k to be further considered may be seen as sectors of $\mathbb{C}^*$ itself; otherwise, they must be considered as sectors of the universal cover $\widetilde{\mathbb{C}}$ of $\mathbb{C}^*$. In general, depending on the problem, we may assume that $k > 1/2$ after performing a change of variable (ramification) $x = t^p$ with a large enough $p \in \mathbb{N}$.

Recall that, unless otherwise specified, we denote by $\Delta, \Delta', \dots$ open sectors in $\mathbb{C}^*$ or $\widetilde{\mathbb{C}}$ and that the notation $\Delta' \Subset \Delta$ means that the closure of the sector Δ' in $\mathbb{C}^*$ or $\widetilde{\mathbb{C}}$ lies in Δ (cf. Def. 1.0.2, p. 2).

---

[2] Maurice Gevrey, 1884-1957, was a French mathematician who introduced the now called "Gevrey classes" in 1918 after he defended his thesis *Sur les équations aux dérivées partielles du type parabolique* at Université de Paris in 1913.

1.2.1 Gevrey Series

Definition 1.2.1 (Gevrey Series of Order s or of Level k)

A series $\sum_{n\geq 0} a_n x^n$ *is of* Gevrey type of order s *(in short,* s-Gevrey) *or* of level $k = 1/s$ *if there exist constants* $C > 0$ *and* $A > 0$ *such that the coefficients* a_n *satisfy for all* n

$$\boxed{|a_n| \leq C(n!)^s A^n}$$

The constants C and A do not depend on n.

It is equivalent to say that a series $\sum_{n\geq 0} a_n x^n$ is s-Gevrey if the series $\sum_{n\geq 0} a_n x^n/(n!)^s$ converges.

Notation 1.2.2 We denote by $\mathbb{C}[[x]]_s$ the set of s-Gevrey series.

Observe that the spaces $\mathbb{C}[[x]]_s$ are filtered as follows:

$$\mathbb{C}\{x\} = \mathbb{C}[[x]]_0 \subset \mathbb{C}[[x]]_s \subset \mathbb{C}[[x]]_{s'} \subset \mathbb{C}[[x]]_\infty = \mathbb{C}[[x]]$$

for all s, s' satisfying $0 < s < s' < +\infty$.

Comments 1.2.3 (on the Examples of Section 1.1.2, p. 3)

▷ A convergent series (cf. Exa. 1.1.3, p. 3) is a 0-Gevrey series.

▷ The Euler series $\widetilde{\mathrm{E}}(x)$ (cf. Exa. 1.1.4, p. 4) is 1-Gevrey and hence s-Gevrey for any $s > 1$. It is s-Gevrey for no $s < 1$.

▷ The hypergeometric series ${}_3F_0(1/z)$ (cf. Exa. 1.1.6, p. 7) is 2-Gevrey. It is s-Gevrey for no $s < 2$.

▷ The series $\widetilde{h}(z)$ (cf. Exa. 1.1.7, p. 8) is 1-Gevrey. Indeed, since $|h_n| \geq n!$ for all n, it is at least 1-Gevrey; and it is at most 1-Gevrey since its Borel transform at infinity converges.

▷ From the fact that $|\ell_{2n+1}| \geq 2^n n!$ and $|\ell_{2n}| \geq 2^n(n-1)!$ we know that, if the series $\tilde{\ell}(z)$ (cf. Exa. 1.1.8, p. 10) is of Gevrey type then it is at least 1/2-Gevrey. From the fact that its Borel transform is convergent it is of Gevrey type and at most 1-Gevrey. Note however that its Borel transform is an entire function and consequently, $\tilde{\ell}(z)$ could be less than 1-Gevrey.

Proposition 1.2.4 $\mathbb{C}[[x]]_s$ *is a differential subalgebra of* $\mathbb{C}[[x]]$ *stable under composition.*

Proof. $\mathbb{C}[[x]]_s$ is clearly a vector subspace of $\mathbb{C}[[x]]$. We have to prove that it is stable under product, derivation and composition.

▷ *Stability of* $\mathbb{C}[[x]]_s$ *under product.* — Consider two s-Gevrey series $\sum_{n\geq 0} a_n x^n$ and $\sum_{n\geq 0} b_n x^n$ satisfying, for all n and for positive constants A, B, C and K, the estimates

$$|a_n| \leq C(n!)^s A^n \quad \text{and} \quad |b_n| \leq K(n!)^s B^n.$$

Their product is the series $\sum_{n\geq 0} c_n x^n$ where $c_n = \sum_{p+q=n} a_p b_q$. Then,

$$|c_n| \leq CK \sum_{p+q=n} (p!)^s (q!)^s A^p B^q \leq CK(n!)^s (A+B)^n. \quad \text{Hence the result.}$$

▷ *Stability of* $\mathbb{C}[[x]]_s$ *under derivation.* — Given an s-Gevrey series $\sum_{n\geq 0} a_n x^n$ satisfying $|a_n| \leq C(n!)^s A^n$ for all n, its derivative $\sum_{n\geq 0} b_n x^n$ satisfies

$$|b_n| = (n+1)\,|a_{n+1}| \leq (n+1)C((n+1)!)^s A^{n+1} \leq C'(n!)^s A'^n$$

for convenient constants $A' > A$ and $C' \geq C$. Hence the result.

▷ *Stability of* $\mathbb{C}[[x]]_s$ *under composition* [Gev18]. — Let $\widetilde{f}(x) = \sum_{p\geq 1} a_p x^p$ and $\widetilde{g}(y) = \sum_{n\geq 0} b_n y^n$ be two s-Gevrey series. The composition $\widetilde{g}\circ\widetilde{f}(x) = \sum_{n\geq 0} c_n x^n$ provides a well-defined power series in x. From the hypothesis, there exist constants $h,k,a,b > 0$ such that, for all p and n, the coefficients of the series $\widetilde{f}$ and $\widetilde{g}$ satisfy respectively $|a_p| \leq h(p!)^s a^p$ and $|b_n| \leq k(n!)^s b^n$.

Faà di Bruno's formula allows us to write

$$n!\,c_n = \sum_{\underline{m}\in I_n} N(\underline{m})\,|\underline{m}|\,!\,b_{|\underline{m}|} \prod_{j=1}^{n} \left(j!\,a_j\right)^{m_j}$$

where I_n stands for the set of non-negative n-tuples $\underline{m} = (m_1, m_2, \ldots, m_n)$ satisfying the condition $\sum_{j=1}^{n} jm_j = n$, where $|\underline{m}| = \sum_{j=1}^{n} m_j$ and the coefficient $N(\underline{m})$ is a positive integer depending neither on $\widetilde{f}$ nor on $\widetilde{g}$. Using the Gevrey hypothesis and the condition $\sum_{j=1}^{n} jm_j = n$, we can then write

$$n!\,|c_n| \leq ka^n \sum_{\underline{m}\in I_n} N(\underline{m})\,|\underline{m}|\,!^{1+s}\,(hb)^{|\underline{m}|} \left(\prod_{j=1}^{n} j!^{m_j}\right)^{1+s}.$$

As clearly $|\underline{m}| \leq n$ and $N(\underline{m}) \leq N(\underline{m})^{1+s}$, with $B = \max(hb, 1)$, we obtain

$$n!\,|c_n| \leq k\,(aB)^n \sum_{\underline{m}\in I_n} \left(N(\underline{m})\,|\underline{m}|\,! \prod_{j=1}^{n} j!^{m_j}\right)^{1+s}$$

and then, from the inequality $\sum_{i=1}^{K} X_i^{1+s} \leq \left(\sum_{i=1}^{K} X_i\right)^{1+s}$ for non-negative s and X_i's, the estimate

$$n!\,|c_n| \leq k\,(aB)^n \left(\sum_{\underline{m}\in I_n} N(\underline{m})\,|\underline{m}|\,! \prod_{j=1}^{n} j!^{m_j}\right)^{1+s}.$$

Now, applying Faà di Bruno's formula to the case of the series $\widetilde{f}(x) = x/(1-x)$ and $\widetilde{g}(x) = 1/(1-x)$, implying thus $\widetilde{g}\circ\widetilde{f}(x) = 1 + x/(1-2x)$, we get the relation

$$\sum_{\underline{m}\in I_n} N(\underline{m})\,|\underline{m}|\,! \prod_{j=1}^{n} (j!)^{m_j} = \begin{cases} 2^{n-1} n! & \text{when } n \geq 1 \\ 1 & \text{when } n = 0; \end{cases}$$

hence, a fortiori,

$$\sum_{\underline{m}\in I_n} N(\underline{m})\,|\underline{m}|\,! \prod_{j=1}^{n} (j!)^{m_j} \leq 2^n\, n!$$

and we can conclude that

$$|c_n| \leq k\,(n!)^s\,(2^{1+s}\,aB)^n$$

for all $n \in \mathbb{N}$, which ends the proof. □

These results can be extended to more general expressions than sums, products and composed series. As an example, consider Gevrey series in several variables be defined as follows.

Definition 1.2.5 *A series* $\widetilde{g}(y_1,\dots,y_r) = \sum_{n_1,\cdots,n_r \geq 0} b_{n_1,\dots,n_r} y_1^{n_1} \dots y_r^{n_r}$ *is said to be* $(s_1,\dots,s_r)$-Gevrey *if there exist positive constants* $C, M_1,\dots,M_r$ *such that, for all n-tuple* $(n_1,\dots,n_r)$ *of non-negative integers, the series satisfies an estimate of the form*

$$\boxed{\left|b_{n_1,\dots,n_r}\right| \leq C(n_1!)^{s_1}\cdots(n_r!)^{s_r} M_1^{n_1}\cdots M_r^{n_r}}$$

It is said to be s-Gevrey *when* $s_1 = \cdots = s_r = s$.

Then, the following proposition holds true:

Proposition 1.2.6 *Let* $\widetilde{f_1}(x), \widetilde{f_2}(x), \dots, \widetilde{f_r}(x)$ *be s-Gevrey series without constant term and let* $\widetilde{g}(y_1,\dots,y_r)$ *be an s-Gevrey series in r variables.*
Then, the series $\widetilde{g}(\widetilde{f_1}(x),\dots,\widetilde{f_r}(x))$ *is an s-Gevrey series.*

Since the expression of the nth derivative of $\widetilde{g}(\widetilde{f_1}(x),\dots,\widetilde{f_r}(x))$ has the same form as in the case of $\widetilde{g}(\widetilde{f}(x))$ the proof is identical to the one for $\widetilde{g}(\widetilde{f}(x))$ and we leave it as an exercise.

The result is, *a fortiori*, true when $\widetilde{g}$ or some of the $\widetilde{f_j}$'s are analytic. The stability of $\mathbb{C}[[x]]_s$ under addition, product or composition can then be seen as a consequence of this proposition.

1.2.2 Algebras of Gevrey Asymptotic Functions

In this section we study asymptotic functions with an extra Gevrey condition.

Definition 1.2.7 (Gevrey Asymptotics of order s**)**

A function $f \in \mathcal{O}(\Delta)$ *is said to be* Gevrey asymptotic of order s *(for short, s-*Gevrey asymptotic*)* to a series $\sum_{n\geq 0} a_n x^n$ on Δ *if for any proper sub-sector* $\Delta' \Subset \Delta$ *there exist constants* $C_{\Delta'} > 0$ *and* $A_{\Delta'} > 0$ *such that, the following estimate holds for all* $N \in \mathbb{N}^*$ *and* $x \in \Delta'$:

$$\boxed{\left|f(x) - \sum_{n=0}^{N-1} a_n x^n\right| \leq C_{\Delta'}(N!)^s A_{\Delta'}^N\,|x|^N} \tag{1.13}$$

A series which is the s-Gevrey asymptotic expansion of a function is said to be an s-Gevrey asymptotic series.

The Borel-Ritt theorem below (Thm. 1.3.1) will show us that any s-Gevrey series is an s-Gevrey asymptotic series and vice-versa after proposition 1.2.10.

Notation 1.2.8 We denote by $\overline{\mathscr{A}}_s(\mathbf{\Delta})$ the set of functions admitting an s-Gevrey asymptotic expansion at 0 on $\mathbf{\Delta}$.

Given an open arc I of S^1, let $\mathbf{\Delta}_I(R)$ denote the sector based on I with radius R. Since there is no possible confusion, we also denote the set of germs of functions admitting an s-Gevrey asymptotic expansion on a sector based on I by

$$\overline{\mathscr{A}}_s(I) = \varinjlim_{R\to 0} \overline{\mathscr{A}}_s\big(\mathbf{\Delta}_I(R)\big).$$

In other words, a germ belongs to $\overline{\mathscr{A}}_s(I)$ if it can be represented by a function $f(x)$ which belongs to $\overline{\mathscr{A}}_s\big(\mathbf{\Delta}_I(R)\big)$ for a certain radius R.

The constants $C_{\mathbf{\Delta}'}$ and $A_{\mathbf{\Delta}'}$ may depend on $\mathbf{\Delta}'$; they do not depend on $N \in \mathbb{N}^*$ and $x \in \mathbf{\Delta}'$. Gevrey asymptotics differ from Poincaré asymptotics by the fact that the dependence of the constant $C_{N,\mathbf{\Delta}'}$ on N (cf. Def. 1.1.1, p. 2) has to be of Gevrey type.

Comments 1.2.9 (On the Examples of Section 1.1.2, p. 3)

The calculations in section 1.1.2, p. 3 show the following Gevrey asymptotic properties:

▷ The Euler function $\mathrm{E}(x)$ is 1-Gevrey asymptotic to the Euler series $\widetilde{\mathrm{E}}(x)$ on any (germ at 0 of) half-plane bisected by a line d_θ with argument θ such that $-\pi < \theta < +\pi$. It is then 1-Gevrey asymptotic to $\widetilde{\mathrm{E}}(x)$ at 0 on the full sector $-3\pi/2 < \arg(x) < +3\pi/2$.

▷ Up to an exponential factor the exponential integral has the same properties on germs of half-planes at infinity.

▷ The generalized hypergeometric series $\widetilde{g}(z)$ of example 1.1.6, p. 7 is 2-Gevrey and we stated that the function $g(z)$ is asymptotic in the rough sense of Poincaré to $\widetilde{g}(z)$ on the half-plane $\Re(z) > 0$ at infinity. We will see (cf. Com. 5.2.6, p. 143) that the function $g(z)$ is actually $1/2$-Gevrey asymptotic to $\widetilde{g}(z)$. Our computations in Sect. 1.1.2, p. 3 do not allow us to state yet such a fact since we did not determine how the constant C depends on N.

▷ The function $h(z)$ of example 1.1.7, p. 8 was proved to be 1-Gevrey asymptotic to the series $\widetilde{h}(z)$ (cf. Estim. (1.10), p. 10) on the right half-plane $\Re(z) > 0$ at infinity.

▷ The function $\ell(z)$ of example 1.1.8, p. 10 satisfies the same estimate (1.10) as $h(z)$ on the sector $\mathbf{\Delta}' = \{-\pi/2 + \delta < \arg(z) < \pi/2 - \delta\}$, for $(0 < \delta < \pi/2)$, with a constant C which can be chosen equal to $C = \mathrm{e}^{-1/2+\mathrm{e}^{1/2}} N! 2^N/(\sin\delta)^{N+1}$. The function $\ell(z)$ is then 1-Gevrey asymptotic to the series $\widetilde{\ell}(z)$ on the right half-plane $\Re(z) > 0$ at infinity.

Proposition 1.2.10 *An s-Gevrey asymptotic series is an s-Gevrey series.*

Proof. Suppose the series $\sum_{n\geq 0} a_n x^n$ is the s-Gevrey asymptotic series of a function f on $\mathbf{\Delta}$. For all N, the result follows from condition (1.13), p. 16 applied twice to

$$a_N x^N = \Big(f(x) - \sum_{n=0}^{N-1} a_n x^n\Big) - \Big(f(x) - \sum_{n=0}^{N} a_n x^n\Big). \qquad \square$$

Proposition 1.2.11 *A function $f \in \overline{\mathscr{A}}(\Delta)$ belongs to $\overline{\mathscr{A}}_s(\Delta)$ if and only if for all $\Delta' \Subset \Delta$ there exist constants $C'_{\Delta'} > 0$ and $A'_{\Delta'} > 0$ such that the following estimate holds for all $N \in \mathbb{N}$ and $x \in \Delta'$:*

$$\boxed{\left|\frac{\mathrm{d}^N f}{\mathrm{d}x^N}(x)\right| \leq C'_{\Delta'}(N!)^{s+1} A'^{N}_{\Delta'}} \tag{1.14}$$

Proof. Let f belong to $\overline{\mathscr{A}}(\Delta)$.

▷ *Prove that condition* (1.14) *implies condition* (1.13), p. 16. — Like in the proof of Prop. 1.1.11, p. 12, write Taylor's formula with integral remainder:

$$f(x) - \sum_{n=0}^{N-1} a_n x^n = \int_0^x \tfrac{1}{(N-1)!}(x-t)^{N-1} f^{(N)}(t)\,\mathrm{d}t = -\tfrac{1}{N!}\int_0^x f^{(N)}(t)\,\mathrm{d}(x-t)^N$$

and conclude that

$$\left|f(x) - \sum_{n=0}^{N-1} a_n x^n\right| \leq \frac{1}{N!} \sup_{t\in\Delta'} \left|\frac{\mathrm{d}^N f}{\mathrm{d}x^N}(t)\right| \cdot |x|^N \leq C'_{\Delta'}(N!)^s A'^{N}_{\Delta'} |x|^N.$$

▷ *Prove that condition* (1.13) *implies condition* (1.14). — Like in the proof of Prop. 1.1.9, p. 11, attach to any $x \in \Delta'$ a circle γ_x centered at x with radius $|x|\delta$, the constant δ being chosen so small that γ_x be contained in Δ and apply Cauchy's integral formula:

$$\frac{\mathrm{d}^N f}{\mathrm{d}x^N}(x) = \frac{N!}{2\pi\mathrm{i}} \int_{\gamma_x} f(t) \frac{\mathrm{d}t}{(t-x)^{N+1}} = \frac{N!}{2\pi\mathrm{i}} \int_{\gamma_x} \Big(f(t) - \sum_{n=0}^{N-1} a_n t^n\Big) \frac{\mathrm{d}t}{(t-x)^{N+1}}$$

since the Nth derivative of a polynomial of degree $N-1$ is 0. Hence,

$$\begin{aligned}\left|\frac{\mathrm{d}^N f}{\mathrm{d}x^N}(x)\right| &\leq \frac{N!}{2\pi} C_{\Delta'}(N!)^s A^N_{\Delta'} \left|\int_{\gamma_x} \frac{|t|^N}{|t-x|^{N+1}}\,\mathrm{d}t\right| \\ &\leq \frac{1}{2\pi} C_{\Delta'}(N!)^{s+1} A^N_{\Delta'} \frac{|x|^N (1+\delta)^N}{|x|^{N+1}\delta^{N+1}} 2\pi\delta|x| \\ &= C_{\Delta'}(N!)^{s+1} A'^{N}_{\Delta'} \quad \text{with} \quad A'_{\Delta'} = A_{\Delta'}\Big(1 + \frac{1}{\delta}\Big).\end{aligned}$$

□

Proposition 1.2.12 (Differential Algebra and Taylor Map)

The set $\overline{\mathscr{A}}_s(\Delta)$ is a differential $\mathbb{C}$-algebra and the Taylor map T_Δ restricted to $\overline{\mathscr{A}}_s(\Delta)$ induces a morphism of differential algebras

$$T = T_{s,\Delta} : \overline{\mathscr{A}}_s(\Delta) \longrightarrow \mathbb{C}[[x]]_s$$

with values in the algebra of s-Gevrey series.

Proof. Let $\mathit{\Delta}' \Subset \mathit{\Delta}$. Suppose f and g belong to $\overline{\mathscr{A}}_s(\mathit{\Delta})$ and satisfy on $\mathit{\Delta}'$

$$\left|\frac{\mathrm{d}^N f}{\mathrm{d}x^N}(x)\right| \leq C(N!)^{s+1}A^N \quad \text{and} \quad \left|\frac{\mathrm{d}^N g}{\mathrm{d}x^N}(x)\right| \leq C'(N!)^{s+1}A'^N.$$

The product fg belongs to $\overline{\mathscr{A}}(\mathit{\Delta})$ (cf. Prop. 1.1.9, p. 11) and its derivatives satisfy

$$\left|\frac{\mathrm{d}^N (fg)}{\mathrm{d}x^N}(x)\right| \leq \sum_{p=0}^{N} C_N^p \frac{\mathrm{d}^p f}{\mathrm{d}x^p}(x)\frac{\mathrm{d}^{N-p} g}{\mathrm{d}x^{N-p}}(x) \leq CC'(N!)^{s+1}(A+A')^N.$$

The fact that the range $T_{s,\mathit{\Delta}}\big(\overline{\mathscr{A}}_s(\mathit{\Delta})\big)$ be included in $\mathbb{C}[[x]]_s$ follows from proposition 1.2.10, p. 17. □

Observe now the effect of a change of variable $x = t^r$, $r \in \mathbb{N}^*$. Clearly, if a series $\widetilde{f}(x)$ is Gevrey of order s (level k) then the series $\widetilde{f}(t^r)$ is Gevrey of order s/r (level kr). What about the asymptotics?

Let $\mathit{\Delta} =]\alpha,\beta[\times]0,R[$ be a sector in $\widetilde{\mathbb{C}}$ (the α and β directions are not given modulo 2π) and let $\mathit{\Delta}_{/r} =]\alpha/r,\beta/r[\times]0,R^{1/r}[$ so that as the variable t runs over $\mathit{\Delta}_{/r}$ the variable $x = t^r$ runs over $\mathit{\Delta}$. From definition 1.2.7, p. 16 we can state:

Proposition 1.2.13 (Gevrey Asymptotics in an Extension of the Variable)
The following two assertions are equivalent:

(i) *the function $f(x)$ is s-Gevrey asymptotic to the series $\widetilde{f}(x)$ on $\mathit{\Delta}$;*

(ii) *the function $g(t) = f(t^r)$ is s/r-Gevrey asymptotic to $\widetilde{g}(t) = \widetilde{f}(t^r)$ on $\mathit{\Delta}_{/r}$.*

Way back, that is, given an s'-Gevrey series $\widetilde{g}(t)$, the series $\widetilde{f}(x) = \widetilde{g}(x^{1/r})$ exhibits, in general, fractional powers of x. To keep working with series of integer powers of x one may use rank reduction as follows [Lod01]. One can uniquely decompose the series $\widetilde{g}(t)$ as a sum

$$\widetilde{g}(t) = \sum_{j=0}^{r-1} t^j \widetilde{g}_j(t^r)$$

where the terms $\widetilde{g}_j(t^r)$ are entire power series in t^r. Set $\omega = e^{2\pi i/r}$ and $x = t^r$. The series $\widetilde{g}_j(x)$ are given, for $j = 0,\ldots,r-1$, by the relations

$$rt^j \widetilde{g}_j(t^r) = \sum_{\ell=0}^{r-1} \omega^{\ell(r-j)} \widetilde{g}(\omega^\ell t).$$

For $j = 0,\ldots,r-1$, let $\mathit{\Delta}^j_{/r}$ denote the sector

$$\mathit{\Delta}^j_{/r} =](\alpha+2j\pi)/r,(\beta+2j\pi)/r[\times]0,R^{1/r}[$$

so that as t runs through $\mathit{\Delta}_{/r} = \mathit{\Delta}^0_{/r}$ then $\omega^j t$ runs through $\mathit{\Delta}^j_{/r}$ and $x = t^r$ runs through $\mathit{\Delta}$.

From the previous relations and proposition 1.2.13 we can state:

Corollary 1.2.14 (Gevrey Asymptotics and Rank Reduction)
The following two assertions are equivalent:

(i) *for* $\ell = 0, \dots, r-1$ *the series* $\widetilde{g}(t)$ *is an* s'*-Gevrey asymptotic series on* $\mathcal{S}_{\ell/r}$ *(in the variable* t*);*

(ii) *for* $j = 0, \dots, r-1$ *the r-rank reduced series* $\widetilde{g}_j(x)$ *is an* $s'r$*-Gevrey asymptotic series on* $\mathcal{S}$ *(in the variable* $x = t^r$*).*

These results allows us to limit the study of Gevrey asymptotics to small values of s ($s \le s_0$) or to large ones ($s \ge s_1$) at convenience.

1.2.3 Flat s-Gevrey Asymptotic Functions

In this section we address the following problem:

To characterize the functions that are
both s-Gevrey asymptotic and flat on a given sector $\mathcal{S}$.

To this end, we first introduce the notion of exponential flatness.

Definition 1.2.15 *A function* f *is said to be* exponentially flat of order k (or k-exponentially flat) on a sector $\mathcal{S}$ *if, for any proper subsector* $\mathcal{S}' \Subset \mathcal{S}$ *of* $\mathcal{S}$*, there exist constants* K *and* $A > 0$ *such that the following estimate holds for all* $x \in \mathcal{S}'$:

$$\boxed{|f(x)| \le K \exp\left(-A/|x|^k\right)} \tag{1.15}$$

The constants K and A may depend on $\mathcal{S}'$.

Notation 1.2.16 We denote the set of k-exponentially flat functions on $\mathcal{S}$ by

$$\overline{\mathcal{A}}^{\le -k}(\mathcal{S}).$$

Proposition 1.2.17 *Let* $\mathcal{S}$ *be an open sector. The functions which are s-Gevrey asymptotically flat on* $\mathcal{S}$ *are the k-exponentially flat functions,* i.e.,

$$\boxed{\overline{\mathcal{A}}_s(\mathcal{S}) \cap \overline{\mathcal{A}}^{<0}(\mathcal{S}) = \overline{\mathcal{A}}^{\le -k}(\mathcal{S})} \qquad \textit{(recall } s = 1/k\textit{)}.$$

Proof. ▷ *Let* $f \in \overline{\mathcal{A}}_s(\mathcal{S}) \cap \overline{\mathcal{A}}^{<0}(\mathcal{S})$ *and prove that* $f \in \overline{\mathcal{A}}^{\le -k}(\mathcal{S})$. From the hypothesis we know that, for all $\mathcal{S}' \Subset \mathcal{S}$, there exist $A > 0, C > 0$ such that the estimate

$$|f(x)| \le C N^{N/k} (A|x|)^N = C \exp\left(\frac{N}{k} \ln\left(N (A|x|)^k\right)\right)$$

holds for all N and all $x \in \mathcal{S}'$ (in condition (1.13) one can replace $N!^{1/k}$ by $N^{N/k}$ with, however, new constants A and C).

For x fixed, we look for a lower bound of the right-hand side of this estimate as N runs over $\mathbb{N}$. The derivative $\varphi'(N) = \ln\big(N(A|x|)^k\big) + 1$ of the function $\varphi(N) = N\ln\big(N(A|x|)^k\big)$ seen as a function of a real variable $N > 0$ vanishes at $N_0 = 1/\big(e(A|x|^k)\big)$ and φ reaches its minimal value $\varphi(N_0) = -1/\big(e(A|x|)^k\big)$ at that point. In general, N_0 is not an integer and the minimal value of $\varphi(N)$ on integer values of N is larger than $\varphi(N_0)$. However, taking into account the monotonicity of φ from N_0 to N_0+1 and the fact that there is at least one integer number between these two values we can assert that

$$\inf_{N\in\mathbb{N}} \varphi(N) \leq \varphi(N_0+1) = \varphi(N_0)\Big(1+\frac{1}{N_0}B(1-(1+N_0)\ln\big(1+\frac{1}{N_0}\big)\Big)\Big).$$

Substituting $1/\big(e(A|x|)^k\big)$ for N_0 in the last factor we obtain a function $\psi(x)$ which is bounded on $\mathcal{S}'$. It follows that $|f(x)| \leq C'\exp\big(-a/|x|^k\big)$ with $a = 1/(keA^k) > 0$ independent of $x \in \mathcal{S}'$. This proves that f belongs to $\overline{\mathcal{A}}^{\leq -k}(\mathcal{S})$.

▷ *Let $f \in \overline{\mathcal{A}}^{\leq -k}(\mathcal{S})$ and prove that $f \in \overline{\mathcal{A}}_s(\mathcal{S}) \cap \overline{\mathcal{A}}^{<0}(\mathcal{S})$.* The hypothesis is now: for all $\mathcal{S}' \Subset \mathcal{S}$, there exist $A > 0, C > 0$ such that an estimate

$$|f(x)| \leq C\exp\Big(-\frac{A}{|x|^k}\Big)$$

holds for all $x \in \mathcal{S}'$. Hence, $|f(x)| \cdot |x|^{-N} \leq C\exp\big(-A/|x|^k\big)|x|^{-N}$ for all N.

For N fixed, look for an upper bound of the right-hand side of this estimate as $|x|$ runs over $\mathbb{R}^+$. Let $\psi(|x|) = \exp\big(-A/|x|^k\big)|x|^{-N}$. Its logarithmic derivative

$$\frac{\psi'(|x|)}{\psi(|x|)} = -\frac{N}{|x|} + \frac{Ak}{|x|^{k+1}}$$

vanishes for $Ak/|x|^k = N$ and ψ reaches its maximal value at that point. It follows that $\max_{|x|>0}\psi(|x|) = \exp\big(-N/k\big)\big(N/(Ak)\big)^{N/k}$ and there exists constants $a = (eAk)^{-1/k}$ and $C > 0$ such that, for all $N \in \mathbb{N}$ and $x \in \mathcal{S}'$ the function f satisfies $|f(x)| \leq CN^{N/k}(a|x|)^N$. Hence, f belongs to $\overline{\mathcal{A}}_s(\mathcal{S}) \cap \overline{\mathcal{A}}^{<0}(\mathcal{S})$. □

1.3 The Borel-Ritt Theorem

With any asymptotic function $f \in \overline{\mathcal{A}}(\mathcal{S})$ over a sector $\mathcal{S}$ the Taylor map $T_{\mathcal{S}}$ associates a formal series $\widetilde{f} = T_{\mathcal{S}}(f)$. We address now the converse problem: is any formal series the Taylor series of an asymptotic function over a given sector $\mathcal{S}$? The theorem below states that the answer is yes for any open sector $\mathcal{S}$ with finite radius in $\mathbb{C}^*$ or $\widetilde{\mathbb{C}}$ in Poincaré asymptotics. In case the series is s-Gevrey an s-Gevrey asymptotic function always exists when the opening of the sector Δ is small enough and, we will see on examples that it might not exist when the opening of Δ is large (cf. Com. 5.1.5, p. 137). Notice that the Taylor series of a function $f \in \overline{\mathscr{A}}(\mathbb{C}^*)$ is necessarily convergent by the removable singularity theorem of Riemann. And thus, when the asymptotic series is divergent, the sector Δ cannot be a full neighborhood of 0 in $\mathbb{C}^*$.

Theorem 1.3.1 (Borel-Ritt) *Let $\Delta \neq \mathbb{C}^*$ be an open sector of $\mathbb{C}^*$ or of the Riemann surface of logarithm $\widetilde{\mathbb{C}}$ with finite radius R.*

(i) (Poincaré asymptotics) *The Taylor map $T_\Delta : \overline{\mathscr{A}}(\Delta) \to \mathbb{C}[[x]]$ is onto.*

(ii) (Gevrey asymptotics) *Suppose Δ has opening $|\Delta| \leq \pi/k$. Then, the Taylor map $T_{s,\Delta} : \overline{\mathscr{A}}_s(\Delta) \to \mathbb{C}[[x]]_s$ is onto. Recall $s = 1/k$.*

Proof. (i) *Poincaré asymptotics.* — Various proofs exist. The one presented here can be found in [Mal95]. For simplicity, we begin with the case of a sector in $\mathbb{C}^*$.

▷ Case when Δ lies in $\mathbb{C}^*$. Modulo rotation it is sufficient to consider the case when $\Delta = \Delta_{-\pi,+\pi}(R)$ is the disc of radius R slit on the negative real axis.

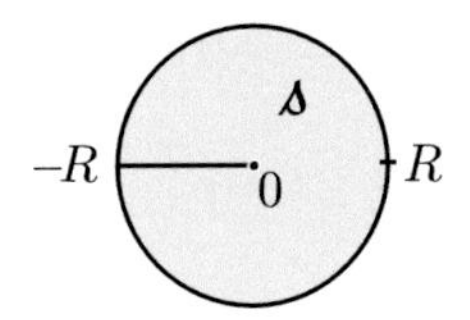

Fig. 1.6

Given any series $\sum_{n\geq 0} a_n x^n \in \mathbb{C}[[x]]$ we look for a function $f \in \overline{\mathscr{A}}(\Delta)$ with Taylor series $T_\Delta f = \sum_{n\geq 0} a_n x^n$. To this end, one introduces functions $\beta_n(x) \in \mathscr{O}(\Delta)$ satisfying the two conditions

$$(1): \quad \sum_{n\geq 0} a_n \beta_n(x) x^n \in \mathscr{O}(\Delta) \quad \text{and} \quad (2): \quad T_\Delta \beta_n(x) \equiv 1 \ \text{ for all } \ n \geq 0.$$

Such functions exist: consider, for instance, the functions $\beta_0 \equiv 1$ and, for $n \geq 1$, $\beta_n(x) = 1 - \exp\big(-b_n/\sqrt{x}\big)$ with a positive constant b_n and $\sqrt{x}$ the principal determination of the square root.

In view to condition (1), observe that since $1 - \mathrm{e}^z = -\int_0^z \mathrm{e}^t\,\mathrm{d}t$ then $|1 - \mathrm{e}^z| < |z|$ for $\Re(z) < 0$. This implies $|\beta_n(x)| \leq b_n/\sqrt{|x|}$ for all $x \in \Delta$ and $n \geq 1$ and then,

$$\big|a_n \beta_n(x) x^n\big| \leq |a_n| b_n |x|^{n-1/2} \leq |a_n| b_n R^{n-1/2}.$$

Now, choose b_n such that the series $\sum_{n\geq 1} |a_n| b_n R^{n-1/2}$ be convergent. Then, the series $\sum_{n\geq 0} a_n \beta_n(x) x^n$ converges normally on Δ and its sum $f(x) = \sum_{n\geq 0} a_n \beta_n(x) x^n$ is holomorphic on Δ.

To prove condition (2), consider any proper sub-sector $\Delta' \Subset \Delta$ of Δ and $x \in \Delta'$.

Then, for any $N \geq 1$, we can write

$$\Big|f(x) - \sum_{n=0}^{N-1} a_n x^n\Big| \leq \Big|\sum_{n=0}^{N-1} a_n(\beta_n(x)-1)x^n\Big| + |x|^N \sum_{n\geq N} \big|a_n\beta_n(x)x^{n-N}\big|.$$

The first summand is a finite sum of terms all asymptotic to 0 and then, is bounded by $C'|x|^N$, for a convenient positive constant C'.

The second summand is bounded by

$$|x|^N\Big(2|a_N| + \sum_{n\geq N+1} |a_n| b_n R^{n-1/2-N}\Big).$$

Choosing $C = C' + 2|a_N| + \sum_{n\geq N+1} |a_n| b_n R^{n-1/2N}$ provides a positive constant C (independent of x but depending on N and Δ') such that

$$\Big|f(x) - \sum_{n=0}^{N} a_n x^n\Big| \leq C|x|^N \quad \text{for all } x \in \Delta'.$$

This ends the proof in this case.

▷ General case when Δ lies in $\widetilde{\mathbb{C}}^*$. — It is again sufficient to consider the case of a sector of the form $\Delta = \{x \in \widetilde{\mathbb{C}}^* \,;\, |\arg(x)| < k\pi, 0 < |x| < R\}$ where $k \in \mathbb{N}^*$. The same proof can be applied after replacing $\sqrt{x}$ by a convenient power x^α of x so that $\Re(x^\alpha)$ is positive for all x in Δ and taking $b_n \equiv 1$ for all $n \leq \alpha$.

(ii) *Gevrey asymptotics*

Let $\widetilde{f}(x) \in \mathbb{C}[[x]]_s$ be an s-Gevrey series which, up to a polynomial, we may assume to be of the form $\widetilde{f}(x) = \sum_{n\geq k} a_n x^n$. It is sufficient to consider a sector Δ of opening π/k (as always, $k = 1/s$) and by means of a rotation, we can then assume that Δ is an open sector bisected by the $\theta = 0$ direction with opening π/k; we denote by R its radius. We must find a function $f \in \overline{\mathcal{A}}_s(\Delta)$, s-Gevrey asymptotic to $\widetilde{f}$ over Δ.

The proof used here is based on the Borel and the Laplace transforms which will be at the core of the Borel-Laplace summation in section 5.3, p. 144.

Since $\widetilde{f}(x)$ is an s-Gevrey series (cf. Def. 1.2.1, p. 14) its k-Borel transform[3]

$$\widehat{f}(\xi) = \sum_{n\geq k} \frac{a_n}{\Gamma(n/k)} \xi^{n-k}$$

is a convergent series[4] and we denote by $\varphi(\xi)$ its sum.

The adequate Laplace transform to "invert" the k-Borel transform (as a function $\varphi(\xi)$, not as a series $\widehat{f}(\xi)$) in the $\theta = 0$ direction would be the k-Laplace

[3] See Sect. 5.3.1, p. 145. The k-Borel transform of a series $\sum_{n\geq k} a_n x^n$ is the usual Borel transform of the series $\sum_{n\geq k} a_n X^{n/k}$ with respect to the variable $X = x^k$ and expressed in the variable $\xi = \zeta^s$.

[4] Although, when k is not an integer, the series $\widehat{f}(\xi)$ is not a series in integer powers of ξ it becomes so after factoring by ξ^{-k}. We mean here that the power series $\xi^k\widehat{f}(\xi)$ is convergent.

transform

$$\mathscr{L}_k(\varphi)(x) = \int_0^{+\infty} \phi(\zeta)\,\mathrm{e}^{-\zeta/x^k}\,\mathrm{d}\zeta$$

where $\zeta = \xi^k$ and $\phi(\zeta) = \varphi(\zeta^{1/k})$. However, although the series $\widehat{f}(\xi)$ is convergent, its sum $\varphi(\xi)$ cannot be analytically continued along $\mathbb{R}^+$ up to infinity in general.

To circumvent this problem we choose $b > 0$ belonging to the disc of convergence of $\widehat{f}(\xi)$ and we consider a truncated k-Laplace transform

$$\boxed{f^b(x) = \int_0^{b^k} \phi(\zeta)\,\mathrm{e}^{-\zeta/x^k}\,\mathrm{d}\zeta} \tag{1.16}$$

instead of the full Laplace transform $\mathscr{L}_k(\varphi)(x)$. Lemma 1.3.2 below shows that the function $f = f^b$ answers the question. $\square$

Lemma 1.3.2 (Truncated Laplace Transform) *Let $\widetilde{f}(x)$ be an s-Gevrey series and let $\varphi(\xi)$ denote the sum of its k-Borel transform $\widehat{f}(\xi)$ (with $s = 1/k$ as usually). Let $b > 0$ belong to the disc of convergence of $\widehat{f}(\xi)$, and $\boldsymbol{\Delta}$ be an open sector bisected by $\theta = 0$ with opening π/k.*

The truncated k-Laplace transform $f^b(x)$ of $\varphi(\xi)$ in the $\theta = 0$ direction given by formula (1.16) *is s-Gevrey asymptotic to $\widetilde{f}(x)$ on $\boldsymbol{\Delta}$.*

Proof. Given $0 < \delta < \pi/2$ and $R' < R$ where R denotes the radius of $\boldsymbol{\Delta}$, consider the proper sub-sector of $\boldsymbol{\Delta}$ defined by $\boldsymbol{\Delta}_\delta = \{x;\, |\arg(x)| < \pi/(2k) - \delta/k \text{ and } |x| < R'\}$. For $x \in \boldsymbol{\Delta}_\delta$ we can write

$$f^b(x) - \sum_{n=k}^{N-1} a_n x^n = \int_0^{b^k} \sum_{n\geq k} \frac{a_n}{\Gamma(n/k)} \zeta^{(n/k)-1}\,\mathrm{e}^{-\zeta/x^k}\,\mathrm{d}\zeta - \sum_{n=k}^{N-1} \frac{a_n}{\Gamma(n/k)} \int_0^{+\infty} \zeta^{(n/k)-1}\,\mathrm{e}^{-\zeta/x^k}\,\mathrm{d}\zeta.$$

Since $\Re(x^k) > 0$ then, $\left|\zeta^{(n/k)-1}\,\mathrm{e}^{-\zeta/x^k}\right| \leq b^{n-k}$ for all $\zeta \in [0, b^k]$. It results that the series $\sum_{n\geq k} \frac{a_n}{\Gamma(n/k)} \zeta^{(n/k)-1}\,\mathrm{e}^{-\zeta/x^k}$ converges normally on $[0, b^k]$ and we can permute sum and integral.

Hence, we can write

$$f^b(x) - \sum_{n=k}^{N-1} a_n x^n = \sum_{n\geq N} \frac{a_n}{\Gamma(n/k)} \int_0^{b^k} \zeta^{(n/k)-1}\,\mathrm{e}^{-\zeta/x^k}\,\mathrm{d}\zeta - \sum_{n=k}^{N-1} \frac{a_n}{\Gamma(n/k)} \int_{b^k}^{+\infty} \zeta^{(n/k)-1}\,\mathrm{e}^{-\zeta/x^k}\,\mathrm{d}\zeta.$$

However, $\left|\zeta/b^k\right|^{(n/k)-1} \leq \left|\zeta/b^k\right|^{(N/k)-1}$ both when $|\zeta| \leq b^k$ and $n \geq N$ and when $|\zeta| \geq b^k$ and $n < N$ and then,

$$\begin{aligned}\Big|f^b(x) - \sum_{n=k}^{N-1} a_n x^n\Big| &\leq \sum_{n\geq N} \frac{|a_n|}{\Gamma(n/k)} \int_0^{b^k} b^{n-N} |\zeta|^{(N/k)-1} \mathrm{e}^{-\zeta \Re(1/x^k)}\, \mathrm{d}\zeta \\ &\quad + \sum_{n=k}^{N-1} \frac{|a_n|}{\Gamma(n/k)} \int_{b^k}^{+\infty} \text{idem} \\ &\leq \sum_{n\geq k} \frac{|a_n|}{\Gamma(n/k)} b^{n-N} \int_0^{+\infty} |\zeta|^{(N/k)-1} \mathrm{e}^{-\zeta \sin(\delta)/|x|^k}\, d\zeta \\ &= \sum_{n\geq k} \frac{|a_n|}{\Gamma(n/k)} b^{n-N} \frac{|x|^N}{(\sin\delta)^{N/k}} \int_0^{+\infty} u^{(N/k)-1} \mathrm{e}^{-u}\, \mathrm{d}u \\ &= \sum_{n\geq k} \frac{|a_n|}{\Gamma(n/k)} b^{n-N} \frac{|x|^N}{(\sin\delta)^{N/k}} \Gamma(N/k) = C\Gamma(N/k) A^N |x|^N\end{aligned}$$

where $A = \frac{1}{b(\sin\delta)^{1/k}}$ and $C = \sum_{n\geq k} \frac{|a_n|}{\Gamma(n/k)} b^n < +\infty$. The constants A and C depend on Δ_δ and on the choice of b but are independent of x. This achieves the proof. □

Comment 1.3.3 (On the Euler Series (Exa. 1.1.4, p. 4))

The proof of the Borel-Ritt theorem provides infinitely many functions asymptotic to the Euler series $\widetilde{\mathrm{E}}(x) = \sum_{n\geq 0}(-1)^n n!\, x^{n+1}$ at 0 on the sector $\Delta = \{x\,;\, |\arg(x)| < 3\pi/2\}$. For instance, the following family provides infinitely many such functions:

$$F_a(x) = \sum_{n\geq 0} (-1)^n n! \Big(1 - \exp(-a/(n!^2 x^{1/3}))\Big) x^{n+1}, \quad a > 0.$$

We saw in example 1.1.4 that the Euler function $\mathrm{E}(x) = \int_0^{+\infty} \frac{\mathrm{e}^{-\xi/x}}{1+\xi}\, \mathrm{d}\xi$ is both solution of the Euler equation and asymptotic to the Euler series on Δ.

We claim that it is the unique function with these properties. Indeed, suppose E_1 be another such function. Then, the difference $\mathrm{E}(x) - \mathrm{E}_1(x)$ would be both asymptotic to the null series 0 on Δ and solution of the homogeneous associated equation $x^2 y' + y = 0$. However, the equation $x^2 y' + y = 0$ admits no such solution on Δ but 0. Hence, $\mathrm{E} = \mathrm{E}_1$ and the infinitely many functions given by the proof of the Borel-Ritt theorem do not satisfy the Euler equation in general.

Taking into account propositions 1.1.9, p. 11, 1.2.12, p. 18 and 1.2.17, p. 20 we can reformulate the Borel-Ritt theorem 1.3.1, p. 22 as follows.

Corollary 1.3.4

The set $\overline{\mathscr{A}}^{<0}(\Delta)$ of flat functions on Δ and the set $\overline{\mathscr{A}}^{\leq -k}(\Delta)$ of k-exponentially flat functions on Δ are differential ideals of $\overline{\mathscr{A}}(\Delta)$ and $\overline{\mathscr{A}}_s(\Delta)$ respectively. The sequences

$$0 \to \overline{\mathscr{A}}^{<0}(\Delta) \longrightarrow \overline{\mathscr{A}}(\Delta) \xrightarrow{T_\Delta} \mathbb{C}[[x]] \to 0$$

and, when $|\boldsymbol{\Delta}| \leq \pi/k$,

$$0 \to \overline{\mathscr{A}}^{\leq -k}(\boldsymbol{\Delta}) \longrightarrow \overline{\mathscr{A}}_s(\boldsymbol{\Delta}) \xrightarrow{T_{s,\boldsymbol{\Delta}}} \mathbb{C}[[x]]_s \to 0$$

are exact sequences of morphisms of differential algebras.

The Borel-Ritt theorem implies the classical Borel theorem in the real case providing thus a new proof of it.

Corollary 1.3.5 (Classical Borel Theorem)

Any formal power series $\sum_{n\geq 0} a_n x^n \in \mathbb{C}[[x]]$ *is the Taylor series at 0 of a* $\mathscr{C}^\infty$*-function of a real variable x.*

Proof. Apply the Borel-Ritt theorem on a sector $\boldsymbol{\Delta}'$ containing $\mathbb{R}^+$ and on a sector $\boldsymbol{\Delta}''$ containing $\mathbb{R}^-$. The two functions so obtained glue together at 0 into a $\mathscr{C}^\infty$-function in a neighborhood of 0 in $\mathbb{R}$. □

1.4 The Cauchy-Heine Theorem

In this section we are given:

▷ a sector $\dot{\boldsymbol{\Delta}} = \boldsymbol{\Delta}_{\alpha,\beta}(R)$ with vertex 0 in $\mathbb{C}^*$;

▷ a point x_0 in $\dot{\boldsymbol{\Delta}}$ and the straight path $\gamma =]0, x_0]$ in $\dot{\boldsymbol{\Delta}}$;

▷ a function $\varphi \in \overline{\mathscr{A}}^{<0}(\dot{\boldsymbol{\Delta}})$ flat at 0 on $\dot{\boldsymbol{\Delta}}$.

Definition 1.4.1 *One defines the* Cauchy-Heine integral *associated with* φ *and* x_0*, to be the function*

$$\boxed{f(x) = \frac{1}{2\pi \mathrm{i}} \int_\gamma \frac{\varphi(t)}{t-x}\,\mathrm{d}t}$$

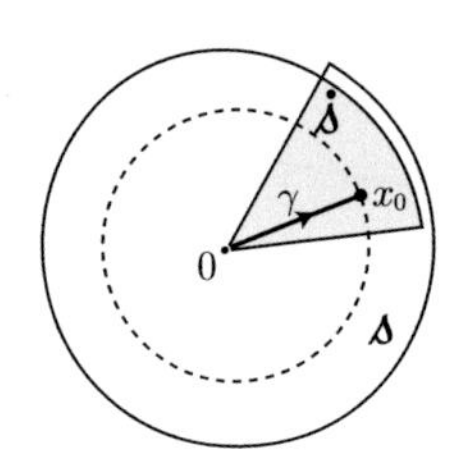

Fig. 1.7

Denote by:

▷ $\boldsymbol{\Delta} = \boldsymbol{\Delta}_{\alpha,\beta+2\pi}(R)$ a sector with vertex 0 in the Riemann surface of log with overlap $\dot{\boldsymbol{\Delta}}$;

▷ θ_0 the argument of x_0 satisfying $\alpha < \theta_0 < \beta$;

▷ $\mathscr{D}_\gamma = \boldsymbol{\Delta}_{\theta_0,\theta_0+2\pi}(|x_0|)$ the disc of radius $|x_0|$ slit along γ;

▷ $\dot{\boldsymbol{\Delta}}' = \dot{\boldsymbol{\Delta}} \cap \{|x| < |x_0|\} = \boldsymbol{\Delta}_{\alpha,\beta}(|x_0|)$;

▷ $\boldsymbol{\Delta}' = \boldsymbol{\Delta} \cap \{|x| < |x_0|\} = \boldsymbol{\Delta}_{\alpha,\beta+2\pi}(|x_0|)$.

The Cauchy-Heine integral determines a well-defined and analytic function f on $\mathscr{D}_\gamma$. By Cauchy's theorem, Cauchy-Heine integrals associated with different points x_0 and x_1 in $\dot{\boldsymbol{\Delta}}$ differ by $\frac{1}{2\pi\mathrm{i}}\int_{\widehat{x_0x_1}} \frac{\varphi(t)}{t-x}\,\mathrm{d}t$, that is, by an analytic function on a neighborhood of 0.

Theorem 1.4.2 (Cauchy-Heine) *With notations and conditions as before and especially, φ flat on $\dot{\boldsymbol{\Delta}}$, the Cauchy-Heine integral $f(x) = \frac{1}{2\pi\mathrm{i}}\int_\gamma \frac{\varphi(t)}{t-x}\,\mathrm{d}t$ has the following properties:*

1. *The function f can be analytically continued from $\mathscr{D}_\gamma$ to $\boldsymbol{\Delta}'$; we also use the term Cauchy-Heine integral when referring to this analytic continuation which we denote again by f.*
2. *The function f belongs to $\overline{\mathscr{A}}(\boldsymbol{\Delta}')$.*
3. *Its Taylor series at 0 on $\boldsymbol{\Delta}'$ reads $T_{\boldsymbol{\Delta}'}f(x) = \sum_{n\geq 0} a_n x^n$ with*

$$a_n = \frac{1}{2\pi\mathrm{i}}\int_\gamma \frac{\varphi(t)}{t^{n+1}}\,\mathrm{d}t.$$

4. *Its variation* $\mathrm{var} f(x) = f(x) - f(x\mathrm{e}^{2\pi\mathrm{i}})$ *is equal to $\varphi(x)$ for all $x \in \dot{\boldsymbol{\Delta}}'$.*
5. *If, in addition, φ belongs to $\overline{\mathscr{A}}^{\leq -k}(\dot{\boldsymbol{\Delta}})$ then, f belongs to $\overline{\mathscr{A}}_s(\boldsymbol{\Delta}')$ with the above Taylor series, that is, if φ is k-exponentially flat on $\dot{\boldsymbol{\Delta}}$ then, f is s-Gevrey asymptotic to the above series $\sum_{n\geq 0} a_n x^n$ on $\boldsymbol{\Delta}'$ (recall $s = 1/k$).*

Proof. The five items can be proved as follows.

1. — Consider, for instance, the function f for values of x to the left[5] of γ. To analytically continue this "branch" of the function f to the right of γ it suffices to deform the path γ by pushing it to the right keeping its endpoints 0 and x_0 fixed. This allows us to go up to the boundary $\arg(x) = \alpha$ of $\boldsymbol{\Delta}'$. We can similarly continue the "branch" of the function f defined for values of x on the right of γ up to the boundary $\arg(x) = \beta + 2\pi$ of $\boldsymbol{\Delta}'$.

2–3. — We have to prove that, for all subsector $\boldsymbol{\Delta}'' \Subset \boldsymbol{\Delta}'$, the function f satisfies the asymptotic estimates of definition 1.1.1, p. 2.

[5] Around 0, one goes to the left when going counterclockwise, and one goes to the right when going clockwise.

▷ Suppose first that $\overline{\Delta''} \cap \gamma = \emptyset$.

Writing

$$\frac{1}{t-x} = \sum_{n=0}^{N-1} \frac{x^n}{t^{n+1}} + \frac{x^N}{t^N(t-x)}$$

as in example 1.1.4, p. 4, we get

$$f(x) = \sum_{n=0}^{N-1} a_n x^n + \frac{x^N}{2\pi\mathrm{i}} \int_\gamma \frac{\varphi(t)}{t^N(t-x)}\,\mathrm{d}t.$$

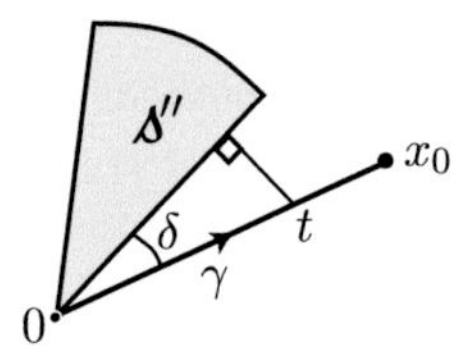

Fig. 1.8

Given $x \in \Delta''$, then $|t-x| \geq \operatorname{dist}(t, \Delta'') = |t|\sin(\delta)$ for all $t \in \gamma$ and so

$$\Big|f(x) - \sum_{n=0}^{N-1} a_n x^n\Big| \leq C|x|^N \tag{1.17}$$

where the constant $C = \frac{1}{2\pi}\Big|\int_\gamma \frac{|\varphi(t)|}{|t|^{N+1}\sin(\delta)}\,\mathrm{d}t\Big|$ is finite (the integral converges since φ is flat at 0 on γ) and depends on N and Δ'', but is independent of $x \in \Delta''$.

▷ Suppose now that $\overline{\Delta''} \cap \gamma \neq \emptyset$. Push homotopically γ into a path made of the union of a segment $\gamma_1 =]0, x_1]$ and a curve γ_2, say a circular arc, from x_1 to x_0 without meeting $\overline{\Delta''}$ as shown in Fig. 1.9. The integral splits into two parts $f_1(x)$ and $f_2(x)$.

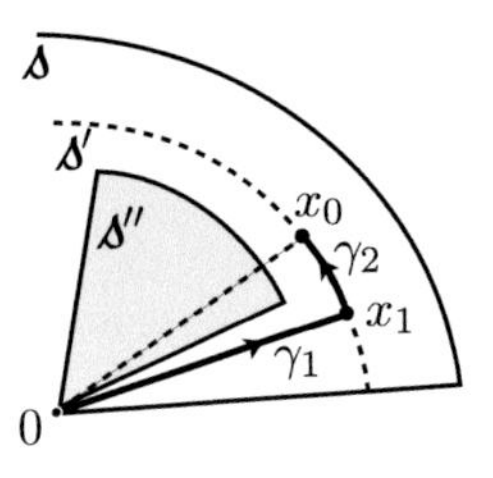

Fig. 1.9

The term $f_1(x)$ belongs to the previous case and is then asymptotic to

$$\sum_{n\geq 0} \frac{1}{2\pi \mathrm{i}} \int_{\gamma_1} \frac{\varphi(t)}{t^{n+1}} \mathrm{d}t\, x^n \quad \text{on } \boldsymbol{\Delta}''.$$

The term $f_2(x)$ defines an analytic function on the disc $|x| < |x_0|$ and is asymptotic to its Taylor series $\sum_{n\geq 0} \frac{1}{2\pi \mathrm{i}} \int_{\gamma_2} \frac{\varphi(t)}{t^{n+1}} \mathrm{d}t\, x^n$. Hence, the result.

4. — Given $x \in \dot{\boldsymbol{\Delta}}'$ compute the variation of f at x.

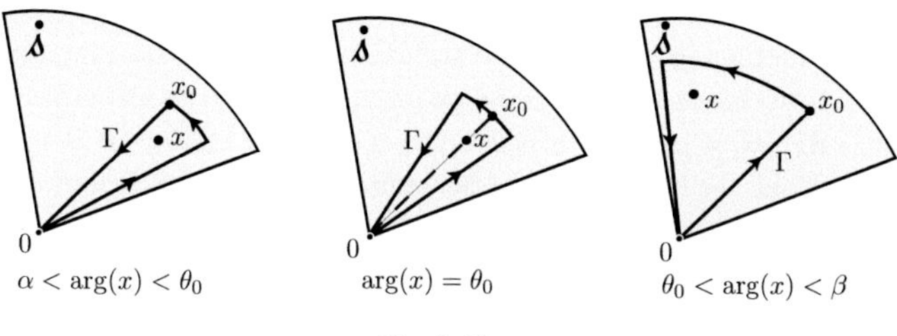

Fig. 1.10

Recall that $x \in \dot{\boldsymbol{\Delta}}'$ means that x belongs to the first sheet of $\boldsymbol{\Delta}'$. So, as explained in the proof of point 1, to evaluate $f(x)$ we might have to push homotopically the path γ to the right into a path γ'. When x lies to the left of γ we can keep $\gamma' = \gamma$. To evaluate $f(x\mathrm{e}^{2\pi\mathrm{i}})$ we might have to push homotopically the path γ to the left into a path γ'' taking $\gamma'' = \gamma$ when x lies to the right of γ.

The concatenation of γ' and $-\gamma''$ generates a path Γ in $\dot{\boldsymbol{\Delta}}$ enclosing x and since the function $\varphi(t)/(t-x)$ is meromorphic on $\dot{\boldsymbol{\Delta}}$ we obtain by the Cauchy's Residue theorem:

$$\operatorname{var} f(x) = f(x) - f(x\mathrm{e}^{2\pi\mathrm{i}}) = \frac{1}{2\pi\mathrm{i}} \int_{\Gamma} \frac{\varphi(t)}{t-x} \mathrm{d}t = \operatorname{Res}\Big(\frac{\varphi(t)}{t-x}, t = x\Big) = \varphi(x) \quad (1.18)$$

5. — Given $\boldsymbol{\Delta}'' \Subset \boldsymbol{\Delta}'$ suppose that the function φ satisfies

$$\big|\varphi(x)\big| \leq K \exp\big(-A/|x|^k\big) \quad \text{on } \boldsymbol{\Delta}''.$$

Consider the case when $\overline{\boldsymbol{\Delta}''} \cap \gamma = \emptyset$. Then, the constant C in estimate (1.17), p. 28 satisfies

$$C \leq \frac{K}{2\pi} \Big| \int_{\gamma} \frac{\exp(-A/|t|^k)}{|t|^{N+1}} \mathrm{d}t \Big| \leq C' A^{-N/k} \Gamma(N/k)$$

with a constant $C' > 0$ independent of N. The case when $\overline{\boldsymbol{\Delta}''} \cap \gamma \neq \emptyset$ is treated similarly by deforming the path γ as in points 2–3. Hence, $f(x)$ is s-Gevrey asymptotic to the series $\sum_{n\geq 0} a_n x^n$ on $\boldsymbol{\Delta}'$. □

Comments 1.4.3 (On the Euler Function (Exa. 1.1.4, p. 4))

Set $\dot{\mathbf{\Delta}} = \mathbf{\Delta}_{\alpha,\beta}(\infty)$ with $\alpha = -3\pi/2$ and $\beta = -\pi/2$ and $\mathbf{\Delta} = \mathbf{\Delta}_{\alpha,\beta+2\pi}(\infty)$. Let $\mathrm{E}(x)$ denote the Euler function as in example1.1.4 and, given θ, let d_θ denote the half line issuing from 0 with the θ direction.

▷ The variation of the Euler function E on $\dot{\mathbf{\Delta}}$ is given by

$$\begin{aligned}\operatorname{var} \mathrm{E}(x) &= \int_{d_{-\pi+\varepsilon}} \frac{\mathrm{e}^{-\xi/x}}{1+\xi}\,\mathrm{d}\xi - \int_{d_{\pi-\varepsilon}} \frac{\mathrm{e}^{-\xi/x}}{1+\xi}\,\mathrm{d}\xi \quad (\varepsilon \text{ small enough})\\ &= -2\pi\mathrm{i}\operatorname{Res}\Big(\frac{\mathrm{e}^{-\xi/x}}{1+\xi}, \xi=-1\Big) \quad \text{(Cauchy's residue Thm.)}\\ &= -2\pi\mathrm{i}\,\mathrm{e}^{1/x}.\end{aligned}$$

▷ Choose the function $\varphi(x) = -2\pi\mathrm{i}\,\mathrm{e}^{1/x}$ and a point $x_0 \in \dot{\mathbf{\Delta}}$, say, $x_0 = -r$ real negative. The function $\varphi(x)$ is 1-exponentially flat on $\dot{\mathbf{\Delta}}$ and we can apply the Cauchy-Heine theorem to $\varphi(x)$.

Denote $\mathbf{\Delta}' = \mathbf{\Delta} \cap \{|x| < |x_0|\}$ and $\dot{\mathbf{\Delta}}' = \mathbf{\Delta}' \cap \{|x| < |x_0|\}$. The Cauchy-Heine theorem provides a function f which, like the Euler function, belongs to $\overline{\mathscr{A}}_1(\mathbf{\Delta}')$ with variation $\varphi(x)$ on $\dot{\mathbf{\Delta}}'$.

We claim that E and f differ by an analytic function near 0. Indeed, the Taylor series of f on $\mathbf{\Delta}'$ reads $\sum_{n\geq 0} a_{n,x_0} x^n$ with coefficients

$$a_{n,x_0} = -\int_0^{x_0} \frac{\mathrm{e}^{1/t}}{t^{n+1}}\,\mathrm{d}t = (-1)^{n-1}\int_{1/r}^{+\infty} u^{n-1}\,\mathrm{e}^{-u}\,\mathrm{d}u$$

while the Taylor coefficients a_n of the Euler function E are given by $a_0 = 0$ and for $n \geq 1$ by $a_n = \lim_{r\to+\infty} a_{n,x_0}$. Since a_{0,x_0} has no limit as r tends to $+\infty$ we consider, instead of f, the function

$$f(x) - a_{0,x_0} = -\int_0^{-r}\Big(\frac{1}{t-x} - \frac{1}{t}\Big)\mathrm{e}^{1/t}\,\mathrm{d}t = \int_{1/r}^{+\infty} \frac{x\mathrm{e}^{-u}}{1+ux}\,\mathrm{d}u.$$

Suppose $x = |x|\,\mathrm{e}^{i\theta}$. Then, the Euler function at x can be defined by the integral

$$\mathrm{E}(x) = \int_{d_\theta} \frac{\mathrm{e}^{-\xi/x}}{1+\xi}\,\mathrm{d}\xi = \int_0^{+\infty} \frac{x\mathrm{e}^{-u}}{1+xu}\,\mathrm{d}u$$

and

$$E(x) - f(x) = -a_{0,x_0} + x\int_0^{1/r} \frac{\mathrm{e}^{-u}}{1+ux}\,\mathrm{d}u$$

which is an analytic function on the disc $|x| < r = |x_0|$.

This property will follow from a general argument of uniqueness given by Watson's lemma 5.1.3, p. 136. Indeed, the function $f(x) - a_{0,x_0} + x\int_0^{1/r} \frac{\mathrm{e}^{-u}}{1+ux}\,\mathrm{d}u$ has the Euler series as asymptotic expansion and is 1-Gevrey asymptotic to the Euler series $\widetilde{\mathrm{E}}(x)$ on $\mathbf{\Delta}'$. Then, it must be equal to the Euler function $\mathrm{E}(x)$ which has the same properties.

▷ *Stokes phenomenon.* — The Euler function $\mathrm{E}(x)$ is also solution of the homogeneous linear differential equation

$$\mathscr{E}_0(y) \equiv x^3\frac{\mathrm{d}^2y}{\mathrm{d}x^2} + (x^2+x)\frac{\mathrm{d}y}{\mathrm{d}x} - y = 0$$

deduced from the Euler equation (1.1), p. 4 by dividing x and then, differentiating once. Since the equation has no singular point but 0 (and infinity) the Cauchy-Lipschitz theorem allows one to analytically continue the Euler function along any path which avoids 0 and then in particular, outside of the sector $-3\pi/2 < \arg(x) < +3\pi/2$. However, when crossing the lateral boundaries of this sector the Euler function $\mathrm{E}(x)$ stops being asymptotic to the Euler series at 0; it even stops having an asymptotic expansion since, from the variation formula above (cf. also the end of Exa. 1.1.4, p. 4), one has now to take into account an exponential term which is unbounded. This phenomenon, known under the name of *Stokes phenomenon*, is at the core of the meromorphic classification of linear differential equations (cf. Sect. 3.5, p. 90).

Let us end this chapter by mentioning the recent book by A. Fruchard and R. Schäfke on composite asymptotic expansions devoted to specific kinds of asymptotic expansions for functions of two variables [FS13]. There, the Gevrey analysis is widely considered with much analogy with the present study.

Chapter 2
Sheaves and Čech Cohomology with an Insight into Asymptotics

Abstract We state the definition and general properties of sheaves from presheaves to espaces étalés. We introduce thoroughly the sheaves we encounter in the next chapters. These are mostly sheaves on the circle S^1 of directions about 0 in $\mathbb{C}$. We reformulate the Borel-Ritt theorem in terms of sheaves. In the second part of the chapter we introduce the Čech cohomology for sheaves of groups, both in the abelian and the non abelian case; and we reformulate the Borel-Ritt and the Cauchy-Heine theorems in the cohomological language.
For more precisions on sheaves and cohomology we refer to the classical literature (cf. [God58], [Ten75], [Ive86] for instance).

2.1 Presheaves and sheaves

Sheaves are the adequate tool to handle objects defined by local conditions without having to make explicit how large is the domain of validity of the conditions. They are mainly used as a bridge from local to global properties. It is convenient to start with the weaker concept of presheaves which we usually denote with an overline.

2.1.1 Presheaves

Let us start with the definition of presheaves with values in the category of sets and continue with the case of various sub-categories (for the definition of a category, see for instance [God58, Sect. 1.7]).

Definition 2.1.1 (Presheaf) *A presheaf (of sets) $\overline{\mathscr{F}}$ over a topological space X called the* base space *is defined by the following data:*

(i) *to any open subset U of X there is a set $\overline{\mathscr{F}}(U)$ whose elements are called* sections of $\overline{\mathscr{F}}$ on U*;*

M. Loday-Richaud, *Divergent Series, Summability and Resurgence II*,
Lecture Notes in Mathematics, DOI 10.1007/978-3-319-29075-1_2

(ii) *to any couple of open subsets* $V \subseteq U$ *there is a map* $\rho_{V,U} : \mathscr{F}(U) \to \mathscr{F}(V)$ *called* restriction map *which satisfies the following two conditions:*

$$\rho_{U,U} = \mathrm{id}_U \quad \textit{for all open set } U,$$
$$\rho_{W,V} \circ \rho_{V,U} = \rho_{W,U} \quad \textit{for all open sets } W \subseteq V \subseteq U.$$

In the language of categories, a presheaf of sets over X is then a *contravariant functor* from the category of open subsets of X into the category of sets.

Unless otherwise specified, we assume that X does not reduce to a single element.

The names "section" and "restriction map" take their origin in example 2.1.2 below which, with the notion of *espace étalé* (cf. Def. 2.1.13, p. 38), will become a reference example.

Example 2.1.2 (A Fundamental Example)

Suppose we are given a topological space F and a continuous map $\pi : F \to X$. One associates with F and π a presheaf $\mathscr{F}$ as follows: for all open set U in X one defines $\mathscr{F}(U)$ as the set of the *sections* of π on U, i.e., the continuous maps $s : U \to F$ such that $\pi \circ s = \mathrm{id}_U$. The restriction maps $\rho_{V,U}$ for $V \subseteq U$ are defined by $\rho_{V,U}(s) = s_{|V}$.

Example 2.1.3 (Constant Presheaf)

Given any set (or group, vector space, etc...) C, the constant presheaf $\mathscr{C}_X$ over X is defined by $\mathscr{C}_X(U) = C$ for all open set U in X and the maps $\rho_{V,U} = \mathrm{id}_C : C \to C$ as restriction maps.

Example 2.1.4 (An Exotic Example)

Given any marked set with more than one element, say $X = (\mathbb{C}, 0)$, one defines a presheaf $\mathscr{G}$ over X as follows: $\mathscr{G}(X) = X$ and $\mathscr{G}(U) = \{0\}$ when $U \neq X$; all the restriction maps are equal to null maps except $\rho_{X,X}$ which is the identity on X.

Below, we consider presheaves with values in a category $\mathscr{C}$ equipped with an algebraic structure. We assume moreover that, in $\mathscr{C}$, there exist products, the terminal objects are the singletons, the isomorphisms are the bijective morphisms. These conditions are satisfied by the sheaves to be considered further.

Definition 2.1.5 *A presheaf over* X with values in a category $\mathscr{C}$ *is a presheaf of sets satisfying the following two conditions:*

(iii) *For all open set* U *of* X *the set* $\mathscr{F}(U)$ *is an object of the category* $\mathscr{C}$*;*

(iv) *For any couple of open sets* $V \subseteq U$ *the map* $\rho_{V,U}$ *is a morphism in* $\mathscr{C}$.

In the next chapters, we will mostly be dealing with presheaves or sheaves of modules, in particular, of abelian groups or vector spaces, and presheaves or sheaves of differential $\mathbb{C}$-algebras, i.e., presheaves or sheaves with values in a category of modules, abelian groups, or vector spaces and presheaves or sheaves with values in the category of differential $\mathbb{C}$-algebras.

Definition 2.1.6 (Morphism of Presheaves)

Given $\overline{\mathscr{F}}$ and $\overline{\mathscr{G}}$ two presheaves over X with values in a category $\mathscr{C}$, a morphism $f : \overline{\mathscr{F}} \to \overline{\mathscr{G}}$ is a collection, for all open sets U of X, of morphisms

$$f(U) : \overline{\mathscr{F}}(U) \longrightarrow \overline{\mathscr{G}}(U)$$

in the category $\mathscr{C}$ which are compatible with the restriction maps, i.e., *such that the diagrams*

$$\begin{array}{ccc} \overline{\mathscr{F}}(U) & \xrightarrow{f(U)} & \overline{\mathscr{G}}(U) \\ \rho_{V,U} \big\downarrow & & \big\downarrow \rho'_{V,U} \\ \overline{\mathscr{F}}(V) & \xrightarrow{f(V)} & \overline{\mathscr{G}}(V) \end{array}$$

commute ($\rho_{V,U}$ and $\rho'_{V,U}$ denote the restriction maps in $\overline{\mathscr{F}}$ and $\overline{\mathscr{G}}$ respectively).

Definition 2.1.7 *A morphism f of presheaves is said to be* injective *or* surjective *when all morphisms $f(U)$ are injective or surjective.*

The morphisms of presheaves from $\overline{\mathscr{F}}$ into $\overline{\mathscr{G}}$ form a set, precisely, they form a subset of $\prod_{U \subseteq X} \mathrm{Hom}(\overline{\mathscr{F}}(U), \overline{\mathscr{G}}(U))$. Composition of morphisms in the category $\mathscr{C}$ induces composition of morphisms of presheaves over X with values in $\mathscr{C}$. It follows that presheaves over X with values in $\mathscr{C}$ form themselves a category.

When $\mathscr{C}$ is abelian, the category of presheaves over X with values in $\mathscr{C}$ is also abelian. In particular, one can talk of an *exact sequence of presheaves*

$$\cdots \to \overline{\mathscr{F}}_{j-1} \xrightarrow{f_j} \overline{\mathscr{F}}_j \xrightarrow{f_{j+1}} \overline{\mathscr{F}}_{j+1} \to \cdots$$

which means that the following sequence is exact for all open set U:

$$\cdots \to \overline{\mathscr{F}}_{j-1}(U) \xrightarrow{f_j(U)} \overline{\mathscr{F}}_j(U) \xrightarrow{f_{j+1}(U)} \overline{\mathscr{F}}_{j+1}(U) \to \cdots .$$

The category of modules over a given ring, hence also the category of abelian groups and the category of vector spaces, are abelian. All three admit the trivial object $\{0\}$ as terminal object.

The category of rings, and in particular, the category of differential $\mathbb{C}$-algebras, is not abelian. Although the quotient of a ring $\mathscr{A}$ by a subring $\mathscr{J}$ is not a ring in general, this becomes true when $\mathscr{J}$ is an ideal and allows one to consider short exact sequences $0 \to \overline{\mathscr{J}} \to \overline{\mathscr{A}} \to \overline{\mathscr{A}} / \overline{\mathscr{J}} \to 0$ of presheaves of rings or of differential $\mathbb{C}$-algebras.

Definition 2.1.8 (Stalk) *Given a presheaf $\overline{\mathscr{F}}$ over X and $x \in X$, the stalk of $\overline{\mathscr{F}}$ at x is the direct limit*

$$\boxed{\overline{\mathscr{F}}_x = \varinjlim_{U \ni x} \overline{\mathscr{F}}(U)}$$

the limit being taken on the filtrant set of the open neighborhoods of x in X ordered by inclusion. The elements of $\overline{\mathscr{F}}_x$ are called germs of sections of $\overline{\mathscr{F}}$ at x.

Let us first recall what is understood by the terms *direct limit* and *filtrant*.

▷ The *direct limit*

$$E = \varinjlim_{\alpha \in I} (E_\alpha, f_{\beta,\alpha})$$

of a direct family $(E_\alpha, f_{\beta,\alpha} : E_\alpha \longrightarrow E_\beta$ for $\alpha \leq \beta)$ (i.e., it is required that the set of indices I be ordered and right filtrant which means that given $\alpha, \beta \in I$ there exists $\gamma \in I$ greater than both α and β; moreover, the morphisms must satisfy $f_{\alpha,\alpha} = \mathrm{id}_\alpha$ and $f_{\gamma,\beta} \circ f_{\beta,\alpha} = f_{\gamma,\alpha}$ for all $\alpha \leq \beta \leq \gamma$) is the quotient of the sum $F = \bigsqcup_{\alpha \in I} E_\alpha$ of the spaces E_α by the equivalence relation $\mathscr{R}$: for $x \in E_\alpha$ and $y \in E_\beta$, one says that

$$x \mathscr{R} y \;\; \textit{if there exists } \gamma \textit{ such that } \gamma \geq \alpha,\ \gamma \geq \beta \;\; \textit{and} \;\; f_{\gamma,\alpha}(x) = f_{\gamma,\beta}(y).$$

In the case of a stalk here considered, the maps $f_{\beta,\alpha}$ are the restriction maps $\rho_{V,U}$.

▷ *Filtrant* means here that, given any two neighborhoods of x, there exists a neighorhood smaller than both of them. Their intersection, for example, provides such a smaller neighborhood.

Thus, a germ φ at x is an equivalence class of sections under the following equivalence relation. Let $s \in \overline{\mathscr{F}}(U)$ and $t \in \overline{\mathscr{F}}(V)$ be two sections of $\overline{\mathscr{F}}$ over open neighborhoods U and V of x in X respectively. The sections s and t are said to be equivalent if there is an open neighborhood $W \subseteq U \cap V$ of x contained both in U and V such that $\rho_{W,U}(s) = \rho_{W,V}(t)$.

By abuse of language and for simplicity, we allow us to say "the germ φ at x" when φ is an element of $\overline{\mathscr{F}}(U)$ with $U \ni x$ identifying so the element φ in the equivalence class to the equivalence class itself.

Warning 1. Given $s \in \overline{\mathscr{F}}(U)$ and $t \in \overline{\mathscr{F}}(V)$ one should be aware of the fact that the equality of the germs $s_x = t_x$ for all $x \in U \cap V$ does not imply the equality of the sections themselves on $U \cap V$.

Counter-Example 2.1.9

A counter-example is given by taking the sections $s \equiv 0$ and $t \equiv 1$ whose germs are everywhere 0 in example 2.1.4, p. 34.

Warning 2. It is worth to notice that a consistent collection of germs for all $x \in U$ does not imply the existence of a section $s \in \overline{\mathscr{F}}(U)$ inducing the given germs at each $x \in U$. *Consistent* means here that any section $v \in \overline{\mathscr{F}}(V)$ representing a given germ at x induces the neighboring germs: there exists an open sub-neighborhood $V' \subseteq V \subseteq U$ of x where the given germs are all represented by v.

Counter-Example 2.1.10

A counter-example is given by the constant presheaf $\overline{\mathscr{C}}_X$ when X is disconnected.

Consider, for instance, $X = \mathbb{R}^*, C = \mathbb{R}$ and the collection of germs $s_x = 0$ for $x < 0$ and $s_x = 1$ for $x > 0$. There exists no constant section over $\mathbb{R}^*$ which induces 0 by restriction to $\mathbb{R}^-$ and 1 by restriction to $\mathbb{R}^+$.

The presheaf $\overline{\mathscr{A}}$ defined in section 2.1.5, p. 41 will provide another example.

Such inconveniences are circumvented by restricting the notion of presheaf to the stronger notion of sheaf given below.

2.1.2 Sheaves

Definition 2.1.11 (Sheaf)

A presheaf $\overline{\mathscr{F}}$ over X is a sheaf *(we denote it then by $\mathscr{F}$) if, for all open set U of X, the following two properties hold:*

1. *If two sections s and σ of $\mathscr{F}(U)$ agree on an open covering $\mathscr{U} = \{U_j\}_{j\in J}$ of U* (i.e., *if they satisfy $\rho_{U_j,U}(s) = \rho_{U_j,U}(\sigma)$ for all j) then $s = \sigma$.*

2. *Given any consistent family of sections $s_j \in \mathscr{F}(U_j)$ on an open covering $\mathscr{U} = \{U_j\}_{j\in J}$ of U there exists a section $s \in \mathscr{F}(U)$ gluing all the s_j's* (i.e., *such that for all j, $\rho_{U_j,U}(s) = s_j$).*

Consistent *means here that, for all i, j, the restrictions of s_i and s_j agree on $U_i \cap U_j$,* i.e., $\rho_{U_i\cap U_j,U_i}(s_i) = \rho_{U_i\cap U_j,U_j}(s_j)$.

Example 2.1.12

The presheaf $\mathscr{F}$ of example 2.1.2, p. 34 is a sheaf. In example 2.1.4, p. 34 condition 1 fails. In the case of the constant presheaf over a disconnected base space X (cf. Exa. 2.1.3, p. 34) and in the case of the presheaf $\overline{\mathscr{A}}$ defined in section 2.1.5, p. 41, condition 2 fails.

It follows from the axioms of sheaves that $\mathscr{F}(\emptyset)$ is a terminal object.
Thus, $\mathscr{F}(\emptyset) = \{0\}$ when $\mathscr{F}$ is a sheaf of modules, abelian groups and vector spaces or of differential $\mathbb{C}$-algebras.

Suppose $\mathscr{F}$ is a sheaf of modules. Then, the restriction maps are linear and condition 1 can be stated :

A section of $\mathscr{F}$ which is zero on all the sets U_j
of an open covering $\mathscr{U} = \{U_j\}$ is the null section.

2.1.3 From Presheaves to Sheaves: Espaces Étalés

With any presheaf $\overline{\mathscr{F}}$ there is a sheaf $\mathscr{F}$ canonically associated as follows. Consider the space $F = \bigsqcup_{x\in X} \overline{\mathscr{F}}_x$ (disjoint union of the stalks of $\overline{\mathscr{F}}$) and endow it with the following topology: a set $\Omega \subseteq F$ is open in F if, for all open set U of X and all section $s \in \overline{\mathscr{F}}(U)$, the set of all elements $x \in U$ such that the germ s_x of s at x belong to Ω is open in X.

Given $s \in \overline{\mathscr{F}}(U)$ where U is an open subset of X, consider the map $\tilde{s} : U \longrightarrow F$ defined by $\tilde{s}(x) = s_x$. Denote by π the projection map $\pi : F \to X,\ s_x \mapsto \pi(s_x) = x$.

The topology on F is the less fine for which $\tilde{s}$ is continuous for all U and s, and the topology induced on the stalks $\overline{\mathscr{F}}_x = \pi^{-1}(x)$ is the discrete topology. The sets $\tilde{s}(U)$ are open in F and the maps $\tilde{s}$ satisfy $\pi(\tilde{s}(x)) = x$ for all $x \in U$. It follows that π is a local homeomorphism.

Definition 2.1.13 (Espace Étalé, Associated Sheaf)

▷ *The topological space F is called the* espace étalé *over X associated with $\overline{\mathscr{F}}$.*

▷ *The* sheaf $\mathscr{F}$ associated with the presheaf $\overline{\mathscr{F}}$ *is the sheaf of continuous sections of $\pi : F \to X$ as defined in example 2.1.2, p. 34.*

Example 2.1.14 (Constant Sheaf)

The espace étalé associated with the constant presheaf $\overline{\mathscr{C}}_X$ in example 2.1.3, p. 34 is the topological space $X \times C$ endowed with the topology product of the given topology on X and of the discrete topology on C. Whereas the sections of $\overline{\mathscr{C}}_X$ are the constant functions over X, the sections of the associated sheaf $\mathscr{C}_X$ are all locally constant functions. The sheaf $\mathscr{C}_X$ is commonly called *the constant sheaf over X with stalk C*. Since there is no possible confusion one calls it too, *the constant sheaf C over X* using the same notation for the sheaf and its stalks.

The maps $i(U)$ given, for all open subsets U of X, by

$$i(U) : \overline{\mathscr{F}}(U) \longrightarrow \mathscr{F}(U), \quad s \longmapsto \tilde{s}$$

define a morphism i of presheaves. These maps may be neither injective (failure of condition 1 in Def. 2.1.11, p. 37. See Exa. 2.1.4, p. 34) nor surjective (failure of condition 2 in Def. 2.1.11. See Exa. 2.1.14, p. 38 or 2.1.25, p. 41). One can check that the morphism i is injective when condition 1 of sheaves (cf. Def. 2.1.11) is satisfied and that it is surjective when both conditions 1 and 2 are satisfied, and so, we can state

Proposition 2.1.15 *The morphism i is an isomorphism of presheaves if and only if $\overline{\mathscr{F}}$ is a sheaf.*

In all cases, the morphism i induces an isomorphism between the stalks $\overline{\mathscr{F}}_x$ and $\mathscr{F}_x$ at any point $x \in X$.

The morphism of presheaves i satisfies the following universal property.

Suppose $\mathscr{G}$ is a sheaf; then, any morphism of presheaves $\overline{\psi} : \overline{\mathscr{F}} \to \mathscr{G}$ can be factored uniquely through the sheaf $\mathscr{F}$ associated with $\overline{\mathscr{F}}$.

In other words, there exists a unique morphism ψ such that the following diagram commutes:

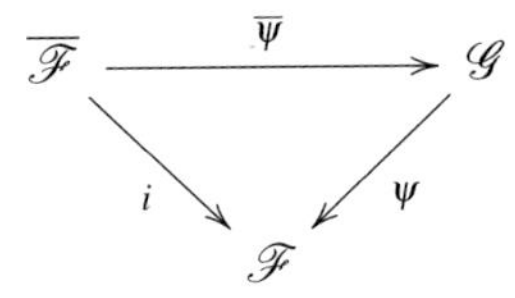

From the fact that,when $\overline{\mathscr{F}}$ is itself a sheaf, the morphism i is an isomorphism of presheaves between $\overline{\mathscr{F}}$ and its associated sheaf $\mathscr{F}$, one can always think of a sheaf

as being the sheaf of the sections of an espace étalé $F \xrightarrow{\pi} X$. From this point of view, it makes sense to consider sections over any subset W of X, open or not, and also to define any section as a collection of germs. Not any collection of germs is allowed. Indeed, if $\varphi \in \overline{\mathscr{F}}(W_x)$ represents the germ s_x on a neighborhood W_x of x then, for the section $s: W \to F$ to be continuous at x, the germs $s_{x'}$ for x' close to x must also be represented by φ. The set $\mathscr{F}(W)$ of the sections of a sheaf $\mathscr{F}$ over a subset W of X is commonly denoted by $\Gamma(W;\mathscr{F})$.

Recall the following definition (cf. end of Sect. 2.1.1, p. 33 and Def. 2.1.11, p. 37).

Definition 2.1.16 (Consistency)

▷ A family of sections $s_j \in \mathscr{F}(W_j)$ *is said to be* consistent *if, when* $W_i \cap W_j$ *is not empty, the restrictions of* s_i *and* s_j *to* $W_i \cap W_j$ *coincide.*

▷ A family of germs *is said to be* consistent *if any germ generates its neighbors.*

One can state:

Proposition 2.1.17 *Given $\mathscr{F}$ a sheaf over X and W any subset of X, open or not, a family of germs $(s_x)_{x\in W}$ is a section of $\mathscr{F}$ over W if and only if it is consistent.*

Definition 2.1.18 *Let $\mathscr{F}$ be the sheaf associated with a presheaf $\overline{\mathscr{F}}$. We define a* local section *of $\mathscr{F}$ to be any section of the presheaf $\overline{\mathscr{F}}$.*

Considering representatives of the germs s_x of a section $s \in \Gamma(W;\mathscr{F})$, proposition 2.1.17 can be reformulated as follows.

Proposition 2.1.19 *Let $\mathscr{F}$ be the sheaf associated with a presheaf $\overline{\mathscr{F}}$ over X and let W be any subset of X, open or not. Sections of $\mathscr{F}$ over W can be seen as consistent collections of local sections $s_j \in \overline{\mathscr{F}}(U_j)$ with U_j open in X and $W \subseteq \bigcup_j U_j$.*

Clearly, such collections are not unique.

When W is not open the inclusion $W \subseteq \bigcup_j U_j$ is proper and the section lives actually on a larger open set (the size of which depends not only on W but both on W and the section).

2.1.4 Morphisms of Sheaves

Definition 2.1.20 (Sheaf Morphism)

A morphism of sheaves is just a morphism of presheaves.

With this definition, proposition 2.1.15, p. 38 has the following corollary.

Corollary 2.1.21 *Let $\mathscr{F}$ be a sheaf and $\mathscr{F}'$ its associated sheaf when considered as a presheaf. Then, $\mathscr{F}$ and $\mathscr{F}'$ are isomorphic sheaves.*

Given two sheaves $\mathscr{F}$ and $\mathscr{F}'$ over X, let $F \xrightarrow{\pi} X$ and $F' \xrightarrow{\pi'} X'$ be their respective espace étalé. From the identification of a sheaf to its espace étalé a morphism $f : \mathscr{F} \to \mathscr{F}'$ of sheaves can be seen as a continuous map, which can also be denoted safely by f, between the associated espaces étalés with the condition that the following diagram commutes:

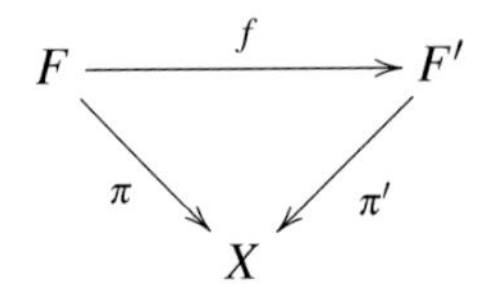

Like presheaves, sheaves with values in a given category $\mathscr{C}$ and their morphisms form a category which is abelian when $\mathscr{C}$ is also abelian. The category of sheaves and the category of espaces étalés with values in a given category $\mathscr{C}$ are equivalent.

Definition 2.1.22 *A morphism $f : \mathscr{F} \to \mathscr{F}'$ of sheaves over X is said to be* injective (resp. surjective, resp. an isomorphism) *if, for any $x \in X$, the stalk map $f_x : \mathscr{F}_x \to \mathscr{F}'_x$ is injective* (resp. *surjective,* resp. *bijective*).

When a morphism $f : \mathscr{F} \to \mathscr{F}'$ is injective then, for all open subset U of X, the map $f(U) : \mathscr{F}(U) \to \mathscr{F}'(U)$ is injective. However, the fact that f be surjective does not imply the surjectivity of the maps $f(U)$ for all U; hence, a surjective morphism of sheaves is not necessarily surjective as a morphism of presheaves, the converse being, of course, true since the functor direct limit is exact.

Example 2.1.23

Take for $\mathscr{F}$ the sheaf of germs of holomorphic functions on $X = \mathbb{C}^*$ and for $\mathscr{F}'$ the subsheaf (see Def. 2.1.24 below) of the non-vanishing functions. The map $f : \varphi \mapsto \exp \circ \varphi$ is a morphism from $\mathscr{F}$ to $\mathscr{F}'$ which is surjective as a morphism of sheaves since the logarithm exists locally on $\mathbb{C}^*$. However, the logarithm is not defined as a univaluate function on all of $\mathbb{C}^*$ and so, the map f is not a surjective morphism of presheaves. For instance, the identical function $\mathrm{Id} : x \mapsto x$ cannot be written in the form $\mathrm{Id} = f(\varphi)$ for any φ in $\mathscr{F}(\mathbb{C}^*)$ or more generally, any φ in $\mathscr{F}(U)$ as soon as U is not simply connected in $\mathbb{C}^*$.

Definition 2.1.24 *A sheaf $\mathscr{F}$ over X is a subsheaf of a sheaf $\mathscr{G}$ over X if, for all open set U, it satisfies the conditions*

- ▷ $\mathscr{F}(U) \subseteq \mathscr{G}(U)$,
- ▷ *the inclusion map $\mathscr{F}(U) \hookrightarrow \mathscr{G}(U)$ commute to the restriction maps.*

The inclusion $j : \mathscr{F} \hookrightarrow \mathscr{G}$ is an injective morphism of sheaves.

2.1.5 Sheaves $\mathscr{A}$ of Asymptotic and $\mathscr{A}_s$ of s-Gevrey Asymptotic Functions over S^1

The sheaves $\mathscr{A}$ and $\mathscr{A}_s$ of asymptotic functions we introduce in this section play a fundamental role in what follows.

$\triangleright$ *Topology of the base space* S^1. — The base space S^1 is the circle of directions from 0. One should consider it as the boundary of the real blow up $\widetilde{\mathbb{C}}$ of 0 in $\mathbb{C}$, that is, as the boundary $S^1 \times \{0\}$ of the space of polar coordinates $(\theta, r) \in S^1 \times [0, \infty[$. For simplicity, we denote S^1 for $S^1 \times \{0\}$.

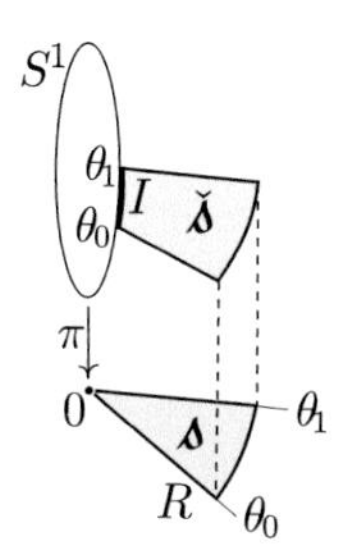

Fig. 2.1

The map $\pi : \widetilde{\mathbb{C}} \to \mathbb{C}$ defined by $\pi(\theta, r) = r e^{i\theta}$ sends S^1 to 0 and $\widetilde{\mathbb{C}} \setminus S^1$ homeomorphically to $\mathbb{C}^*$. A basis of open sets of S^1 is given by the arcs $I =]\theta_0, \theta_1[$ seen as the direct limit of the domains $\breve{\Delta} = I \times]0, R[$ in $\widetilde{\mathbb{C}}$ as R tends to 0. Such domains are identified via π to sectors $\Delta = \{x = r e^{i\theta};\ \theta_0 < \theta < \theta_1 \text{ and } 0 < r < R\}$ of $\mathbb{C}^*$.

$\triangleright$ *The presheaf* $\overline{\mathscr{A}}$ *over* S^1. — Given an open arc $I =]\theta_0, \theta_1[$ we denote by $\Delta_{I,R} = I \times]0, R[$ a sector based on I with radius R. The sections of $\overline{\mathscr{A}}$ over I are given by

$$\overline{\mathscr{A}}(I) = \varinjlim_{R \to 0} \overline{\mathscr{A}}(\Delta_{I,R}).$$

Suppose an element of $\overline{\mathscr{A}}(I)$ is represented by two functions $\varphi \in \overline{\mathscr{A}}(\Delta_{I,R})$ and $\psi \in \overline{\mathscr{A}}(\Delta_{I,R})$ on the same sector $\Delta_{I,R}$. This means that there exists a subsector $\Delta_{I,R'}$ of $\Delta_{I,R}$ on which φ and ψ coincide. By analytic continuation, we conclude that $\varphi = \psi$ on all of $\Delta_{I,R}$.

Choosing as restriction maps the usual restriction of functions, this defines a presheaf of differential $\mathbb{C}$-algebras. The example below shows that such a presheaf is not a sheaf.

Example 2.1.25

Consider the lacunar series (see [Rud87, Hadamard's Thm. 16.6 and Exa. 16.7])

$$f_1(x) = \sum_{n \geq 0} a_n (x-1)^{2^n} \quad \text{with } a_n = \exp(-2^{n/2}).$$

Since $\limsup_{n\to+\infty} |a_n|^{2^{-n}} = 1$ its radius of convergence as a series in powers of $x-1$ is equal to 1. We know from a theorem of Hadamard that its natural domain of holomorphy is the open disc $D = \{x \in \mathbb{C}\,;\, |x-1| < 1\}$. The series of the derivatives of any order (starting from order 0) converge uniformly on the closed disc $\overline{D}$. The function f_1 admits then an asymptotic expansion at 0 on any sector included in $\overline{D}$.

Consider now the arc $I =]-\frac{\pi}{2}, \frac{\pi}{2}[$ of S^1. To any $\theta \in I$ there is a sector $\Delta_\theta = I_\theta \times]0, R_\theta[$ on which f_1 is well defined and belongs to $\overline{\mathscr{A}}(\Delta_\theta)$. However, as θ approaches $\pm\frac{\pi}{2}$ the radius R_θ tends to 0 and there is no sector $\Delta = I \times]0, R[$ with $R > 0$ on which f_1 is even defined. Thus, condition 2 of definition 2.1.11, p. 37 fails on $U = I$.

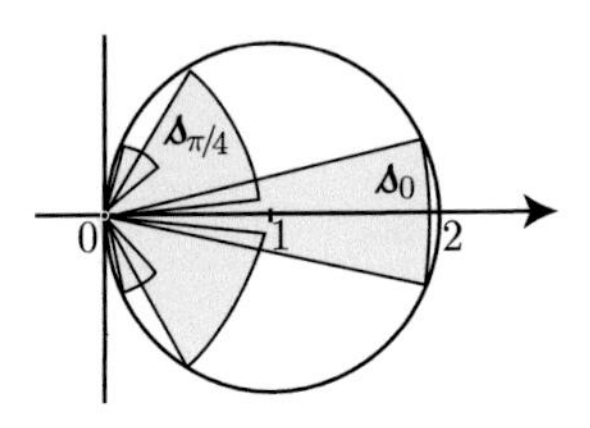

Fig. 2.2

▷ *The sheaf $\mathscr{A}$ over S^1.* — The sheaf $\mathscr{A}$ of asymptotic functions over S^1 is the sheaf associated with the presheaf $\overline{\mathscr{A}}$. A section of $\mathscr{A}$ over an interval I is defined by a collection of asymptotic functions $f_j \in \overline{\mathscr{A}}(\Delta_j)$ on $\Delta_j = I_j \times]0, R_j[$ where $\{I_j\}$ is an open covering of I and $R_j \neq 0$ for all j. The sheaf $\mathscr{A}$ is a sheaf of differential $\mathbb{C}$-algebras.

▷ *The subsheaf $\mathscr{A}^{<0}$ of flat germs.* — Given an open sector $\Delta_{I,R} = I \times]0, R[$ (cf. Notation 1.1.2, p. 3), we define

$$\overline{\mathscr{A}}^{<0}(I) = \varinjlim_{R\to 0} \overline{\mathscr{A}}^{<0}(\Delta_{I,R}).$$

The set $\overline{\mathscr{A}}^{<0}(I)$ is a subset of $\overline{\mathscr{A}}(I)$. Considering the restriction maps $\rho_{J,I}$ of the presheaf $\overline{\mathscr{A}}(I)$ restricted to $\overline{\mathscr{A}}^{<0}(I)$ we obtain a presheaf $I \mapsto \overline{\mathscr{A}}^{<0}(I)$ over S^1. The associated sheaf is denoted by $\mathscr{A}^{<0}$ and is a subsheaf of $\mathscr{A}$ over S^1.

▷ *The Taylor map.* — The Taylor map $T_{\Delta_{I,R}} : \overline{\mathscr{A}}(\Delta_{I,R}) \to \mathbb{C}[[x]]$ induces a map

$$T : \mathscr{A} \to \mathbb{C}[[x]]$$

also called *Taylor map* which is a morphism of sheaves of $\mathbb{C}$-differential algebras with kernel $\mathscr{A}^{<0}$. Thus, $\mathscr{A}^{<0}$ is a subsheaf of ideals of $\mathscr{A}$.

▷ *The sheaf $\mathscr{A}_s$ over S^1.* — Similarly, one defines a presheaf $\overline{\mathscr{A}}_s$ over S^1 by setting

$$\overline{\mathscr{A}}_s(I) = \varinjlim_{R\to 0} \overline{\mathscr{A}}_s(\Delta_{I,R})$$

for the set of (equivalence classes of) s-Gevrey asymptotic functions on a sector based on I. Its associated sheaf is denoted by $\mathscr{A}_s$.

▷ *The sheaf $\mathscr{A}^{\leq -k}$ over S^1.* — One also defines a presheaf by setting

$$\overline{\mathscr{A}}^{\leq -k}(I) = \varinjlim_{R \to 0} \overline{\mathscr{A}}^{\leq -k}(\Delta_{I,R})$$

and $\mathscr{A}^{\leq -k}$ denotes the associated sheaf over S^1. According to proposition 1.2.17, p. 20, the presheaf $\overline{\mathscr{A}}^{\leq -k}$ is a sub-presheaf of $\overline{\mathscr{A}}_s$, and then, $\mathscr{A}^{\leq -k}$ is a subsheaf of $\mathscr{A}_s$, precisely, the subsheaf of s-Gevrey flat germs.

The Taylor map $T : \mathscr{A} \to \mathbb{C}[[x]]$ induces a Taylor map

$$T = T_s : \mathscr{A}_s \longrightarrow \mathbb{C}[[x]]_s$$

which is a morphism of sheaves of $\mathbb{C}$-differential algebras with kernel $\mathscr{A}^{\leq -k}$. Thus, $\mathscr{A}^{\leq -k}$ is a subsheaf of ideals of $\mathscr{A}_s$.

2.1.6 Quotient Sheaves and Exact Sequences

From now, unless otherwise specified, we suppose that all the sheaves or presheaves we consider are sheaves or presheaves of *abelian groups* (or, more generally, sheaves or presheaves with values in an abelian category $\mathscr{C}$). Recall that such sheaves or presheaves and their morphisms form themselves an abelian category which will allow us to talk of exact sequences of sheaves.

Given a sheaf $\mathscr{G}$ with values in $\mathscr{C}$ and a subsheaf $\mathscr{F}$ one defines a presheaf by setting $U \mapsto \mathscr{G}(U)/\mathscr{F}(U)$ for all open set U of the base space X, the restriction maps being induced by those of $\mathscr{G}$.

Condition 1 of sheaves is always satisfied (for a proof see [Mal95, Annexe 1] for instance) while condition 2 fails in general (cf. Exa. 2.1.27, p. 44).

Definition 2.1.26 *One defines the* quotient sheaf $\mathscr{H} = \mathscr{G}/\mathscr{F}$ *to be the sheaf over X associated with the presheaf*

$$U \longmapsto \mathscr{G}(U)/\mathscr{F}(U) \quad \textit{for all open set } U \textit{ of } X$$

with restriction maps induced by those of $\mathscr{G}$.

If $\mathscr{F}$ and $\mathscr{G}$ are sheaves of abelian groups or of vector spaces so is the quotient $\mathscr{H}$. If $\mathscr{G}$ is a sheaf of algebras and $\mathscr{F}$ a subsheaf of ideals, then $\mathscr{H}$ is a sheaf of algebras.

As noticed at the end of section 2.1.3, p. 37, the fact that the quotient presheaf satisfies condition 1 of sheaves (Def. 2.1.11, p. 37) means that the natural map

$$\mathscr{G}(U)/\mathscr{F}(U) \to \mathscr{H}(U)$$

is injective. If condition 2 were also satisfied then this natural map would be surjective. However, this is not true, in general, as shown by the example 2.1.27 below.

Example 2.1.27 (Euler Equation and Quotient Sheaf)
We saw in example 1.1.4, p. 4 that the Euler equation

$$x^2\frac{\mathrm{d}y}{\mathrm{d}x}+y=x \tag{1.1}$$

admits an actual solution $\mathrm{E}(x)=\int_0^{+\infty}\frac{\mathrm{e}^{-\xi/x}}{1+\xi}\,\mathrm{d}\xi$ which, on the sector $-\frac{3\pi}{2}<\arg(x)<+\frac{3\pi}{2}$, is asymptotic to the Euler series $\widetilde{\mathrm{E}}(x)=\sum_{n\geq 0}(-1)^n n!x^{n+1}$.

Consider the homogeneous version of the Euler equation

$$\mathscr{E}_0\, y\equiv x^3\frac{\mathrm{d}^2y}{\mathrm{d}x^2}+(x^2+x)\frac{\mathrm{d}y}{\mathrm{d}x}-y=0.$$

Recall that one obtains the equation $\mathscr{E}_0 y=0$ by dividing equation (1.1) by x and differentiating. In any direction, $\mathscr{E}_0\, y=0$ admits a two dimensional $\mathbb{C}$-vector space of solutions spanned by $\mathrm{e}^{1/x}$ and $\mathrm{E}(x)$.

Following P. Deligne we denote by $\mathscr{V}$ the sheaf over S^1 of the germs of solutions of $(\mathscr{E}_0)$ having an asymptotic expansion at 0 and we denote by $\mathscr{V}_\theta$ the stalk of $\mathscr{V}$ in a θ direction. The sheaf $\mathscr{V}$ is a sheaf of vector spaces and a subsheaf of $\mathscr{A}$ seen as a sheaf of vector spaces. Since $\mathrm{E}(x)$ has an asymptotic expansion in all directions θ satisfying $-3\pi/2<\theta<3\pi/2$ and $\mathrm{e}^{1/x}$ has an asymptotic expansion (equal to 0) on $\Re(x)<0$ we can assert that

$$\dim_{\mathbb{C}}\mathscr{V}_\theta=\begin{cases}2 & \text{if } +\pi/2<\theta<3\pi/2,\\ 1 & \text{if } -\pi/2\leq\theta\leq+\pi/2\cdot\end{cases}$$

Denote by $\mathscr{V}^{<0}=\mathscr{V}\cap\mathscr{A}^{<0}$ the subsheaf of flat germs of $\mathscr{V}$. We observe that $\mathscr{V}(S^1)=\{0\}$ and $\mathscr{V}^{<0}(S^1)=\{0\}$, hence the quotient $\mathscr{V}(S^1)/\mathscr{V}^{<0}(S^1)=\{0\}$.

A global section of the quotient sheaf $\mathscr{V}/\mathscr{V}^{<0}$ is a collection of solutions over an open covering of S^1 which agree on the intersections up to flat solutions. The solution E induces such a global section while $\mathrm{e}^{1/x}$ does not. Thus, the space of global sections $\Gamma(S^1;\mathscr{V}/\mathscr{V}^{<0})$ has dimension $\dim_{\mathbb{C}}\Gamma(S^1;\mathscr{V}/\mathscr{V}^{<0})=1$. This shows that the quotient sheaf $\mathscr{V}/\mathscr{V}^{<0}$ is different from the quotient presheaf. The quotient sheaf $\mathscr{V}/\mathscr{V}^{<0}$ is isomorphic to the constant sheaf $\mathbb{C}$ as a sheaf of $\mathbb{C}$-vector spaces.

Let $f:\mathscr{F}\to\mathscr{G}$ be a morphism of sheaves with values in $\mathscr{C}$ over the same base space X. Let $\rho_{V,U}$ and $\rho'_{V,U}$ denote the restriction maps in $\mathscr{F}$ and $\mathscr{G}$ respectively. One can define the presheaves $\overline{\mathscr{K}er}(f),\overline{\mathscr{I}m}(f)$ and $\overline{\mathscr{C}oker}(f)$ over X with values in $\mathscr{C}$ by setting

▷ for $\overline{\mathscr{K}er}(f):U\longmapsto\ker\big(f(U):\mathscr{F}(U)\to\mathscr{G}(U)\big)$ for all open set $U\subseteq X$ with restriction maps $r_{V,U}=\rho_{V,U}|_{\ker(f(U))}$;

▷ for $\overline{\mathscr{I}m}(f):U\longmapsto f\big(\mathscr{F}(U)\big)$ with restriction maps $r'_{V,U}=\rho'_{V,U}|_{f(\mathscr{F}(U))}$;

▷ for $\overline{\mathscr{C}oker}(f):U\longmapsto\mathscr{G}(U)/f\big(\mathscr{F}(U)\big)$ with restriction maps canonically induced from $\rho'_{V,U}$ on the quotient.

So defined, $\overline{\mathscr{K}er}(f)$ and $\overline{\mathscr{I}m}(f)$ appear as sub-presheaves of $\mathscr{F}$ and $\mathscr{G}$ respectively, $\overline{\mathscr{C}oker}(f)$ as a quotient of $\mathscr{G}$. For a definition by a universal property we refer to the classical literature. One can check that the presheaf $\overline{\mathscr{K}er}(f)$ is actually a sheaf (precisely, a subsheaf of $\mathscr{F}$). Hence, the definition:

Definition 2.1.28 *The sheaves kernel, image and cokernel of a given morphism of sheaves f can be defined as follows.*

▷ *The* kernel $\mathscr{K}er(f)$ *of the sheaf morphism f, is the sheaf defined by*

$$U \longmapsto \ker\big(f(U)\big) \quad \textit{for all open set} \quad U \subseteq X$$

with the restriction maps $\rho_{V,U}|_{\ker(f(U))}$.

▷ *The* image $\mathscr{I}m(f)$ *and the* cokernel $\mathscr{C}oker(f)$ *of the sheaf morphism f, are the sheaves respectively associated with the presheaves $\overline{\mathscr{I}m}(f)$ and $\overline{\mathscr{C}oker}(f)$.*

The sheaves $\mathscr{C}oker(f)$ and $\mathscr{I}m(f)$ are respectively a quotient and a kernel:

$$\mathscr{C}oker(f) = \mathscr{G}/\mathscr{I}m(f), \quad \mathscr{I}m(f) = \mathscr{K}er\big(\mathscr{G} \to \mathscr{C}oker(f)\big)$$

where $\mathscr{G} \to \mathscr{C}oker(f)$ stands for the canonical quotient map.

Definition 2.1.29 *Exactness of sequences of presheaves and of sheaves are defined by the following non-equivalent conditions:*

▷ *A* sequence of presheaves $\overline{\mathscr{F}} \xrightarrow{\bar{f}} \overline{\mathscr{G}} \xrightarrow{\bar{g}} \overline{\mathscr{H}}$ *is said to be* exact *when*

$$\overline{\mathscr{I}m}(\bar{f}(U)) = \overline{\mathscr{K}er}(\bar{g}(U)) \textit{ for all open set } U \subseteq X.$$

▷ *A* sequence of sheaves $\mathscr{F} \xrightarrow{f} \mathscr{G} \xrightarrow{g} \mathscr{H}$ *is said to be* exact *when*

$$\operatorname{Im}(f_x) = \operatorname{Ker}(g_x) \textit{ for all } x \in X.$$

▷ *A sequence* $\cdots \to \mathscr{F}_{n-1} \xrightarrow{f_n} \mathscr{F}_n \xrightarrow{f_{n+1}} \mathscr{F}_{n+1} \to \cdots$ *of presheaves or sheaves is* exact *when each subsequence* $\mathscr{F}_{n-1} \xrightarrow{f_n} \mathscr{F}_n \xrightarrow{f_{n+1}} \mathscr{F}_{n+1}$ *is exact.*

A sequence of sheaves can be seen as a sequence of presheaves. One can show that exactness as a sequence of presheaves implies exactness as a sequence of sheaves the converse being false in general. Precisely, to a short (hence to any) exact sequence of presheaves $0 \to \overline{\mathscr{F}} \xrightarrow{\bar{f}} \overline{\mathscr{G}} \xrightarrow{\bar{g}} \overline{\mathscr{H}} \to 0$ there corresponds canonically the exact sequence of sheaves $0 \to \mathscr{F} \xrightarrow{f} \mathscr{G} \xrightarrow{g} \mathscr{H} \to 0$. Reciprocally, an exact sequence $0 \to \mathscr{F} \xrightarrow{f} \mathscr{G} \xrightarrow{g} \mathscr{H} \to 0$ of sheaves can be seen as a sequence of presheaves but, in general, only the truncated sequence $0 \to \mathscr{F} \xrightarrow{f} \mathscr{G} \xrightarrow{g} \mathscr{H}$ is exact as a sequence of presheaves.

Let $Presh_X$ and Sh_X denote respectively the categories of presheaves and sheaves over X with values in a given abelian category $\mathscr{C}$. In the language of categories the properties above are formulated as follows.

▷ The functor of sheafification $Presh_X \to Sh_X$ is exact.

▷ The functor of inclusion $Sh_X \hookrightarrow Presh_X$ is only left exact.

2.1.7 The Borel-Ritt Theorem Revisited

By construction, $\overline{\mathscr{A}}^{<0}(I)$ and $\overline{\mathscr{A}}^{\leq -k}(I)$ are the kernels of the Taylor maps

$$T_I : \overline{\mathscr{A}}(I) \longrightarrow \mathbb{C}[[x]] \quad \text{and} \quad T_{s,I} : \overline{\mathscr{A}}_s(I) \longrightarrow \mathbb{C}[[x]]_s$$

respectively for any open arc I of S^1. Hence, the sequences

$$0 \rightarrow \overline{\mathscr{A}}^{<0} \longrightarrow \overline{\mathscr{A}} \xrightarrow{\ T\ } \mathbb{C}[[x]] \quad \text{and} \quad 0 \rightarrow \overline{\mathscr{A}}^{\leq -k} \longrightarrow \overline{\mathscr{A}}_s \xrightarrow{\ T_s\ } \mathbb{C}[[x]]_s$$

are exact sequences of presheaves and they generate the exact sequences of sheaves of differential algebras

$$0 \rightarrow \mathscr{A}^{<0} \longrightarrow \mathscr{A} \xrightarrow{\ T\ } \mathbb{C}[[x]] \quad \text{and} \quad 0 \rightarrow \mathscr{A}^{\leq -k} \longrightarrow \mathscr{A}_s \xrightarrow{\ T_s\ } \mathbb{C}[[x]]_s.$$

The Borel-Ritt theorem 1.3.1, p. 22 allows one to complete these sequences into short exact sequences as follows.

Corollary 2.1.30 (Borel-Ritt) *The sequences*

$$0 \rightarrow \mathscr{A}^{<0} \longrightarrow \mathscr{A} \xrightarrow{\ T\ } \mathbb{C}[[x]] \rightarrow 0, \tag{2.1}$$

$$0 \rightarrow \mathscr{A}^{\leq -k} \longrightarrow \mathscr{A}_s \xrightarrow{\ T_s\ } \mathbb{C}[[x]]_s \rightarrow 0 \tag{2.2}$$

are exact sequences of sheaves of differential $\mathbb{C}$-algebras over S^1. Equivalently, the quotient sheaves $\mathscr{A}/\mathscr{A}^{<0}$ and $\mathscr{A}_s/\mathscr{A}^{\leq -k}$ are isomorphic via the Taylor map to the constant sheaves $\mathbb{C}[[x]]$ and $\mathbb{C}[[x]]_s$ respectively, as sheaves of differential $\mathbb{C}$-algebras.

With this approach, the surjectivity of T or T_s means that, given any series and any direction there exist a sector containing the direction and a function asymptotic to the given series on that sector. We cannot not claim that the sector can be chosen to be arbitrarily wide.

Observe that the sequences (2.1) and (2.2), p. 46 are not exact sequences of presheaves over S^1. Indeed, the range of the Taylor map $T : \mathscr{A}(S^1) \rightarrow \mathbb{C}[[x]]$, as well as the range of $T_s : \mathscr{A}_s(S^1) \rightarrow \mathbb{C}[[x]]_s$, consists in convergent series and, consequently, these maps are not surjective.

2.1.8 Change of Base Space: Direct Image, Restriction and Extension by 0

By definition of continuity, for $f : X \rightarrow Y$ continuous and U open in Y, the set $f^{-1}(U)$ is open in X. Thus, we can set the following definition.

Definition 2.1.31 (Direct Image) *Let $f : X \to Y$ be a continuous map. With any sheaf $\mathscr{F}$ over X one can associate a sheaf $f_*\mathscr{F}$ over Y called its* direct image *by setting*

$$f_*\mathscr{F}(U) = \mathscr{F}\left(f^{-1}(U)\right) \quad \textit{for all open set } U \textit{ in } Y,$$

with restriction maps

$$\rho_{*V,U}(s_*) = \rho_{f^{-1}(V),f^{-1}(U)}(s) \quad \textit{for all open sets } V \subseteq U \textit{ in } Y.$$

When $\mathscr{F}$ is a sheaf of abelian groups, vector spaces, etc..., so is its direct image $f_*\mathscr{F}$.

To a morphism $\varphi : \mathscr{F} \to \mathscr{G}$ of sheaves over X there corresponds a morphism of sheaves $\varphi_* : f_*\mathscr{F} \to f_*\mathscr{G}$ over Y defined by

$$s_* \in f_*\mathscr{F}(U) = \mathscr{F}\left(f^{-1}(U)\right) \longmapsto \varphi(s_*) \in \mathscr{G}\left(f^{-1}(U)\right) = f_*\mathscr{G}(U).$$

The functor direct image is left exact. Thus, to an exact sequence

$$0 \to \mathscr{F}' \xrightarrow{u} \mathscr{F}' \xrightarrow{v} \mathscr{F}''$$

there corresponds the exact sequence

$$0 \to f_*\mathscr{F}'' \xrightarrow{u_*} f_*\mathscr{F}' \xrightarrow{v_*} f_*\mathscr{F}''.$$

We suppose now that X is a subspace of Y with inclusion $j : X \hookrightarrow Y$ and that we are given $\mathscr{G}$ a sheaf over Y. The restriction of $\mathscr{G}$ into a sheaf over X is fully natural in terms of espaces étalés. We denote by $\pi : G \to Y$ the espace étalé associated with $\mathscr{G}$.

Definition 2.1.32 (Restriction) *The sheaf $\mathscr{G}$ restricted to X is the sheaf $\mathscr{G}_{|X}$ over X with espace étalé*

$$\pi_{|_{\pi^{-1}(X)}} : \pi^{-1}(X) \to X.$$

The definition makes sense since as $\pi : G \to Y$ is a local homeomorphism so is $\pi_{|_{\pi^{-1}(X)}} : \pi^{-1}(X) \to X$. The restricted sheaf can also be seen as the inverse image of $\mathscr{G}$ by the inclusion map j, a viewpoint which we won't develop here.

As in the previous section we consider now sheaves of abelian groups and we denote by 0 the neutral element. With $X \overset{j}{\hookrightarrow} Y$ let $\mathscr{F}$ and $\mathscr{F}'$ be sheaves over X and Y respectively.

Definition 2.1.33 (Extension)

▷ *A sheaf $\mathscr{F}'$ is an* extension *of a sheaf $\mathscr{F}$ if its restriction $\mathscr{F}'_{|X}$ to X is isomorphic to $\mathscr{F}$.*

▷ *An extension $\mathscr{F}'$ of $\mathscr{F}$ is an* extension by 0 *if, for all $y \in Y \setminus X$, the stalk $\mathscr{F}'_y$ is 0. (Equivalently, $\mathscr{F}'_{|Y \setminus X}$ is the constant sheaf 0.)*

Definition 2.1.34 (Support of a Section)
The support of a section $s \in \Gamma(U;\mathscr{F})$ *is the subset of U where s does not vanish:*

$$\operatorname{supp}(s) = \{x \in U\,;\, s_x \neq 0\}.$$

Example 2.1.35
Let $\mathscr{E}$ be the sheaf of $\mathbb{C}$-vector spaces generated over S^1 by the function $\mathrm{e}^{1/x}$.

The sheaf $\mathscr{E}$ is isomorphic to the constant sheaf with stalk $\mathbb{C}$ over S^1. Let $\mathrm{e}(x)$ be the class of $\mathrm{e}^{1/x}$ in the quotient sheaf $\mathscr{E}/\mathscr{E}^{<0}$ where $\mathscr{E}^{<0} = \mathscr{E} \cap \mathscr{A}^{<0}$. Thus, $\mathrm{e}(x) = 0$ for $\Re(x) < 0$ and the support of e is the arc $-\pi/2 \leq \arg(x) \leq \pi/2$, a closed subset of S^1.

The support $\operatorname{supp}(s)$ is always a closed subset of U, for, if a germ s_x is 0 then, there is an open neighborhood V_x of x on which s_x is represented by the 0 function generating thus the germs 0 on a neighborhood of x.

Recall that a subset X of Y is said to be *locally closed in Y* if any point $x \in X$ admits in Y a neighborhood $V_Y(x)$ such that its intersection $V_Y(x) \cap X$ is closed in $V_Y(x)$. This is equivalent to saying that there exist X_1 open in Y and X_2 closed in Y such that $X = X_1 \cap X_2$.

Definition 2.1.36 (Sheaf $j_!\mathscr{F}$)
Suppose X is locally closed in Y and denote by $j : X \hookrightarrow Y$ the inclusion map of X in Y. Given $\mathscr{F}$ a sheaf of abelian groups over X one defines the sheaf $j_!\mathscr{F}$ over Y by setting, for all open U of Y,

$$j_!\mathscr{F}(U) = \{s \in \Gamma(X \cap U;\mathscr{F})\,;\, \operatorname{supp}(s) \textit{ is closed in } U\}$$

with restriction maps induced by those of $j_*\mathscr{F}$ (of which $j_!\mathscr{F}$ is a subsheaf).

One can check that $j_!\mathscr{F}$ is a sheaf; it is then clearly a subsheaf of $j_*\mathscr{F}$ and there is a canonical inclusion $j_!\mathscr{F} \hookrightarrow j_*\mathscr{F}$. Moreover, $j_!\mathscr{F}$ is an extension of $\mathscr{F}$ by 0. When X is closed in Y then the two sheaves coincide: $j_!\mathscr{F} = j_*\mathscr{F}$. Unlike the functor j_* which is only left exact, the functor $j_!$ is exact.

The extension of sheaves by 0 provides a characterization of locally closed subspaces as follows: *X is locally closed in Y if and only if, for any sheaf $\mathscr{F}$ over X, there is a unique extension of $\mathscr{F}$ to Y by* 0 (cf. [Ten75] Thm. 3.8.6).

Example 2.1.37 ($j_* \neq j_!$)
As an illustration consider the sheaf $\mathscr{E}'$ generated as a sheaf of $\mathbb{C}$-vector spaces by $\mathrm{e}^{1/x}$ over the punctured disc $D^* = \{x \in \mathbb{C}\,;\, 0 < |x| < 1\}$ and consider the inclusion $j : D^* \hookrightarrow \mathbb{C}$.

The direct image $j_*\mathscr{E}'$ of $\mathscr{E}'$ by j is a non-constant sheaf of $\mathbb{C}$-vector spaces. Indeed, for U a connected open set in $\mathbb{C}$, one has $j_*\mathscr{E}'(U) \simeq \mathbb{C}^n$ where n is the number of connected components of $U \cap D^*$.

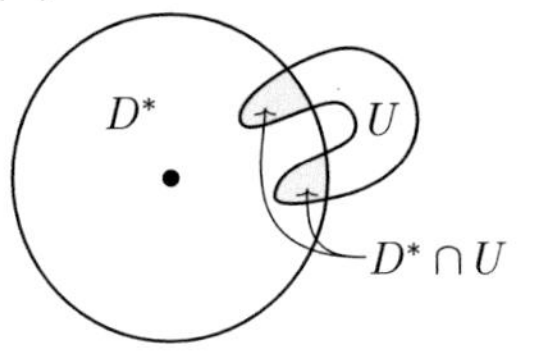

Fig. 2.3. Non connected intersection of two connected open sets

The stalks of $j_*\mathscr{E}'$ are given by

$$j_*\mathscr{E}'_x \simeq \begin{cases} \mathbb{C} & \text{if } x \in \overline{D}^*, \\ 0 & \text{otherwise,} \end{cases}$$

so that, in some way, the direct image $j_*\mathscr{E}'$ spreads $\mathscr{E}'$ out, onto the closure of D^*. Thus, the direct image $j_*\mathscr{E}'$ is an extension of $\mathscr{E}'$ but not an extension by 0.

On the contrary, the sheaf $j_!\mathscr{E}'$ is an extension of $\mathscr{E}'$ by 0. It is well defined since D^* being open in $\mathbb{C}$ is also locally closed in $\mathbb{C}$. This shows that $j_*\mathscr{E}' \neq j_!\mathscr{E}'$ and therefore, that the functors j_* and $j_!$ are different.

2.2 Abelian Čech Cohomology

Let $\mathscr{F}$ be a sheaf over a topological space X. We assume that $\mathscr{F}$ is a *sheaf of abelian groups*. The set $\Gamma(U;\mathscr{F})$ of sections of $\mathscr{F}$ over a $U \subset X$ is then naturally endowed with the structure of an abelian group and $\mathscr{F}(\emptyset) = \{0\}$, the trivial abelian group 0. Unless otherwise specified, all the coverings we consider are coverings by open sets.

2.2.1 *Cohomology of a Covering*

Let $\mathscr{U} = \{U_i\}_{i\in I}$ be an open covering of X.
Denote $U_{i,j} = U_i \cap U_j$, $U_{i,j,k} = U_i \cap U_j \cap U_k$, etc.

Definition 2.2.1 *One defines the* Čech complex of $\mathscr{F}$ associated with the covering $\mathscr{U}$ *to be the differential complex*

$$0 \to \prod_{i_0} \Gamma(U_{i_0};\mathscr{F}) \xrightarrow{d_0} \prod_{i_0,i_1} \Gamma(U_{i_0,i_1};\mathscr{F}) \xrightarrow{d_1}$$
$$\cdots \xrightarrow{d_{n-1}} \prod \Gamma(U_{i_0,\dots,i_n};\mathscr{F}) \xrightarrow{d_n} \prod \Gamma(U_{i_0,\dots,i_{n+1}};\mathscr{F}) \xrightarrow{d_{n+1}} \cdots$$

where, for all n, the map d_n is defined by

$$d_n : f = (f_{i_0,\dots,i_n}) \longmapsto g = (g_{i_0,\dots,i_{n+1}})$$

with

$$g_{i_0,\dots,i_{n+1}} = \sum_{\ell=0}^{n+1} (-1)^\ell f_{i_0,\dots,i_{\ell-1},\widehat{i_\ell},i_{\ell+1},\dots,i_{n+1}}\big|_{U_{i_0,\dots,i_{n+1}}}$$

the hat over i_ℓ indicating that the index i_ℓ is omitted.

Each term of the complex is an abelian group.

The maps d_n are morphisms of abelian groups. Consequently, the image $\operatorname{im} d_n$ and the kernel $\ker d_n$ are abelian groups. For all n, the maps d_n are "differentials" which, in this context, means that $d_n \circ d_{n-1} = 0$ and thus, $\operatorname{im} d_{n-1} \subset \ker d_n$ and the quotients $\ker d_n / \operatorname{im} d_{n-1}$ are abelian groups.

Definition 2.2.2 *One calls*

▷ *n*-cochains of $\mathscr{U}$ (with values) in $\mathscr{F}$ *the elements of the abelian group*

$$\mathscr{C}^n(\mathscr{U};\mathscr{F}) = \prod \Gamma(U_{i_0,\dots,i_n};\mathscr{F}),$$

▷ *n*-cocycles of $\mathscr{U}$ (with values) in $\mathscr{F}$ *the elements of the abelian group*

$$\mathscr{Z}^n(\mathscr{U};\mathscr{F}) = \ker d_n,$$

▷ *n*-coboundaries of $\mathscr{U}$ (with values) in $\mathscr{F}$ *the elements of the abelian group*

$$\mathscr{B}^n(\mathscr{U};\mathscr{F}) = \operatorname{im} d_{n-1},$$

▷ *n*-th Čech cohomology group of $\mathscr{U}$ (with values) in $\mathscr{F}$ *the abelian group*

$$H^n(\mathscr{U};\mathscr{F}) = \mathscr{Z}^n(\mathscr{U};\mathscr{F})/\mathscr{B}^n(\mathscr{U};\mathscr{F}) = \ker d_n/\operatorname{im} d_{n-1}.$$

In particular, $H^0(\mathscr{U};\mathscr{F})$ is isomorphic to $\Gamma(X;\mathscr{F})$, the set of the global sections of $\mathscr{F}$ over X.

Two n-cocycles which induce the same class in $H^n(\mathscr{U};\mathscr{F})$ are said to be cohomologous n-cocycles of $\mathscr{U}$ with values in $\mathscr{F}$ (or cohomologous in $H^n(\mathscr{U};\mathscr{F})$): they differ by an n-coboundary of $\mathscr{U}$ with values in $\mathscr{F}$.

Definition 2.2.3 (Refinement of a Covering) *A covering $\mathscr{V} = \{V_j\}_{j\in J}$ is said to be* finer *than the covering $\mathscr{U} = \{U_i\}_{i\in I}$, and we denote $\mathscr{V} \preceq \mathscr{U}$, if any element in $\mathscr{V}$ is contained in at least one element of $\mathscr{U}$.*
Equivalently, one can say that there exists a map

$$\sigma : J \longrightarrow I \quad \textit{such that } V_j \subset U_{\sigma(j)} \textit{ for all } j \in J.$$

The map σ is called inclusion map *or* simplicial map.

With the simplicial map σ are naturally associated the maps

$$\sigma_n^* : \mathscr{C}^n(\mathscr{U};\mathscr{F}) \longrightarrow \mathscr{C}^n(\mathscr{V};\mathscr{F}), \quad f = (f_{i_0,\dots,i_n}) \longmapsto \sigma_n^* f = (F_{j_0,\dots,j_n})$$

given by

$$F_{j_0,\dots,j_n} = f_{\sigma(j_0),\dots,\sigma(j_n)}|_{V_{j_0,\dots,j_n}}.$$

The family $\sigma^* = (\sigma_n^*)$ defines a morphism of Čech complexes and induces, for all n, a morphism of groups

$$\mathfrak{S}^n(\mathscr{V},\mathscr{U}) : H^n(\mathscr{U};\mathscr{F}) \longrightarrow H^n(\mathscr{V};\mathscr{F}).$$

It turns out that these latter homomorphims are independent of the choice of the simplicial map σ (cf. Prop. 2.3.5, p. 59 where a proof is given in the general non abelian case).

They satisfy, for all n, the following two conditions:

$$\mathfrak{S}^n(\mathscr{U},\mathscr{U}) = \mathrm{Id} \quad \text{for all } \mathscr{U}, \tag{2.3}$$

$$\mathfrak{S}^n(\mathscr{W},\mathscr{V}) \circ \mathfrak{S}^n(\mathscr{V},\mathscr{U}) = \mathfrak{S}^n(\mathscr{W},\mathscr{U}) \quad \text{for all } \mathscr{W} \preceq \mathscr{V} \preceq \mathscr{U}. \tag{2.4}$$

The case $n = 1$ has the following specificity:

Proposition 2.2.4 *The morphism* $\mathfrak{S}^1(\mathscr{V},\mathscr{U}) : H^1(\mathscr{U};\mathscr{F}) \longrightarrow H^1(\mathscr{V};\mathscr{F})$ *is injective.*

We refer to [Ten75, Thm. 4.15, p. 148] and to Prop. 2.3.7, p. 59 which provides a proof in the general non abelian case.

2.2.2 Cohomology of a Space

With a view to define the cohomology of the topological space X with values in the sheaf $\mathscr{F}$ from the cohomology of coverings $\mathscr{U}$ of X we first observe the following facts:

▷ The open coverings $\mathscr{U}$ of X are ordered by the fineness relation making them a right filtrant system[1].

▷ With conditions (2.3) and (2.4) the system $(H^n(\mathscr{U};\mathscr{F}), \mathfrak{S}^n(\mathscr{V},\mathscr{U}))$ becomes a direct system of abelian groups,

These properties suggest to define $H^n(X\,;\mathscr{F})$ as being the direct limit (cf. Def. p. 36) of the groups $H^n(\mathscr{U}\,;\mathscr{F})$ the limit being taken on the open coverings $\mathscr{U}$ ordered with fineness using the maps $\mathfrak{S}^n(\mathscr{V},\mathscr{U})$. However, all together, the open coverings of a topological space X do not constitute a set in general. This difficulty can be circumvented by limiting the considered coverings to those that are indexed by a given convenient set, that is, a set of indices large enough to allow arbitrarily fine coverings. In the cases we consider any countable set of indices is convenient, say, $\mathbb{N}$ or $\mathbb{Z}$. For $X = S^1$, we may consider coverings with a finite number of open sets since there exist finite coverings of S^1 that are arbitrarily fine. From now on, we assume that the coverings are indexed by subsets J of $\mathbb{N}$.

To circumvent this difficulty another trick due to R. Godement consists in considering only coverings (U_x) by open sets indexed by the points $x \in X$ with the condition $x \in U_x$ (cf. [God58, Sect. 5.8, p. 223]). Hence, the following definition:

Definition 2.2.5 *The* n-th Čech cohomology group of the space X (with values) in $\mathscr{F}$ *is the direct limit of the cohomology groups* $H^n(\mathscr{U};\mathscr{F})$*, the limit being taken over open coverings ordered with fineness. One denotes*

$$\boxed{H^n(X;\mathscr{F}) = \varinjlim_{\mathscr{U}} H^n(\mathscr{U};\mathscr{F})}$$

[1] "Right filtrant" means here that to each finite family $\mathscr{U}_1,\ldots,\mathscr{U}_p$ of open coverings of X there is a covering $\mathscr{V}$ finer than all of them.

When X is a manifold and $n > \dim X$ there exists arbitrarily fine coverings without intersections $n+1$ by $n+1$ and then, $H^n(X;\mathscr{F}) = 0$. The canonical isomorphism $H^0(X;\mathscr{F}) \simeq \Gamma(X;\mathscr{F})$ is valid without restriction.

The following two results are useful.

Theorem 2.2.6 (Leray's Theorem) *Given $\mathscr{U}$ an acyclic covering of X which is either closed and locally finite or open then,*

$$H^n(\mathscr{U};\mathscr{F}) = H^n(X;\mathscr{F}) \quad \textit{for all } n.$$

Acyclic means that $H^n(U_i;\mathscr{F}) = 0$ for all $U_i \in \mathscr{U}$ and all $n \geq 1$.

We refer to [God58, Thm. 5.2.4, Cor. p. 209], (case $\mathscr{U}$ closed and locally finite) and to [God58, Thm. 5.4.1, Cor. p. 213] (case $\mathscr{U}$ open).

Theorem 2.2.7 *To any short exact sequence of sheaves of abelian groups over X*

$$0 \to \mathscr{G} \longrightarrow \mathscr{F} \longrightarrow \mathscr{H} \to 0$$

there is a long exact sequence of cohomology

$$\begin{aligned} 0 \to H^0(X;\mathscr{G}) &\longrightarrow H^0(X;\mathscr{F}) \longrightarrow H^0(X;\mathscr{H}) \\ &\xrightarrow{\delta_0} H^1(X;\mathscr{G}) \longrightarrow H^1(X;\mathscr{F}) \longrightarrow H^1(X;\mathscr{H}) \\ &\quad\xrightarrow{\delta_1} H^2(X;\mathscr{G}) \longrightarrow H^2(X;\mathscr{F}) \longrightarrow H^2(X;\mathscr{H}) \\ &\qquad\xrightarrow{\delta_2} \cdots \end{aligned}$$

The maps $\delta_0, \delta_1, \delta_2, \ldots$ are called *coboundary maps*. For their general definition, see the references above.

2.2.3 The Borel-Ritt Theorem and Cohomology

We know from Corollary 2.1.30, p. 46 that the sheaves $\mathscr{A}/\mathscr{A}^{<0}$ and $\mathscr{A}_s/\mathscr{A}^{\leq -k}$ are constant sheaves with stalks $\mathbb{C}[[x]]$ and $\mathbb{C}[[x]]_s$ respectively. Their global sections

$$\Gamma(S^1;\mathscr{A}/\mathscr{A}^{<0}) \equiv H^0(S^1;\mathscr{A}/\mathscr{A}^{<0})$$

and

$$\Gamma(S^1;\mathscr{A}_s/\mathscr{A}^{\leq -k}) \equiv H^0(S^1;\mathscr{A}_s/\mathscr{A}^{\leq -k})$$

are then also respectively isomorphic to $\mathbb{C}[[x]]$ and $\mathbb{C}[[x]]_s$ and we can state the following corollary of the Borel-Ritt theorem.

Corollary 2.2.8 (Borel-Ritt) *The Taylor map induces the following isomorphisms:*

$$H^0(S^1;\mathscr{A}/\mathscr{A}^{<0}) \simeq \mathbb{C}[[x]], \quad H^0(S^1;\mathscr{A}_s/\mathscr{A}^{\leq -k}) \simeq \mathbb{C}[[x]]_s.$$

We can synthesize:

$$\left.\begin{array}{l}\text{formal series}\\ \widetilde{f}(x)=\sum\limits_{n\geq 0} a_n x^n \in \mathbb{C}[[x]]\end{array}\right\} \Longleftrightarrow \left\{\begin{array}{l}\text{(equivalence class of a)}\\ \text{0-cochain } (f_j)_{j\in J} \text{ over } S^1\\ \text{with values in } \mathscr{A} \text{ and}\\ \text{coboundary } (f_j-f_\ell)_{j,\ell\in J}\\ \text{with values in } \mathscr{A}^{<0}\end{array}\right.$$

The components $f_j(x)$ of the 0-cochains are all asymptotic to $\widetilde{f}(x)$.

$$\left.\begin{array}{l}s\text{-Gevrey series}\\ \widetilde{f}(x)=\sum\limits_{n\geq 0} a_n x^n \in \mathbb{C}[[x]]_s\end{array}\right\} \Longleftrightarrow \left\{\begin{array}{l}\text{(equivalence class of a)}\\ \text{0-cochain } (f_j)_{j\in J} \text{ over } S^1\\ \text{with values in } \mathscr{A}_s \text{ and}\\ \text{coboundary } (f_j-f_\ell)_{j,\ell\in J}\\ \text{with values in } \mathscr{A}^{\leq -k}\end{array}\right.$$

The components $f_j(x)$ of the 0-cochains are all s-Gevrey asymptotic to $\widetilde{f}(x)$. From proposition 1.2.17, p. 20 it would actually be sufficient to ask for the coboundary to be with values in $\mathscr{A}^{<0}$. This latter equivalence will be improved in corollary 5.2.2, p. 140.

2.2.4 The Space S^1 and the Cauchy-Heine Theorem

Since, in what follows, we will mostly be dealing with sheaves over $X = S^1$, it is worth developing this case further. With $X = S^1$ things are often made simpler by the fact that S^1 is a manifold of dimension 1. On another hand, one has to take into account the fact that S^1 has a non-trivial π_1.

Definition 2.2.9 (Good Covering) *An open covering $\mathscr{I} = (I_j)_{j\in J}$ of S^1 is said to be a* good covering *if the following conditions are satisfied:*

- ▷ *$\mathscr{I}$ is finite with $|J| = p$ elements,*
- ▷ *its elements I_j are connected* (i.e., *open arcs of S^1*),
- ▷ *it has thickness ≤ 2* (i.e., *no 3-by-3 intersections*),
- ▷ *when $p = 2$ its two open arcs I_1 and I_2 are proper arcs of S^1 so that $I_1 \cap I_2$ consists of two disjoint open arcs which we denote by $\dot{I}_1$ and $\dot{I}_2$; when $p \geq 3$ its*

open arcs I_j can be indexed by the cyclic group $\mathbb{Z}/p\mathbb{Z}$ so that $\dot{I}_j := I_j \cap I_{j+1} \neq \emptyset$ and $I_j \cap I_\ell = \emptyset$ as soon as $|\ell - j| > 1$ modulo p.

The definition implies that open arcs of a good covering are not properly nested. The family of arcs $\dot{I}_j = I_j \cap I_{j+1}$ is sometimes called the *nerve of the covering $\mathscr{I}$*.

The case $p = 1$, that is, the case of coverings of S^1 by just one arc, is worth to consider. These unique arcs cannot be proper arcs of S^1: one has to introduce overlapping arcs, that is, arcs of the universal covering of S^1 of length $> 2\pi$. Such coverings are widely used to make proofs simpler by using the additivity of 1-cocycles. A typical example is given by the Cauchy-Heine theorem (Thm. 1.4.2, p. 27 and Cor. 2.2.15 below).

Definition 2.2.10 (Elementary good covering) *An open covering $\mathscr{I} = \{I\}$ with only one overlapping open arc $I =]\alpha, \beta + 2\pi[$ and nerve $\dot{I} =]\alpha, \beta[\subsetneq S^1$ is called an* elementary good covering.

Remark 2.2.11 From the topology on S^1 defined page 41 an arc of S^1 appears as an equivalent class of sectors with vertex 0, same opening and arbitrary small radius in $\mathbb{C}^*$. Arcs of length more that 2π correspond to equivalence classes of sectors in the universal covering $\widetilde{\mathbb{C}}$ of $\mathbb{C}^*$.

Abusively, we will talk of covering of S^1 by sectors, replacing once more the equivalence class by one of its representative.

Example 2.2.12 (The Euler series as a 0-cochain)

The Euler series $\widetilde{f}(x)$, which belongs to $\mathbb{C}[[x]]_1$, can be seen as a 0-cochain as follows. Consider the covering $\mathscr{I} = \{I_1, I_2\}$ of S^1 consisting of the two arcs

$$I_1 =]-3\pi/2, +\pi/2[\quad \text{and} \quad I_2 =]-\pi/2, +3\pi/2[.$$

These intersect over the two arcs

$$\dot{I}_1 = \{x \in S^1\,;\, \Re(x) < 0\} \quad \text{and} \quad \dot{I}_2 = \{x \in S^1\,;\, \Re(x) > 0\}.$$

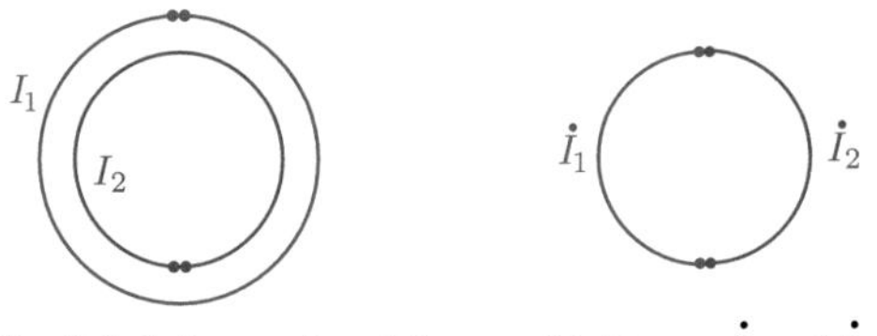

Fig. 2.4. *Left*: arcs I_1 and I_2 — *Right*: arcs $\dot{I}_1$ and $\dot{I}_2$

The corresponding 0-cochain to consider is the pair $(f_1(x), f_2(x))$ consisting of the restrictions of the Euler function $f(x)$ to I_1 and I_2 respectively. Both $f_1(x)$ and $f_2(x)$ are sections of $\mathscr{A}_1$. The coboundary $(\dot{f}_1, \dot{f}_2)$ is given by

$$\dot{f}_1(x) = f_1(x) - f_2(x) = 2\pi \mathrm{i}\, \mathrm{e}^{1/x} \text{ on } \dot{I}_1 \quad \text{and} \quad \dot{f}_2(x) = f_2(x) - f_1(x) = 0 \text{ on } \dot{I}_2$$

and has values in $\mathscr{A}^{\leq -1}$.

Since the component $\dot{f}_2$ is trivial one could also consider a branch covering consisting of the unique arc $I =]-3\pi/2, +3\pi/2[$ overlapping on $\dot{I} = \{x \in S^1\,;\, \Re(x) < 0\}$.

The (branched) 0-cochain $f(x)$ is 1-Gevrey asymptotic to $\widetilde{f}(x)$ on I and its coboundary $f^+(x) - f^-(x) = 2\pi\mathrm{i}\exp(1/x)$ is a section of $\mathscr{A}^{\leq -1}$ over $\dot{I}$.

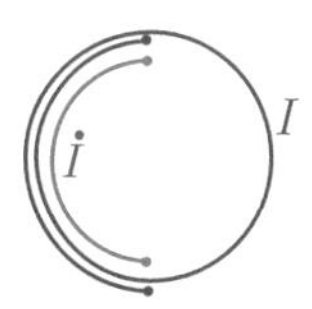

Fig. 2.5. Arcs I and $\dot{I}$

Given a good covering $\mathscr{I} = \{I_j\}$ of S^1, a 1-cochain is a family

$$f_{j,\ell} \in \Gamma(I_j \cap I_\ell; \mathscr{F}) \quad \text{for } j \text{ and } \ell \in \mathbb{Z}/p\mathbb{Z}.$$

The 1-cocycle conditions $\dot{f}_{j,k} + \dot{f}_{k,\ell} = \dot{f}_{j,\ell}$ on $I_j \cap I_k \cap I_\ell$ for all j,k,ℓ are empty since so are the 3-by-3 intersections; consequently, any 1-cochain is a 1-cocycle. Taking into account the necessary conditions $f_{j,j} = 0$ and $f_{k,j} = -f_{j,k}$ on 1-cocycles, a 1-cocycle can thus be seen as any collection $(\dot{f}_j \in \Gamma(\dot{I}_j; \mathscr{F}))$ for $j \in \mathbb{Z}/p\mathbb{Z}$.

By linearity, a 1-cocycle $(\dot{f}_j)_{j\in\mathbb{Z}/p\mathbb{Z}}$ can be decomposed into a sum $\sum_{j\in\mathbb{Z}/p\mathbb{Z}} \dot{\varphi}_j$ where $\dot{\varphi}_j$ is the 1-cocycle over the covering $\mathscr{I}$ having all trivial components (equal to 0, the neutral element) but the jth equal to $\dot{f}_j$. Fix j and consider the elementary good covering $\mathscr{I}_j$ whose nerve is $\dot{I}_j$ and the 1-cocycle $\dot{f}_j \in \Gamma(\dot{I}_j, \mathscr{F})$. The covering $\mathscr{I}$ is finer than $\mathscr{I}_j$. We identify the 1-cocycles $\dot{\varphi}_j$ and $\dot{f}_j$ and we say that the 1-cocycle $\dot{\varphi}_j$ can be lifted into the elementary 1-cocycle $\dot{f}_j$.

Proposition 2.2.13 *There exist arbitrarily fine good coverings of S^1*

Henceforth, when $\mathscr{F}$ is a sheaf over S^1, to determine $H^1(S^1; \mathscr{F})$ it suffices to consider good coverings.

Example 2.2.14 (Euler Equation and Cohomology)

We consider the elementary good covering $\mathscr{I} = \{I\}$ of S^1 defined by the overlapping interval $I =]-3\pi/2, +3\pi/2[$ with self-intersection $\dot{I} =]-3\pi/2, -\pi/2[$ and we consider the sheaf $\mathscr{V}$ of asymptotic solutions of the Euler equation (cf. Exa. 2.1.27, p. 44). A 1-cocycle of $\mathscr{I}$ in $\mathscr{V}$ is a section $\dot{\varphi}(x) = af(x) + b\mathrm{e}^{-1/x}$ over $\dot{I}$ with arbitrary constants a and b in $\mathbb{C}$. There is no 1-cocycle condition. The 0-cochains are of the form $cf(x)$ over I, with $c \in \mathbb{C}$ an arbitrary constant and they generate the 1-coboundaries $c\big(f(x\mathrm{e}^{2\pi\mathrm{i}}) - f(x)\big) = 2\pi\mathrm{i}c\mathrm{e}^{-1/x}$. The cohomological class of $\dot{\varphi}$ is then uniquely represented by $af(x)$ for $-3\pi/2 < \arg(x) < -\pi/2$. Hence, $H^1(\mathscr{I}; \mathscr{V})$ is a vector space of dimension one, isomorphic to $\mathbb{C}$.

Given $\mathscr{J}$ a covering of S^1 finer than $\mathscr{I}$ we saw (cf. Prop. 2.2.4, p. 51) that the map

$$\mathfrak{S}_{\mathscr{J},\mathscr{I}} : H^1(\mathscr{I}; \mathscr{V}) \to H^1(\mathscr{J}; \mathscr{V})$$

is injective. Let us check that it is surjective on the example of the covering

$$\mathscr{J} = \{J_1, J_2\} \quad \text{for} \quad J_1 =]-\pi/4, 5\pi/4[\quad \text{and } J_2 =]-5\pi/4, \pi/4[.$$

We set $\dot{J}_1 =]3\pi/4, 5\pi/4[$ and $\dot{J}_2 =]-\pi/4, \pi/4[$.

A 1-cocycle $(\dot{\varphi}_1, \dot{\varphi}_2)$ over the covering $\mathscr{J}$ is cohomologous to $(\dot{\varphi}_1 + \dot{\varphi}_2, 0)$ via the 0-cochain $(0, \dot{\varphi}_2)$ where we keep denoting by $\dot{\varphi}_2$ the continuation of $\dot{\varphi}_2$ to $\dot{J}_2$. However, $\dot{\varphi} = \dot{\varphi}_1 + \dot{\varphi}_2$ can be continued to $\dot{I}$ (we keep denoting by $\dot{\varphi}$ the continuation) and therefore, the 1-cocycle $(\dot{\varphi}_1 + \dot{\varphi}_2, 0)$ lifts up into the 1-cocycle $\dot{\varphi}$ of the covering $\mathscr{I}$.

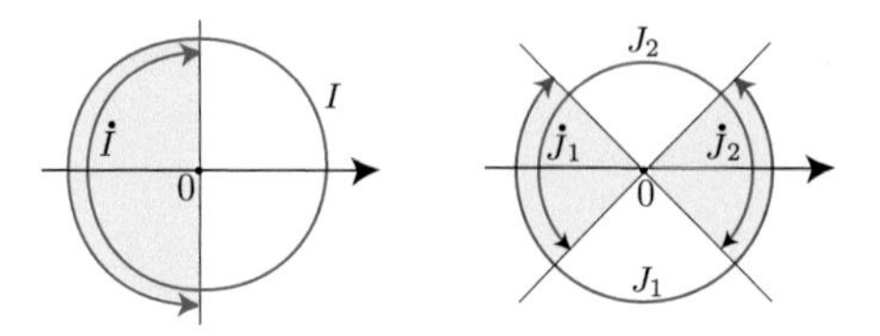

Fig. 2.6. *Left*: covering $\mathscr{I}$ — *Right*: covering $\mathscr{J}$

The proof extends to any good covering $\mathscr{J}$ finer than $\mathscr{I}$ by induction on the number of connected 2-by-2 intersections. We can conclude that $H^1(\mathscr{I};\mathscr{V}) = H^1(S^1;\mathscr{V})$.

The same result can be seen as a consequence of a theorem of Leray (Thm. 2.2.6, p. 52) after showing that $\mathscr{I}$ is acyclic for $\mathscr{V}$.

In the case when $X = S^1$ the long exact sequence of cohomology of theorem 2.2.7, p. 52 reduces to

$$0 \to H^0(S^1;\mathscr{G}) \longrightarrow H^0(S^1;\mathscr{F}) \longrightarrow H^0(S^1;\mathscr{H})$$
$$\xrightarrow{\delta_0} H^1(S^1;\mathscr{G}) \longrightarrow H^1(S^1;\mathscr{F}) \longrightarrow H^1(S^1;\mathscr{H}) \to 0.$$

The coboundary map δ_0 is defined as follows: The sheaf $\mathscr{H}$ is the quotient of $\mathscr{F}$ by $\mathscr{G}$. A 0-cocycle in $H^0(S^1;\mathscr{H})$ is a collection of $f_i \in \Gamma(I_i;\mathscr{F})$ such that $f_i - f_j$ belongs to $\Gamma(I_{i,j};\mathscr{G})$ for all i, j. Denote by $\mathscr{I} = \{I_i\}$ the covering of S^1 by the open arcs I_i for all i. There corresponds the 1-cocycle $(g_{i,j} = f_i - f_j)_{i,j}$ of $\mathscr{I}$ with values in $\mathscr{G}$. Other representatives f_i' of the 0-cocycle on $\mathscr{I}$ yield a cohomologous 1-cocycle $(g'_{i,j} = f'_i - f'_j)_{i,j}$ of $\mathscr{I}$ with values in $\mathscr{G}$. Indeed, for all i, j we can write $g_{i,j} - g'_{i,j} = (f_i - f'_i) - (f_j - f'_j)$ where $(f_i - f'_i)$ and $(f_j - f'_j)$ are defined on I_i and I_j respectively and both have values in $\mathscr{G}$. This defines an element of $H^1\big(\mathscr{I};\mathscr{G}\big)$ and therefore, an element of $H^1(S^1;\mathscr{G})$. We leave to the reader the proof of the fact that this latter element does not depend on the choice of the covering $\mathscr{I}$ on which the initial 0-cocycle of $H^0\big(S^1\,;\mathscr{H}\big)$ is represented.

The Cauchy-Heine theorem (Thm. 1.4.2, p. 27) can be reformulated as a cohomological condition as follows.

Corollary 2.2.15 (Cauchy-Heine)

(i) *The natural map* $H^1(S^1, \mathscr{A}^{<0}) \to H^1(S^1, \mathscr{A})$ *is the null map.*

(ii) *The natural map* $H^1(S^1, \mathscr{A}^{\le -k}) \to H^1(S^1, \mathscr{A}_s)$ *is the null map.*

Proof. (i) It suffices to prove the assertion for any good covering. Given a covering $\mathscr{I}$ of S^1 there is a natural map from $H^1(\mathscr{I};\mathscr{A}^{<0})$ into $H^1(\mathscr{I};\mathscr{A})$ (cohomologous 1-cocycles of $H^1(\mathscr{I};\mathscr{A}^{<0})$ are also cohomologous in $H^1(\mathscr{I};\mathscr{A})$). By linearity, it suffices to consider the case of an elementary good covering $\mathscr{I} = \{I\}$

with self-intersection $\dot{I}$ (cf. Def. 2.2.10, p. 54). The Cauchy-Heine theorem as stated in Thm. 1.4.2, p. 27 says that a 1-cocycle of $H^1(\mathscr{I};\mathscr{A}^{<0})$ is a coboundary in $H^1(\mathscr{I};\mathscr{A})$, that is, it is cohomologous to the trivial 1-cocycle 0 in $H^1(\mathscr{I};\mathscr{A})$.

(ii) Same proof by replacing $\mathscr{A}^{<0}$ by $\mathscr{A}^{\leq -k}$ and $\mathscr{A}$ by $\mathscr{A}_s$. □

Although these maps are zero maps, far from being null spaces, $H^1(S^1,\mathscr{A})$ and $H^1(S^1,\mathscr{A}_s)$ are huge spaces.

2.3 Non-abelian Čech Cohomology

We assume now that $\mathscr{F}$ is a *sheaf of non-abelian groups* over a topological space X and we denote multiplicatively the group law. The set $\Gamma(U;\mathscr{F})$ of sections of $\mathscr{F}$ over a subset $U \subset X$ inherits from $\mathscr{F}$ the structure of a non-abelian group: one defines the product of two sections point wise in $\mathscr{F}$. Unless otherwise specified the term covering still means *covering by open sets*.

For our purpose and especially for the meromorphic classification of systems (cf. Sect. 3.5, p. 90), we need only to consider non-abelian cohomology in degree 0 and degree 1, that is, to consider the cases of H^0 and H^1. Indeed, the base to consider being the space S^1 of dimension 1 all cohomological information is contained in H^0 and H^1. The cohomology sets H^0 and H^1 can be defined similarly as in the abelian case; this is no longer possible for non-abelian cohomology sets in higher degrees which, fortunately, we do not consider here. For this section we refer to [Fre57].

2.3.1 Non-abelian Cohomology in Degree 0

Suppose we are given a covering $\mathscr{U} = (U_i)_{i\in I}$ and denote $U_{i,j} = U_i \cap U_j$ for all i,j. As in the abelian case a 0-*cochain* (of $\mathscr{U}$ with values in $\mathscr{F}$) is a family $(f_i)_{i\in I} \in \prod \Gamma(U_i\,;\mathscr{F})$ of sections of $\mathscr{F}$ over the U_i's. We denote by $\mathscr{C}^0(\mathscr{U}\,;\mathscr{F})$ the set of 0-cochains of $\mathscr{U}$ with values in $\mathscr{F}$. The 0-cochain $(f_i)_{i\in I}$ is said to be a 0-*cocycle* if $f_i = f_j$ on $U_{i,j}$ for all $i,j \in I$. From the definition of a sheaf a 0-cocycle determines a global section of $\mathscr{F}$. Two 0-cocycles are said to be cohomologous if they determine the same global section and one can identify the set of global sections $\Gamma(X\,;\mathscr{F})$ to the set of cohomology classes of 0-cocycles.

Definition 2.3.1 *$H^0(\mathscr{U}\,;\mathscr{F}) \simeq \Gamma(X\,;\mathscr{F})$ is the set of cohomology classes of 0-cocycles of $\mathscr{U}$ with values in $\mathscr{F}$. One also denotes $\Gamma(X\,;\mathscr{F})$ by $H^0(X\,;\mathscr{F})$.*

Via its identification to $\Gamma(X\,;\mathscr{F})$, the 0th set of cohomology $H^0(\mathscr{U}\,;\mathscr{F})$ inherits the structure of a non-abelian group.

Let the covering $\mathscr{V} = (V_j)_{j\in J}$ refine the covering $\mathscr{U}$ (cf. Def. 2.2.3, p. 50) and let $\sigma : J \to I$ such that $V_j \subseteq U_{\sigma(j)}$ be a simplicial map for $\mathscr{U}$ and $\mathscr{V}$. To σ there corresponds a map$\sigma^* : \mathscr{C}^0(\mathscr{U}\,;\mathscr{F}) \to \mathscr{C}^0(\mathscr{V}\,;\mathscr{F})$ defined by $\sigma^*((f_i)_{i\in I}) = (f_{\sigma(j)|V_j})_{j\in J}$.

Recall that the notation $_{|V_j}$ means "restriction to V_j". The map σ^* induces an isomorphism of groups between $H^0(\mathscr{U}\,;\mathscr{F})$ and $H^0(\mathscr{V}\,;\mathscr{F})$ which is compatible with the identification to $\Gamma(X\,;\mathscr{F})$.

2.3.2 Non-abelian Cohomology in Degree 1

As in the abelian case a 1-cochain is a family $(f_{i,j})_{i,j\in J} \in \prod \Gamma(U_{i,j}\,;\mathscr{F})$ of sections of $\mathscr{F}$ over the intersections $U_{i,j} = U_i \cap U_j$. We denote $U_{i,j,k} = U_i \cap U_j \cap U_k$.

Definition 2.3.2 *A 1-cochain $(f_{i,j})_{i,j\in J} \in \Gamma(U_{i,j}\,;\mathscr{F})$ is said to be a* 1-cocycle *if it satisfies, for all $i,j,k \in I$, the* cocycle condition *(also said* Chasles condition*)*

$$\boxed{f_{i,j}\,f_{j,k} = f_{i,k} \quad on \quad U_{i,j,k}.}$$

One has $f_{i,i}(x) = e_x$, the neutral element in the stalk $\mathscr{F}_x$, for all $x \in U_i$ and all $i \in I$ and $f_{i,j}(x)\,f_{j,i}(x) = e_x$ for all $x \in U_{i,j}$ and all $i,j \in I$.

Definition 2.3.3 *Two 1-cochains $(f_{i,j})$ and $(g_{i,j})$ of $\mathscr{U}$ with values in $\mathscr{F}$ are said to be* cohomologous *if there exists a 0-cochain (h_i) of $\mathscr{U}$ with values in $\mathscr{F}$ which conjugates $(f_{i,j})$ and $(g_{i,j})$, that is, which satisfies for all $i,j \in I$ the relation of cohomology*

$$\boxed{f_{i,j} = h_i^{-1}\,g_{i,j}\,h_j \quad on \quad U_{i,j}.} \tag{2.5}$$

The relation of cohomology is an equivalence relation; it is compatible with the cocycle condition so that we can talk of cohomologous 1-cocycles and set the following definition.

Definition 2.3.4 *The* first cohomology set of $\mathscr{U}$ with values in $\mathscr{F}$ *is the set $H^1(\mathscr{U}\,;\mathscr{F})$ of the cohomology classes of 1-cocycles of $\mathscr{U}$ with values in $\mathscr{F}$.*

When the group $\mathscr{F}$ is abelian $H^1(\mathscr{U}\,;\mathscr{F})$ coincide with the first abelian cohomology group (denoted the same way with no harm).

It is worth to notice however that when the group law of $\mathscr{F}$ is not abelian a product of 1-cocycles is not a 1-cocycle in general and the set $H^1(\mathscr{U}\,;\mathscr{F})$ *does not inherit from $\mathscr{F}$ the structure of a group*. Nevertheless, $H^1(\mathscr{U}\,;\mathscr{F})$ earns a special element called *neutral element* defined as the cohomology class of the neutral 1-cocycle $(f_{i,j})$ where $f_{i,j} = e_x$ for all $x \in U_{i,j}$ and all $i,j \in I$.

Let $\mathscr{V} = (V_j)_{j\in J} \preceq \mathscr{U}$ be a covering finer than $\mathscr{U}$ and $\sigma : J \to I$ a simplicial map, meaning that σ satisfies the condition $V_j \subseteq U_{\sigma(j)}$ for all j. The associated map $\sigma^* : \mathscr{C}^1(\mathscr{U}\,;\mathscr{F}) \to \mathscr{C}^1(\mathscr{V}\,;\mathscr{F})$ is compatible with the cocycle and the cohomology conditions.

It results that the map σ^* induces an application

$$\mathfrak{S}(\mathscr{V},\mathscr{U}) : H^1(\mathscr{U};\mathscr{F}) \longrightarrow H^1(\mathscr{V};\mathscr{F}).$$

The map $\mathfrak{S}(\mathscr{V},\mathscr{U})$ sends the neutral element of $H^1(\mathscr{U};\mathscr{F})$ to the neutral element of $H^1(\mathscr{V};\mathscr{F})$.

Proposition 2.3.5 *As in the abelian case, the map $\mathfrak{S}(\mathscr{V},\mathscr{U})$ does not depend on the choice of the simplicial map σ.*

Proof. Let $(f_{i,j})$ be a 1-cocycle of $\mathscr{U}$ and $\sigma,\tau : J \to I$ be two simplicial maps for $\mathscr{U}$. We must prove that the 1-cocycles $(g_{i,j}) = (f_{\sigma(i),\sigma(j)}\big|_{V_{i,j}})$ and $(h_{i,j}) = (f_{\tau(i),\tau(j)}\big|_{V_{i,j}})$ are cohomologous. To this end, consider the 0-cochain $\varphi_i = f_{\tau(i),\sigma(i)}\big|_{V_i}$ and observe that the cocycle condition implies

$$g_{i,j} \equiv f_{\sigma(i),\sigma(j)}\big|_{V_{i,j}} = f_{\sigma(i),\tau(i)} f_{\tau(i),\tau(j)} f_{\tau(j),\sigma(j)}\big|_{V_{i,j}} = \varphi_i^{-1} h_{i,j} \varphi_j.$$

This achieves the proof. □

One can check, like in the abelian case, the relations

$$\begin{cases} \mathfrak{S}(\mathscr{U},\mathscr{U}) = \mathrm{id} \quad \text{for all} \quad \mathscr{U} \\ \mathfrak{S}(\mathscr{W},\mathscr{V})\mathfrak{S}(\mathscr{V},\mathscr{U}) = \mathfrak{S}(\mathscr{W},\mathscr{U}) \quad \text{for all} \quad \mathscr{W} \preceq \mathscr{V} \preceq \mathscr{U} \end{cases}$$

and again, the family of sets $H^1(\mathscr{U}\,;\mathscr{F})$ and applications $\mathfrak{S}(\mathscr{V},\mathscr{U})$ is a direct system of sets (see Def. 2.1.8, p. 35,and below). We can then state the following definition.

Definition 2.3.6 *The first set of (non-abelian) Čech cohomology of the space X with values in $\mathscr{F}$ is the direct limit*

$$\boxed{H^1(X\,;\mathscr{F}) = \varinjlim_{\mathscr{U}} H^1(\mathscr{U};\mathscr{F})}$$

the limit being taken over coverings $\mathscr{U}$ ordered with fineness.

Non-abelian 1-cohomology satisfies the same important property of injectivity as in the abelian case.

Proposition 2.3.7 *The map $\mathfrak{S}(\mathscr{V},\mathscr{U}) : H^1(\mathscr{U};\mathscr{F}) \longrightarrow H^1(\mathscr{V};\mathscr{F})$ is injective.*

Proof. Let $f = (f_{i,j})$ and $g = (g_{i,j})$ be two 1-cocycles of $\mathscr{U}$ with values in $\mathscr{F}$ and denote $\sigma^*(f) = (F_{i,j})$ and $\sigma^*(g) = (G_{i,j})$. Suppose that the 1-cocycles $\sigma^*(f)$ and $\sigma^*(g)$ of $\mathscr{V}$ are cohomologous, that is, that there exists a 0-cochain $\phi = (\phi_\alpha)$ such that $F_{\alpha,\beta} = \phi_\alpha^{-1} G_{\alpha,\beta}\, \phi_\beta$ on $V_\alpha \cap V_\beta$ for all α,β. We must prove that there exists a 0-cochain $\varphi = (\varphi_i)$ of $\mathscr{U}$ with values in $\mathscr{F}$ conjugating f and g.

One can choose φ as follows: for $x \in U_i$, choose α such that x belongs to V_α and set

$$\varphi_i(x) = g_{i,\sigma(\alpha)}(x)\,\phi_\alpha(x)\,f_{\sigma(\alpha),i}(x).$$

The value thus obtained for $\varphi_i(x)$ is independent of the choice of α. Indeed, let β be another possible choice. Then, we have

$$\begin{aligned} g_{i,\sigma(\beta)}(x)\,\phi_\beta(x)\,f_{\sigma(\beta),i}(x) &= g_{i,\sigma(\beta)}\,G_{\alpha,\beta}^{-1}\,\phi_\alpha\,F_{\alpha,\beta}\,f_{\sigma(\beta),i}(x) \\ &= \left(g_{i,\sigma(\beta)}\,g_{\sigma(b),\sigma(\alpha)}\right)\,\phi_\alpha\,\left(f_{\sigma(\alpha),\sigma(\beta)}\,f_{\sigma(\beta),i}(x)\right) \\ &= g_{i,\sigma(\alpha)}(x)\,\phi_\alpha(x)\,f_{\sigma(\alpha),i}(x). \end{aligned}$$

It follows from the relation

$$\varphi_i^{-1}\,g_i\,\varphi_j(x) = f_{i,\sigma(\alpha)}\,\phi_\alpha^{-1}\left(g_{\sigma(\alpha),i}\,g_{i,j}\,g_{j,\sigma(\alpha)}\right)\phi_\alpha\,f_{\sigma(\alpha),j}(x)$$

where $g_{\sigma(\alpha),i}\,g_{i,j}\,g_{j,\sigma(\alpha)} = e_x$ that $\varphi_i^{-1}\,g_i\,\varphi_j(x) = f_{i,j}(x)$ for all $x \in U_i$. Hence the cocycles f and g are cohomologous as 1-cocycles of $\mathscr{U}$ with values in $\mathscr{F}$. □

From this proposition it follows that one can identify any element of $H^1(\mathscr{U}\,;\mathscr{F})$ to an element of $H^1(X\,;\mathscr{F})$ and see $H^1(X\,;\mathscr{F})$ as the union of $H^1(\mathscr{U}\,;\mathscr{F})$ for all coverings $\mathscr{U}$ of X. In particular, any element of $H^1(X\,;\mathscr{F})$ is represented in a unique way in $H^1(\mathscr{U}\,;\mathscr{F})$ for $\mathscr{U}$ fine enough.

Remark 2.3.8 Note that, for short and by abuse of language, instead of saying that the two 1-cocycles f and g are cohomologous as 1-cocycles of $\mathscr{U}$ with values in $\mathscr{F}$ one also says that f and g are cohomologous in $H^1(\mathscr{U}\,;\mathscr{F})$.

2.3.3 Exact Sequences

Let $\mathscr{F}$ be a sheaf of groups over X and $\mathscr{G}$ a subsheaf of groups.

We recall that the quotient sheaf $\mathscr{F}/\mathscr{G}$ is, by definition, the sheaf associated with the presheaf $U \mapsto \mathscr{F}(U)/\mathscr{G}(U)$, i.e., the classes of elements of $\mathscr{F}(U)$ under the action of $\mathscr{G}(U)$ to the right. One defines similarly the quotient $\mathscr{G}\backslash\mathscr{F}$ by taking the classes of elements of $\mathscr{F}(U)$ under the action of $\mathscr{G}(U)$ to the left. As for groups themselves if $\mathscr{G}$ is not normal in $\mathscr{F}$ the quotient $\mathscr{F}/\mathscr{G}$ is not a sheaf of groups.

Definition 2.3.9 *By analogy with the case of linear morphisms a sequence of sets*

$$G \xrightarrow{f} F \xrightarrow{g} H$$

where H contains a neutral element e is said to be exact if $g^{-1}(e) = f(G)$.

The coboundary map $\delta : H^0(X\,;\mathscr{F}/\mathscr{G}) \longrightarrow H^1(X\,;\mathscr{G})$ exists also in the case of sheaves of non-abelian groups and is defined as follows.

Let $c \in H^0(X\,;\mathscr{F}/\mathscr{G})$. By definition, c is a collection of compatible sections of the presheaf $U \mapsto \mathscr{F}(U)/\mathscr{G}(U)$. This means that there exists a covering $\mathscr{U} = (U_i)_{i\in I}$ of X and sections $c_i \in \mathscr{F}(U_i)$ such that $c_{i,j} = c_i^{-1}c_j$ belongs to $\mathscr{G}(U_{i,j})$ for all i and j, thus defining c in the form $c = (c_i)_{i\in I}$. The $c_{i,j}$'s provide us with a 1-cochain and actually, a 1-cocycle $(c_{i,j})_{i,j\in I}$ of $\mathscr{U}$ with values in $\mathscr{G}$. The cohomology class of $(c_{i,j})$ in $H^1(\mathscr{U}\,;\mathscr{G})$ does not depend on the choice of the c_i's to represent c. Indeed, let $c'_i \in \mathscr{F}(U_i), i \in I$ be other representatives of c. Setting $b_i = c_i^{-1}c'_i$ the new 1-cocycle $(c'_{i,j})$ satisfies $c'_{i,j} = c'^{-1}_i c'_j = b_i^{-1}c_i^{-1}c_j b_j = b_i^{-1}c_{i,j}b_j$. Since b_i belongs to $\mathscr{G}(U_i)$ for all i this proves that the two 1-cocycles are cohomologous in $H^1(\mathscr{U}\,;\mathscr{G})$; denote by $\delta_{\mathscr{U}}(c)$ their common cohomology class.

Finally, let us prove that the image of $\delta_{\mathscr{U}}$ in $H^1(X\,;\mathscr{G})$ does not depend on the choice of $\mathscr{U}$. To this end, consider a covering $\mathscr{V} = (V_j)_{j\in J}$ finer than $\mathscr{U}$ and a simplicial map $\sigma : J \to I$. We can represent the 0-cochain c in the form $c = (\gamma_j)_{j\in J}$ where $\gamma_j = c_{\sigma(j)|V_j}$. Denote by $\delta_{\mathscr{V}}(c)$ the cohomology class built from the γ_j's in $H^1(\mathscr{V}\,;\mathscr{G})$ like $\delta_{\mathscr{U}}(c)$ has been built from the c_j's in $H^1(\mathscr{U}\,;\mathscr{G})$. It follows that $\delta_{\mathscr{V}}(c) = \mathfrak{S}(\mathscr{V},\mathscr{U})\,\delta_{\mathscr{U}}(c)$ which proves that the 0-cochains $(c_i)_{i\in I}$ and $(\gamma_j)_{j\in J}$ induce the same cohomology class in $H^1(X\,;\mathscr{G})$.

This achieves the construction of the coboundary map δ. □

Theorem 2.3.10 *With a short exact sequence of sheaves of sets with neutral element e:*

$$e \longrightarrow \mathscr{G} \xrightarrow{\ i\ } \mathscr{F} \xrightarrow{\ p\ } \mathscr{F}/\mathscr{G} \longrightarrow e$$

there is the long exact sequence of cohomology sets:

$$e \longrightarrow H^0(X\,;\mathscr{G}) \xrightarrow{i_0^*} H^0(X\,;\mathscr{F}) \xrightarrow{p_0^*} H^0(X\,;\mathscr{F}/\mathscr{G}) \xrightarrow{\delta} H^1(X\,;\mathscr{G}) \xrightarrow{i_1^*} H^1(X\,;\mathscr{F}).$$

If, moreover, $\mathscr{G}$ is normal in $\mathscr{F}$ one can continue the exact sequence to the right with the extra term $\xrightarrow{p_1^*} H^1(X\,;\mathscr{F}/\mathscr{G})$.

Proof. We have to prove the following non trivial three points.

▷ *Prove the equality* $\mathfrak{I}(p_0^*) = \ker(\delta)$.

Let $f \in H^0(X\,;\mathscr{F})$: there exists a covering $\mathscr{U} = (U_i)$ of X and sections f_i of $\mathscr{F}$ over U_i such that $f_i^{-1}(x)f_j(x) = e_x$ for all i,j and $x \in U_{i,j}$. The image $c = p_0^*(f)$ is defined by the 0-cochain $(c_i) = (f_i \mod \mathscr{G})$ in $H^0(\mathscr{U}\,;\mathscr{F}/\mathscr{G})$. Applying the coboundary map we obtain the neutral 1-cocycle $\delta(c_i) = (f_i^{-1}(x)f_j(x) = e_x)$ for all i,j and $x \in U_{i,j}$. This proves the inclusion $\mathfrak{I}(p_0^*) \subseteq \ker(\delta)$.

Conversely, let $c \in H^0(X\,;\mathscr{F}/\mathscr{G})$ be such that $\delta(c)$ is the neutral element e of $H^1(X\,;\mathscr{G})$. By definition of c, there exist a covering $\mathscr{U}$ and a 0-cocycle (c_i) of $\mathscr{U}$ with values in $\mathscr{F}$, such that $c_{i,j} = c_i^{-1}c_j$ belongs to $\Gamma(U_{i,j}\,;\mathscr{G})$ for all i,j. Moreover, $\delta\big((c_i)_{i\in I}\big)$ is the 1-cocycle $(c_{i,j})$ in $H^1(\mathscr{U}\,;\mathscr{G})$. However, from the injectivity of the maps $\mathfrak{S}(\mathscr{V},\mathscr{U})$ the hypothesis $\delta(c) = e$ means that, for $\mathscr{U}$ fine enough, the 0-cocycle $(c_{i,j})$ itself is trivial in $H^1(\mathscr{U};\mathscr{G})$. After refining $\mathscr{U}$ if necessary, we can then assert that there exists a 0-cochain (b_i) of $\mathscr{U}$ with values in $\mathscr{G}$ such that $c_{i,j} = b_i^{-1} b_j$ for all j. The 0-cochain $(\gamma_i = c_i b_i^{-1})$ defines an element of $H^0(X;\mathscr{F})$ satisfying $p_0^*(\gamma_i) = c$. Hence, the converse inclusion $\ker(\delta) \subseteq \Im(p_0^*)$ and the result.

▷ *Prove the equality* $\Im(\delta) = \ker(i_1^*)$.

As seen above, an element $\gamma = \delta(c) \in H^1(X;\mathscr{G})$ is given by a 1-cocycle $(c_{i,j})$ of a covering $\mathscr{U} = (U_i)$ with values in $\mathscr{G}$ satisfying $c_{i,j} = c_i^{-1} c_j$ and $c_i \in \Gamma(U_i;\mathscr{F})$ for all i, j. This latter condition means that $(c_{i,j})$ seen as a 1-cocycle of $\mathscr{U}$ with values in $\mathscr{F}$ is a 1-coboundary. Hence, its image $i_1^*(c_{i,j})$ is trivial in $H^1(X;\mathscr{F})$. This proves that $\Im(\delta) \subseteq \ker(i_1^*)$.

Conversely, given $\gamma \in \ker(i_1^*)$ we must prove that γ belongs to $\Im(\delta)$. By hypothesis, there exist a covering $\mathscr{U}$ and a 1-cocycle $(c_{i,j})$ of $\mathscr{U}$ with values in $\mathscr{G}$ which defines γ and is trivial as a 1-cocycle with values in $\mathscr{F}$: there exists a 0-cochain $c = (c_i)$ in $H^0(\mathscr{U};\mathscr{F})$ such that $c_{i,j} = c_i^{-1} c_j$ for all i, j. The image $p_0^*(c)$ defines a 0-cochain of $H^0(X;\mathscr{F}/\mathscr{G})$ such that $\delta(p_0^*(c)) = (c_{i,j})$. Hence, the converse inclusion $\ker(i_1^*) \subseteq \Im(\delta)$ and the result.

▷ *Prove the equality* $\Im(i_1^*) = \ker(p_1^*)$.

When $\mathscr{G}$ is normal in $\mathscr{F}$ the quotient sheaf $\mathscr{F}/\mathscr{G}$ is a sheaf of groups and it makes sense to consider the first cohomology sets $H^1(\mathscr{U};\mathscr{F}/\mathscr{G})$ and $H^1(X;\mathscr{F}/\mathscr{G})$ for $\mathscr{U}$ a covering of X (cf. Def. 2.3.4, p. 58 and 2.3.6, p. 59) since, indeed, the cohomology relation (2.5), p. 58 makes sense.

By definition of the quotient $\mathscr{F}/\mathscr{G}$, if a 1-cocycle in $H^1(X;\mathscr{F})$ has values in $\mathscr{G}$ then it is trivial as a 1-cocycle in $H^1(X;\mathscr{F}/\mathscr{G})$. Hence, the inclusion $\Im(i_1^*) \subseteq \ker(p_1^*)$.

Conversely, let $c \in H^1(X;\mathscr{F})$ belong to the kernel of p_1^*. This means that there exit a covering $\mathscr{U} = (U_i)_{i\in I}$ of X and a 1-cocycle $(c_{i,j})$ representing c in $H^1(\mathscr{U};\mathscr{F})$ which is trivial as a 1-cocycle in $H^1(\mathscr{U};\mathscr{F}/\mathscr{G})$. It results that there exists a 0-cochain $(c_i) \in \prod \Gamma(U_i;\mathscr{F}/\mathscr{G})$ such that $c_{i,j} \mod \mathscr{G} = c_i^{-1} c_j$ for all i, j. The 0-cochain (c_i) is defined by a collection of local sections with values in $\mathscr{F}$ related to each other by local sections with values in $\mathscr{G}$. One can thus find a covering $\mathscr{U}' = (U'_\alpha)_{\alpha\in A}$ of X finer than $\mathscr{U}$, for which the 0-cochain (c_i) is given in the form of a 0-cochain (c'_α) where $c'_\alpha \in \Gamma(U'_\alpha;\mathscr{F})$ and $c'^{-1}_\alpha c'_\beta \in \Gamma(U'_{\alpha,\beta};\mathscr{G})$ for all α, β. On the covering $\mathscr{U}'$ the cohomology class c is represented by a 1-cocycle $(c'_{\alpha,\beta}) = \mathfrak{S}^1(\mathscr{U}',\mathscr{U})((c_{i,j}))$ with values in $\mathscr{F}$. The 1-cocycle $(c'_\alpha c'_{\alpha,\beta} c'^{-1}_\beta)$ is cohomologous to $(c'_{\alpha,\beta})$ in $H^1(\mathscr{U}';\mathscr{F})$ and has values in $\mathscr{G}$. Hence, the inclusion $\ker(p_1^*) \subseteq \Im(i_1^*)$ and the result.

This achieves the proof of the theorem. □

Chapter 3
Linear Ordinary Differential Equations: Basic Facts and Infinitesimal Neighborhoods at an Irregular Singular Point

Abstract With a view to discussing the theories of summability on solutions of differential equations we include a chapter on the theory of linear differential equations. Only part of the results are proved; otherwise, references to the classical literature are given. We begin by discussing the link between equations, systems and $\mathscr{D}$-modules introducing formal and meromorphic equivalence in each case. We state and discuss the theorem of formal classification, making explicit a normal form. We sketch the calculation of the formal invariants from Newton polygons in the case of equations.
We include the main asymptotic existence theorem both in its classical and sheaf forms and prove that the sheaf form implies the classical one.
We show how the meromorphic classification of systems is related to the Stokes phenomenon. The linear Stokes phenomenon is fully described by the Malgrange-Sibuya theorem, itself improved by the Stokes cocycle theorem. We provide Sibuya's proof of the Malgrange-Sibuya theorem. We make explicit the link between the Stokes phenomenon and the summation of a formal fundamental solution.
We end the chapter with the definition of infinitesimal neighborhoods suited to describing the singularities of linear differential equations. The adequacy of such neighborhoods to characterize the summability properties of the formal solutions of a given differential equation is discussed in chapters 5 and 7 (Defs. 5.4.1, p. 180 and 7.7.1, p. 231).

We denote by

$$D = b_n(x)\frac{\mathrm{d}^n}{\mathrm{d}x^n} + b_{n-1}(x)\frac{\mathrm{d}^{n-1}}{\mathrm{d}x^{n-1}} + \cdots + b_0(x) \quad \text{where } b_n(x) \not\equiv 0 \tag{3.1}$$

a linear differential operator of order n with analytic coefficients at $x = 0$.

Unless otherwise specified, we assume that the coefficients $b_n, b_{n-1}, \dots, b_0$ do not vanish simultaneously at $x = 0$. When the coefficients $b_n, b_{n-1}, \dots, b_0$ are polynomials in x their maximal degree is called the *degree of* D.

M. Loday-Richaud, *Divergent Series, Summability and Resurgence II*,
Lecture Notes in Mathematics, DOI 10.1007/978-3-319-29075-1_3

3.1 Equation versus System

With the differential equation $Dy = 0$ above, setting $Y = {}^t\left[y\ y'\ \cdots\ y^{(n-1)}\right]$ where the symbol ^{t} is used here for denoting the transposed matrix, one associates its companion system $\Delta Y = 0$ defined by the n-dimensional order one differential operator

$$\Delta = \frac{\mathrm{d}}{\mathrm{d}x} - B(x) \quad \text{where } B(x) = \begin{bmatrix} 0 & 1 & \cdots & 0 \\ \vdots & \ddots & \ddots & \vdots \\ \vdots & & 0 & 1 \\ -\dfrac{b_0}{b_n} & \cdots & \cdots & -\dfrac{b_{n-1}}{b_n} \end{bmatrix}.$$

Some questions (such as the effective calculation of the formal invariants) are easier to solve on equations than on systems and vice-versa (such as the formal and meromorphic classifications and the related Stokes phenomenon). Thus, it is worth to address the converse problem of determining if and how one can put a given system in companion form. A solution is given by the cyclic vector lemma below. See also volume I [MS16, Thm. 2.21] and [BCL03] for more general statements and references.

Definition 3.1.1 (Gauge Transformation)
Given a system $\Delta Y \equiv \frac{dY}{\mathrm{d}x} - B(x)Y = 0$ of dimension n with meromorphic coefficients a gauge transformation *is a linear change of the unknown variables $Z = TY$ with T invertible in a sense to be made precise.*

In the case when T belongs to $\mathrm{GL}(n, \mathbb{C}\{x\}[1/x])$ the gauge transformation T is said to be a meromorphic gauge transformation.
In the case when T belongs to $\mathrm{GL}(n, \mathbb{C}[[x]][1/x])$ it is said to be a formal (meromorphic) gauge transformation.

A gauge transformation $Z = TY$ changes the system $\Delta Y = 0$ into the differential system ${}^T\!\Delta Z = 0$ with

$${}^T\!\Delta = T\Delta T^{-1} = \frac{\mathrm{d}}{\mathrm{d}x} - \frac{dT}{\mathrm{d}x}T^{-1} - TBT^{-1}. \tag{3.2}$$

When T is meromorphic (resp. formal), so is ${}^T\!\Delta$; however, ${}^T\!\Delta$ may be meromorphic for some formal T. We can now answer the question above.

Proposition 3.1.2 ((Deligne's) Cyclic Vector Lemma)
To any system $\Delta Y = 0$ with meromorphic coefficients there is a meromorphic gauge transformation $Z = TY$ such that the transformed system ${}^T\!\Delta = 0$ is in companion form.

The formulation in terms of cyclic vectors (cf. Rem. 3.2.6, p. 68) is due to P. Deligne [Deli70, Lem II.1.3] although more algorithmic proofs already existed [Cop36], [Jac37]. The companion form is obtained by differential elimination. Despite the fact that the program is short and simple it is not (at least, not yet) available in computer algebra systems such as Mathematica or Maple (see [BCL03] for a sketched algorithm and references; see also [Ram84, Thm. 1.6.16]). As a consequence of the cyclic vector lemma, theoretical properties can be proved equally on equations or systems (as long as these properties remain unchanged under meromorphic gauge transformations). To perform calculations one could, in principle, by using the algorithm, go from equations to systems and conversely at convenience. Actually, these algorithms are usually very "expensive" and used sparingly.

3.2 The Viewpoint of $\mathscr{D}$-Modules

The notion of differential module, or equivalently, of $\mathscr{D}$-module generalizes the notion of order one differential system in an abstract setting free of coordinates. In this point of view, the gauge transformations and the meromorphic or formal equivalence arise naturally.

Suppose we are given a differential field (K,∂): a field K equipped with a derivation ∂, that is, with an additive map $\partial : K \to K$ satisfying the Liebniz rule

$$\partial(ab) = \partial(a)b + a\partial(b) \quad \text{for all } a,b \in K.$$

Precisely, for our purpose, we suppose that K is either the field $\mathbb{C}\{x\}[1/x]$ of meromorphic series at 0 or the field $\mathbb{C}[[x]][1/x]$ of the formal ones. The derivation is $\partial = \mathrm{d}/\mathrm{d}x$. The constant subfield C of K, i.e., the set of the elements $f \in K$ satisfying $\partial f = 0$, is $C = \mathbb{C}$ and the C-vector space of derivations of K has dimension 1 and generator ∂.

3.2.1 $\mathscr{D}$-Modules and First Order Differential Systems

Definition 3.2.1 *A* differential module[1] (M,∇) *of rank n over K is a K-vector space M of dimension n equipped with a map*

$$\nabla : M \longrightarrow M,$$

called connection, *which satisfies the two conditions:*

(i) ∇ *is additive;*

(ii) ∇ *satisfies the* Leibniz rule $\nabla(fm) = \partial f \cdot m + f\nabla(m)$ *for all $f \in K$ and $m \in M$.*

We observe that ∇ is also $\mathbb{C}$-linear. Indeed, when $f \in \mathbb{C}$ is a constant the Leibniz rule reads $\nabla(fm) = f\nabla(m)$. The link with differential systems is as follows.

[1] In French, one says "un vectoriel à connexion".

Choose a K-basis $\underline{e} = [e_1\ e_2\ \cdots\ e_n]$ of M and let

$$[\varepsilon_1\ \varepsilon_2\ \cdots\ \varepsilon_n] = -[e_1\ e_2\ \cdots\ e_n]B \quad \text{with } B \in \mathrm{gl}(n,K)$$

be its image by ∇ (the minus sign is introduced to fit the usual notations for systems and has no special meaning). The connection ∇ is fully determined by the matrix B. Indeed, let $y = \sum_{j=1}^{n} y_j e_j$ be any element of M. In matrix notation, we write $y = \underline{e}Y$ where $Y = {}^t[y_1\ \cdots\ y_n]$ is the column matrix of the components of y in the basis $\underline{e}$. Then, applying the Leibniz rule, we see that ∇y is uniquely determined by

$$\nabla y = \underline{e}(\partial Y - BY).$$

Thus, with the connection ∇ and the K-basis $\underline{e}$ is naturally associated the differential operator $\Delta = \partial - B$ of order one and dimension n.

Definition 3.2.2 *Let (M_1, ∇_1) and (M_2, ∇_2) be differential modules.*

(i) *A* morphism of differential modules *from (M_1, ∇_1) to (M_2, ∇_2) is a K-linear map $\mathcal{T} : M_1 \to M_2$ which commutes to the connections ∇_1 and ∇_2,* i.e., *such that the following diagram commutes:*

$$\begin{array}{ccc} M_1 & \xrightarrow{\mathcal{T}} & M_2 \\ \nabla_1 \downarrow & & \downarrow \nabla_2 \\ M_1 & \xrightarrow{\mathcal{T}} & M_2 \end{array}$$

(ii) *A morphism $\mathcal{T}$ is an* isomorphism *if $\mathcal{T}$ is bijective.*

Denote by n_1 and n_2 the rank of (M_1, ∇_1) and (M_2, ∇_2) respectively. Choose K-basis $\underline{e_1}$ and $\underline{e_2}$ of M_1 and M_2 and denote by Δ_1 and Δ_2 the differential system operators associated with of ∇_1 and ∇_2 in the basis $\underline{e_1}$ and $\underline{e_2}$ respectively. Denote by T the matrix of $\mathcal{T}$ in these basis. The definition says that $\mathcal{T}$ is a morphism if T satisfies the relation

$$\Delta_2 T = T\Delta_1.$$

It says that $\mathcal{T}$ is an isomorphism if, in addition, $n_1 = n_2$ and the matrix T is invertible so that the condition may be written

$$\Delta_1 = T^{-1}\Delta_2 T$$

and is also valid for T^{-1} in the form $\Delta_1 T^{-1} = T^{-1}\Delta_2$; hence, the commutation of the diagram with $T : M_1 \to M_2$ replaced by $T^{-1} : M_2 \to M_1$. We recognize the formula linking the operators Δ_1 and Δ_2 under the gauge transformation $\mathcal{T}$ (cf. Def. ref-gauge). Suppose $M_1 = M_2 =: M$. An invertible K-morphism $\mathcal{T}$ is just a change of K-basis in M. Therefore, to the connection ∇ there are the infinitely many system operators $T^{-1}\Delta T$ associated with all $T \in \mathrm{GL}(n,K)$ and it is natural to set the following definition.

Definition 3.2.3 *Two differential operators $\Delta_1 = \partial - B_1$ and $\Delta_2 = \partial - B_2$ are said to be* K-equivalent *if there exists a gauge transformation T in* $\mathrm{GL}(n,K)$ *such that*

$$\boxed{\Delta_1 = T^{-1}\Delta_2 T}\,.$$

When $K = \mathbb{C}\{x\}[1/x]$ is the field of meromorphic series the systems are said to be meromorphically equivalent.

When $K = \mathbb{C}[[x]][1/x]$ is the field of formal meromorphic series they are said to be formally equivalent *or* formally meromorphically equivalent.

In modern language, we should say K-similar but the old denomination K-equivalent is still in common use.

The condition is clearly an equivalence relation: indeed, any system operator Δ satisfies $\Delta = I^{-1}\Delta I$; if $\Delta_1 = T^{-1}\Delta_2 T$ then $\Delta_2 = S^{-1}\Delta_1 S$ with $S = T^{-1}$; if $\Delta_1 = T^{-1}\Delta_2 T$ and $\Delta_2 = S^{-1}\Delta_3 S$ then $\Delta_1 = (ST)^{-1}\Delta_3(ST)$. With this definition, a differential module can be identified to an equivalence class of systems.

Denote by $\mathscr{D} = K[\partial]$ the ring of differential operators on K, i.e., the ring of polynomials in ∂ with coefficients in K satisfying the non-commutative rule

$$\partial x = x\partial + 1.$$

Let us show how a differential module can be identified to a $\mathscr{D}$-module, i.e., a module over the ring $\mathscr{D}$ in the classical sense. To this end, we go to a dual approach as follows.

Consider $\mathscr{D}^n$ as a left $\mathscr{D}$-module and denote by $\underline{\varepsilon} = [\varepsilon_1 \cdots \varepsilon_n]$ its canonical $\mathscr{D}$-basis. Given a n-dimensional system operator $\Delta = \partial - B$ with coefficients in K we make it act linearly on $\mathscr{D}^n$ to the right by setting

$$\sum_{j=1}^{n} P_j\varepsilon_j \longmapsto [P_1 \cdots P_n]\Delta = [P_1\partial \cdots P_n\partial] - [P_1 \cdots P_n]B.$$

The cokernel $\mathscr{D}^n/\mathscr{D}^n\Delta$ has a natural structure of left $\mathscr{D}$-module (but no natural structure of right-module over $\mathscr{D}$) and has rank n (its dimension as K-vector space). Denote by $M \equiv \mathscr{D}^n/\mathscr{D}^n\Delta$ this K-vector space of dimension n. The images in the cokernel of the n elements $\varepsilon_1,\ldots,\varepsilon_n$ — which we keep denoting by $\varepsilon_1,\ldots,\varepsilon_n$ — of the canonical $\mathscr{D}$-basis $\underline{\varepsilon}$ form a K-basis of M. On another hand, the operator ∂ acting on M to the left defines a connection on M: indeed, it acts additively and satisfies the Liebniz rule. The question remains to determine which class of systems it represents. From the relation $\partial - B = 0$ in M we deduce that, for all $j = 1,\ldots,n$, the components of $\partial\varepsilon_j$ in the basis $\underline{\varepsilon}$ are given by the jth row of the matrix B. Hence,

$$\partial[\varepsilon_1\varepsilon_2\cdots\varepsilon_n] = [\varepsilon_1\varepsilon_2\cdots\varepsilon_n]\,{}^tB.$$

And we can conclude that the system operator associated with the connection ∂ is the adjoint $\Delta^* = \partial + {}^tB$ of Δ.
We can state:

Proposition 3.2.4 *Given a differential system operator $\Delta = \partial - B$ with coefficients B in K the pair $(M = \mathscr{D}^n/\mathscr{D}^n\Delta, \partial)$ defines a differential module of rank n over K with connection $\partial = \nabla^*$ adjoint to Δ.*

From now on, we may talk of the differential module $\mathscr{D}^n/\mathscr{D}^n\Delta$, the connection $\nabla = \partial$ being understood. With this result we can identify left $\mathscr{D}$-modules and differential modules equipped with a K-basis. Observe, in particular, that a morphism or an isomorphism

$$\phi : \mathscr{D}^n/\mathscr{D}^n\Delta_1 \longrightarrow \mathscr{D}^n/\mathscr{D}^n\Delta_2$$

in the sense of definition 3.2.2, p. 66 is a morphism or an isomorphism of $\mathscr{D}$-modules in the classical sense and reciprocally.

Proposition 3.2.5 *Two system operators $\Delta_1 = \partial - B_1$ and $\Delta_2 = \partial - B_2$ with coefficients in K are K-equivalent if and only if the $\mathscr{D}$-modules $\mathscr{D}^n/\mathscr{D}^n\Delta_1$ and $\mathscr{D}^n/\mathscr{D}^n\Delta_2$ are isomorphic.*

Proof. We have to prove that two differential systems $\Delta_1 Y = 0$ and $\Delta_2 Y = 0$ on one hand and their adjoints $\Delta_1^* Y = 0$ and $\Delta_2^* Y = 0$ on the other hand are simultaneously K-equivalent. To this end, consider fundamental solutions $\mathscr{Y}_1$ and $\mathscr{Y}_2$ of $\Delta_1 Y = 0$ and $\Delta_2 Y = 0$ respectively in any convenient extension of K (for instance, the formal fundamental solutions given by Thm. 3.3.1, p. 72). The systems $\Delta_1 Y = 0$ and $\Delta_2 Y = 0$ are equivalent if and only if there exists a gauge transformation $T \in \mathrm{GL}(n, K)$ such that $\Delta_1 = T^{-1}\Delta_2 T$ or equivalently $\mathscr{Y}_2 = T\mathscr{Y}_1$. This latter relation is equivalent to the relation ${}^t\mathscr{Y}_2^{-1} = {}^tT^{-1}\,{}^t\mathscr{Y}_1^{-1}$. Hence the result since ${}^t\mathscr{Y}_1^{-1}$ and ${}^t\mathscr{Y}_2^{-1}$ are fundamental solutions of the adjoints equations $\Delta_1^* Y = 0$ and $\Delta_2^* Y = 0$ respectively. □

Remark 3.2.6 ; Let us end this section with a remark on the cyclic vector lemma (Prop. 3.1.2, p. 64). In a differential module (M, ∇) of rank n one calls *cyclic vector* any vector $e \in M$ such that the n vectors $e, \nabla e, \ldots, \nabla^{n-1}e$ form a K-basis of M. In such a basis, the matrix of the connection ∇ reads in the form

$$B_\nabla = \begin{bmatrix} 0 & \cdots & 0 & a_{n-1} \\ 1 & & \vdots & \vdots \\ \vdots & \ddots & \vdots & \vdots \\ 0 & \cdots & 1 & a_0 \end{bmatrix}.$$

Let Δ be a system of dimension n with coefficients in K and $\mathscr{D}^n/\mathscr{D}^n\Delta$ the associated $\mathscr{D}$-module . In a cyclic basis $\underline{e}$ the system Δ admits $-{}^tB_\nabla$ as matrix which is a companion form (cf. Sect. 3.1, p. 64) but, *stricto sensu*, the minus signs in the sup-diagonal of 1's. One can cancel these minus signs by taking the basis $(e, -\partial e, \ldots, (-1)^{n-1}\partial^{n-1}e)$.

3.2.2 $\mathscr{D}$-Modules and Order n Differential Operators

The aim of this section is to describe the K-equivalence of order n linear differential operators with coefficients in K. Consider a single linear differential operator

$$D = \partial^n + b_{n-1}(x)\partial^{n-1} + \cdots + b_0(x), \quad b_0, \ldots, b_{n-1} \in K.$$

The operator D acts linearly on $\mathscr{D}$ by multiplication to the right. Its cokernel $\mathscr{D}/\mathscr{D}D$ has a natural structure of left $\mathscr{D}$-module. The pair $(\mathscr{D}/\mathscr{D}D, \partial)$ defines a differential module of rank n. Again, by abuse, we talk of the differential module $\mathscr{D}/\mathscr{D}D$, the connection ∂ being understood.

Proposition 3.2.7 *Let Δ be the companion system operator of D* (cf. *Sect. 3.1, p. 64*)*. Then, the $\mathscr{D}$-modules $\mathscr{D}/\mathscr{D}D$ and $\mathscr{D}^n/\mathscr{D}^n\Delta$ are isomorphic.*

Proof. Consider the maps

$$U : \mathscr{D}^n \longrightarrow \mathscr{D}, \ (\delta_1 \cdots \delta_n) \longmapsto \delta_1 + \delta_2\partial + \cdots + \delta_n\partial^{n-1}$$

and the projection over the last component $V : \mathscr{D}^n \to \mathscr{D}, \ (\delta_1 \cdots \delta_n) \longmapsto \delta_n$.

The maps U and V are $\mathscr{D}$-linear; the diagram

$$\begin{array}{ccc} \mathscr{D}^n & \xrightarrow{\cdot\Delta} & \mathscr{D}^n \\ \downarrow V & & \downarrow U \\ \mathscr{D} & \xrightarrow{\cdot D} & \mathscr{D} \end{array}$$

commutes and it can be completed into the commutative diagram with exact rows

$$\begin{array}{ccccccccc} 0 & \longrightarrow & \mathscr{D}^n & \xrightarrow{\cdot\Delta} & \mathscr{D}^n & \longrightarrow & \mathscr{D}^n/\mathscr{D}^n\Delta & \longrightarrow & 0 \\ & & \downarrow V & & \downarrow U & & \downarrow u & & \\ 0 & \longrightarrow & \mathscr{D} & \xrightarrow{\cdot D} & \mathscr{D} & \longrightarrow & \mathscr{D}/\mathscr{D}D & \longrightarrow & 0. \end{array}$$

The quotient map u does exist. It is left $\mathscr{D}$-linear and surjective since U is also left $\mathscr{D}$-linear and surjective. On the other hand, the modules $\mathscr{D}^n/\mathscr{D}^n\Delta$ and $\mathscr{D}/\mathscr{D}D$ have equal ranks. Therefore, u is an isomorphism of K-vector spaces and in particular, is injective. □

From propositions 3.2.7, p. 69 and 3.2.5, p. 68 we can set the following definition.

Definition 3.2.8 (Equivalent Operators)
*Two linear differential operators D_1 and D_2 of order n are said to be K-*equivalent *if the $\mathscr{D}$-modules $\mathscr{D}/\mathscr{D}D_1$ and $\mathscr{D}/\mathscr{D}D_2$ are isomorphic.*

Let us now make explicit the equivalence of order n linear differential operators in the spirit of definition 3.2.3, p. 66.

Recall that $\mathcal{D} = K[\partial]$ is a non commutative ring with non-commutation relations generated by $\partial x = x\partial + 1$. In the ring $\mathcal{D}$ there is an euclidian division on the right and on the left. Consequently, any left or right ideal is principal and any two differential operators have a greatest common divisor on the left (denoted by lgcd) and on the right (rgcd) as well as a least common multiple on the left (llcm) and on the right (rlcm). These gcd and lcm are uniquely determined by adding the condition that they are monic polynomials, which we do.

The counterpart for a differential operator $D \in \mathcal{D}$ of a gauge transformation for a system involves a transformation $\mathcal{T}_A$, with $A \in \mathcal{D}$, of the form

$$\mathcal{T}_A(D) = \mathrm{llcm}(D,A)A^{-1}.$$

By this, we mean that we take the lcm of D and A on the left and we divide it by A on the right (this is possible since, by definition, A can be factored on the right in any llcm involving A). In other words, $\mathcal{T}_A(D)$ is the factor of smallest degree we must multiply A on the left to obtain a left multiple of D. Notice that such a factor is unique due to the uniqueness of $\mathrm{llcm}(D,A)$ as a monic polynomial.

Proposition 3.2.9 *The differential operators D_1 and $D_2 \in \mathcal{D}$ are K-equivalent if and only if there exists $A \in \mathcal{D}$ prime to D_2 to the right such that*

$$\boxed{D_1 = \mathcal{T}_A(D_2)}\,.$$

We may notice that, as A and D_2 are prime, the operators D_2 and $\mathcal{T}_A(D_2)$ have the same order.

Proof. By definition, the K-equivalence of D_1 and D_2 means that there is an isomorphism of $\mathcal{D}$-modules

$$\varphi : \mathcal{D}/\mathcal{D}D_1 \longrightarrow \mathcal{D}/\mathcal{D}D_2.$$

As a morphism of $\mathcal{D}$-modules the map φ is well defined by

$$\varphi(1+\mathcal{D}D_1) = A + \mathcal{D}D_2.$$

For any $L \in \mathcal{D}$, one has then $\varphi(L+\mathcal{D}D_1) = LA+\mathcal{D}D_2$. Since $\varphi(D_1) = 0$ there exists $L_1 \in \mathcal{D}$ such that $D_1A = L_1D_2$. Conversely, any A such that there is an L_1 satisfying $D_1A = L_1D_2$ determines a morphism of $\mathcal{D}$-modules from $\mathcal{D}/\mathcal{D}D_1$ into $\mathcal{D}/\mathcal{D}D_2$ by setting $\varphi(1+\mathcal{D}D_1) = A+\mathcal{D}D_2$.

The injectivity of φ means that the condition $\varphi(L) = 0$, i.e., $LA = PD_2$ for a certain $P \in \mathcal{D}$, implies $L = QD_1$ with $Q \in \mathcal{D}$. Hence, to any relation $LA = PD_2$ there is $Q \in \mathcal{D}$ such that $PD_2 = QD_1A$, that is to say, any left common multiple of A and D_2 is a left multiple of D_1A. Otherwise said, D_1A is the llcm of A and D_2 and then,

$$D_1 = \mathcal{T}_A(D_2).$$

Let us now express the surjectivity of φ. This amount to the fact that there exists $L \in \mathcal{D}$ such that $\varphi(L+\mathcal{D}D_1) = 1+\mathcal{D}D_2$, which means that there is $P \in \mathcal{D}$ such that $LA + PD_2 = 1$. This is a Bézout relation for A and D_2 on the right which means that A and D_2 are prime on the right. □

3.3 Formal Meromorphic Classification

We denote by $\widetilde{K} = \mathbb{C}[[x]][1/x]$ the field of all meromorphic series at 0 either convergent or not and by $K = \mathbb{C}\{x\}[1/x]$ the subfield of the convergent ones. All linear differential systems or equations we consider have convergent meromorphic coefficients, i.e., coefficients in K although this section would be valid for systems or equations with coefficients in $\widetilde{K}$ as well.

The formal classification of linear differential systems is the classification under $\widetilde{K}$-equivalence (cf. Def 3.2.3, p. 66 and Prop. 3.2.9, p. 70) whereas the meromorphic[2] classification is the classification under K-equivalence. The former one, which we consider now, is solved by building a *normal form* (cf. Thm. 3.3.1, p. 72), which although not unique, clearly contains all formal invariants. A system belongs to a formal class when there is a formal gauge transformation which changes the selected normal form into that system; this means (cf. Def. 3.1.1, p. 64) that there exists an invertible matrix $T(x)$ with formal meromorphic entries for which the linear change of variable $Z = TY$ changes the normal form into the given system. The entries of T satisfy a linear meromorphic differential system which depends on the normal form and on the given system and which is singular at 0 in general. It turns out that, given a normal form and a system in its formal class, the gauge transformations connecting the two are not unique in general.

In this section, we introduce the main theoretical results on the formal classes of systems and we state some definitions attached to normal forms. Their transcription in terms of equations is straightforward by restricting the results to the first row of a formal fundamental solution of its companion system. In the case of equations we sketch the practical algorithms based on Newton polygons to compute the formal invariants. There exist similar algorithms for systems but they are far more complicated (See, for instance, the appendix and references in [BCL03]).

3.3.1 The System Case

Consider an order one linear differential system with meromorphic coefficients and dimension n

$$\Delta Y \equiv \frac{\mathrm{d}Y}{\mathrm{d}x} - B(x)Y = 0 \quad \text{with} \quad B(x) \in g\ell(n,K). \tag{3.3}$$

[2] We use the term *meromorphic* in the sense of convergent meromorphic. Otherwise, we specify *formal meromorphic* or just *formal*.

Recall (cf. Sect. 3.1, p. 64) that a gauge transformation $Z = TY$ changes the differential system $\mathrm{d}Y/\mathrm{d}x - B(x)Y = 0$ into the differential system $\mathrm{d}Z/\mathrm{d}x - {}^{T}B(x)Z = 0$ with

$$\boxed{{}^{T}B = \frac{\mathrm{d}T}{\mathrm{d}x}T^{-1} + TBT^{-1}} \tag{3.4}$$

When $T(x)$ is meromorphic $\big($we denote $T \in \mathbb{G} = \mathrm{GL}(n, \mathbb{C}\{x\}[1/x])\big)$ the matrix ${}^{T}B(x)$ is also meromorphic. However, the matrix ${}^{T}B(x)$ may be convergent for some divergent T. We denote by $\widetilde{\mathbb{G}}(B)$ the set of formal meromorphic gauge transformations $T \in \widetilde{\mathbb{G}} \equiv \mathrm{GL}(n, \mathbb{C}[[x]][1/x])$ such that ${}^{T}B(x)$ is convergent. The set $\widetilde{\mathbb{G}}(B)$ contains $\mathbb{G}$. While $\mathbb{G}$ is a group, $\widetilde{\mathbb{G}}(B)$ is not. The *meromorphic class* of the system is its orbit under the gauge transformations in $\mathbb{G}$ while its *formal class* is its (larger) orbit under those in $\widetilde{\mathbb{G}}(B)$.

The formal classification of n-dimensional meromorphic linear differential systems is performed by selecting, in each class, a system of a special form called *a normal form*. There exist algorithms to fully calculate a normal form of any given system (cf. end of Sect. 3.3.3.3, p. 84).

Theorem 3.3.1 (Formal Fundamental Solution and Normal Form)

(i) *To the system* (3.3)*:*

$$\mathrm{d}Y/\mathrm{d}x = B(x)Y$$

there is a formal fundamental solution (i.e., *a matrix of n linearly independent formal solutions) of the form*

$$\boxed{\widetilde{\mathscr{Y}}(x) = \widetilde{F}(x)\, x^{L}\, \mathrm{e}^{Q(1/x)}}$$

where

▷ $Q(1/x) = \bigoplus_{j=1}^{J} q_j(1/x) I_{n_j}$ *(assume the q_j's are distinct) is a diagonal matrix satisfying $Q(0) = 0$; its diagonal entries are polynomials in 1/x or in a fractional power $1/t = 1/x^{1/p}$ of $1/x$; the notation I_{n_j} stands for the identity matrix of dimension n_j and $\sum n_j = n$. The smallest possible number p is called the* degree of ramification *of the system,* $\mathrm{e}^{Q(1/x)}$ *the* irregular part *of $\mathscr{Y}(x)$ and the q_j's the* determining polynomials.

▷ $L \in g\ell(n, \mathbb{C})$ *is a constant matrix called the* matrix of the exponents of formal monodromy.

▷ $\widetilde{F}(x) \in \mathrm{GL}(n, \mathbb{C}[[x]][1/x])$ *is an invertible formal meromorphic matrix.*

(ii) *The matrix* $\mathscr{Y}_0(x) = x^{L}\, \mathrm{e}^{Q(1/x)}$ *is a (formal) fundamental solution of a system*

$$\frac{\mathrm{d}Y}{\mathrm{d}x} = B_0(x)Y$$

with polynomial coefficients in x and $1/x$. The system $\mathrm{d}Y/\mathrm{d}x = B_0(x)Y$ is formally equivalent to the initial system $\mathrm{d}Y/\mathrm{d}x = B(x)Y$ via the formal gauge transformation $\widetilde{F}(x)$ (hence, the relation $B(x) = {}^{\widetilde{F}}B_0(x)$) and it is called a normal form *of the given system $\mathrm{d}Y/\mathrm{d}x = B(x)Y$.*
The fundamental matrix $\mathscr{Y}_0(x)$ is called a normal solution.

It is worth to discuss the nature of formal solutions. Let us do it on the example of the formal exponential e^x. As a formal object e^x is a symbol which satisfies the condition $\mathrm{d}\mathrm{e}^x/\mathrm{d}x = \mathrm{e}^x$. It makes no sense to talk of the value of e^x for $x = 1$ like it makes no sense to talk of the value of a formal series $\sum a_n x^n$ at $x = 1$. They are symbols, not functions. Nonetheless, it may be useful to identify the formal e^x to functions satisfying the same condition, in order for example, to find analytic solutions. In this identification, one may choose any analytic solution of the equation $\mathrm{d}y/\mathrm{d}x = y$: the classical exponential function $\exp(x)$, which satisfies $\exp(0) = 1$, as well as any other solution $c\exp(x)$ of the equation $\mathrm{d}y/\mathrm{d}x = y$ satisfying the initial condition $c\exp(0) = c$ for any $c \neq 0$. This freedom of choices is reflected in the *exponential torus* in differential Galois theory (cf. volume I [MS16, Sect. 2.2.3.2]). Note however, that unless there might be some ambiguity, we denote by e^x equally the formal exponential and the classical analytic one. The notation exp is used exclusively for the analytic exponential.

Let $p \geq 1$ be the degree of ramification of the system and $x = t^p$.

Denote $Q'(1/t) = Q(1/x)$ and $q_j'(1/t) = q_j(1/x)$ so that all $q_j'(1/t)$'s are true polynomials in the variable $1/t$. Notice however that the degree of ramification of a particular $q_j(1/x)$ may be smaller than p (actually any divisor of p including 1 might arise).

The case when the degree of ramification p of $Q(1/x)$ is equal to 1 is referred to as the *case without ramification* or the *unramified case*. The name refers to the fact that the determining polynomials in $Q(1/x)$ contain no roots although there might exist roots of x and logarithms in the formal fundamental solution coming from the factor x^L. The matrix L can be chosen in Jordan form with blocks compatible[3] with the block structure of Q implying thus that Q and L commute, hence also $\mathrm{e}^{Q(1/x)}$ and x^L.

The case when the degree of ramification p is > 1 is referred to as the *case with ramification* or the *ramified case*. The matrix L can no longer be chosen in Jordan form and does not commute with $Q(1/x)$ although it can be chosen so that its Jordan form commutes with $Q(1/x)$. Because the system is meromorphic in x (in particular, without root of x) the matrix $Q(1/x)$ is globally invariant under the action of the Galois group G_p of the ramification $t^p = x$. Choose for instance the generator $\omega = \mathrm{e}^{2\pi\mathrm{i}/p}$ of G_p. This means that, with a diagonal scalar block $q_j'(1/t)I_{n_j}$, the matrix $Q'(1/t)$ contains all distinct blocks in its orbit $q_j'(1/(\omega^\ell t))I_{n_j}$ for all $\ell \in \mathbb{N}$ and the Galois group permutes the blocks of a same orbit. If the degree of ramification of $q_j(1/x)$ is $p' \leq p$ then its orbit contains p' distinct elements.

[3] If $Q = \oplus q_j I_{n_j}$ then $L = \oplus L_j$ where L_j has size n_j and can split into more diagonal blocks.

Turning back to the variable x we can write:

$$Q\big(1/(\mathrm{e}^{2\pi\mathrm{i}}x)\big) = R^{-1}\,Q(1/x)\,R \tag{3.5}$$

where R is a matrix of permutation by blocks. We can order the scalar blocks of Q so that R is a direct sum of circulant matrices (by blocks). The matrix L can be chosen in the form

$$L = U^{-1}JU$$

where J is in Jordan form compatible with the scalar block structure of Q, and U is a matrix of diagonalization of R; hence, J commutes with Q, and U is a direct sum (over the various orbits) of van der Monde matrices by blocks (i.e., van der Monde matrices tensored by unit matrices I_{n_j} according to the size of the blocks in each orbit). Unless R is the identity, the matrices U and $\mathrm{e}^{Q(1/x)}$ do not commute implying thus that x^L and $\mathrm{e}^{Q(1/x)}$ do not commute. For more precision we refer to [BJL79a] and [Lod01].

The proof of theorem 3.3.1, p. 72 proceeds in two steps. The first part is a classification theorem over a finite extension of $\widetilde{K}$, actually the extension by $t^p = x$ although the degree of ramification p is not given a priori. This means the formal classification over a field $\widetilde{K}_p = \mathbb{C}[[t]][1/t]$ for a convenient value of p by considering $x = t^p$ as a function of t. One obtains a formal solution in the form

$$\widetilde{Y}(t) = \widetilde{\phi}(t)\,t^{pJ'}\,\mathrm{e}^{Q'(1/t)}$$

where $Q'(1/t)$ is as before, J' is a constant matrix in Jordan form commuting with Q' and $\widetilde{\phi}(t)$ is an invertible matrix with entries in $\widetilde{K}_p$. In this form, the theorem was first stated by Hukuhara and Turrittin and, for this reason, it is often called *Hukuhara-Turrittin theorem*. Later, it received a complete proof in terms of systems by Wasow [Was76, Thm. 19.1] and in terms of equations by B. Malgrange [DMR07, p. 104, Thm. (4.2.1)]; see also Hsieh-Sibuya's book [HS99, Th. XIII-6-1] and Levelt [Lev75]. The second part of the proof amounts to symmetrize the formal solution $\widetilde{Y}(t)$ in order to fire out the roots appearing in $\widetilde{\phi}(t)$. This was first done by W. Balser, W. Jurkat and D. Lutz [BJL79a]; a simpler, however in the same spirit, proof providing moreover a general expression for a normal form in terms of rank reduced systems, is given in [Lod01].

Let us briefly explain why the scalar blocks of Q (or Q') are globally invariant and permuted under the action of G_p (cf. Eq. (3.5), p. 74). The fundamental solution $\widetilde{Y}(t)$ reads $\widetilde{Y}(t) = \widetilde{\Phi}(t)\,\mathrm{e}^{Q'(1/t)}$ where $\widetilde{\Phi}(t) = \widetilde{\phi}(t)\,t^{pJ'}$ is formal logarithmic (i.e., a polynomial in $\ln(t)$ with coefficients that might contain complex powers of t and formal series in t but no exponential). Multiplying t by ω leaves $x = t^p$ invariant and changes $\widetilde{Y}(t)$ into another formal fundamental solution. There exists thus a constant invertible matrix C such that $\widetilde{\Phi}(\omega t)\,\mathrm{e}^{Q'(1/(\omega t))} = \widetilde{\Phi}(t)\,\mathrm{e}^{Q'(1/t)}C$. Hence, the relation $\widetilde{\Phi}(t)^{-1}\widetilde{\Phi}(\omega t) = \mathrm{e}^{Q'(1/t)}\,C\,\mathrm{e}^{-Q'(1/(\omega t))}$. Decompose the matrix $C = \big[C_{j,\ell}\big]$ into blocks fitting the scalar block structure of $Q'(1/t)$. The right-hand side of this latter relation reads $\left[C_{j,\ell}\,\mathrm{e}^{q'_j(1/t)-q'_\ell(1/(\omega t))}\right]$ and it suffices to write down that this matrix contains no exponential and is invertible to conclude.

A *minimal full set of formal invariants* can be read on any normal solution. Unless otherwise specified we refer to a normal solution $\mathscr{Y}_0(x) = x^L\,\mathrm{e}^{Q(1/x)}$. Observe first that normal solutions are not unique. Indeed, consider for instance the normal solution $\mathscr{Y}_0(x)$ and a n dimensional permutation matrix P, or any matrix, commuting with $Q(1/x)$. A formal fundamental solution of the given system $Y' = B(x)Y$ reads in the form

$$\mathscr{Y}(x)\,P = \big(\widetilde{F}(x)\,P\big)\,\big(P^{-1}\,\mathscr{Y}_0(x)\,P\big)$$

where $\widetilde{F}(x)\,P$ belongs to $\mathrm{GL}(n,\widetilde{K})$ and $P^{-1}\,\mathscr{Y}_0(x)\,P = x^{P^{-1}Q(1/x)P}\,\mathrm{e}^{P^{-1}Q(1/x)P}$ is also a normal solution; $P^{-1}\,\mathscr{Y}_0(x)\,P$ is a formal fundamental solution of $Y' = {}^{P^{-1}}B_0(x)Y$.

In the unramified case (i.e., with ramification degree $p = 1$), a minimal full set of formal invariants is given by the invariants of similarity of L (eigenvalues and size of the corresponding irreducible Jordan blocks) jointly with the determining polynomials. In the ramified case (i.e., with ramification degree $p > 1$) a minimal full set of formal invariants is given by the invariants of similarity of L jointly with one determining polynomial $q_j(1/x)$ in each orbit under the action of the group G_p (cf. [Lod01, Cor. 2.5.2]).

Any normal form is meromorphically equivalent to $\mathrm{d}Y/\mathrm{d}x = B_0(x)\,Y$. That's why, sometimes, one generalizes the definition by calling normal form any system meromorphically equivalent to $\mathrm{d}Y/\mathrm{d}x = B_0(x)\,Y$.

3.3.2 Some Definitions Related to Normal Forms

Let us now state some definitions associated with the formal invariants. We choose a formal fundamental solution

$$\widetilde{\mathscr{Y}}(x) = \widetilde{F}(x)\,x^L\,\mathrm{e}^{Q(1/x)} \quad \text{with} \quad Q = \bigoplus_{j=1}^{J} q_j(1/x)\,I_{n_j} \quad \text{and distinct } q_j\text{'s} \qquad (3.6)$$

of system (3.3): $\mathrm{d}Y/\mathrm{d}x = B(x)\,Y$ and the normal form $Y' = B_0(x)Y$ with the fundamental solution $\mathscr{Y}_0(x) = x^L\,\mathrm{e}^{Q(1/x)}$.

Definition 3.3.2 (Singular points)

(i) *If B is analytic at 0* (i.e., *Poincaré rank $r_0 = -1$) then 0 is said to be an* ordinary point *of the system.*

(ii) *If B has a pole at 0* (i.e., *Poincaré rank $r_0 \geq 0$) then 0 is said to be a* singular point *of the system. If, in addition,*

(a) $Q = 0$ *then the point 0 is said to be* regular singular; *if moreover, $\widetilde{F}(x)\,x^L$ is a power series in non negative integer powers of x the singular point 0 is said to be* apparent.
(b) $Q \neq 0$ *then the point 0 is said to be* irregular singular.

The following proposition provides a characterization of regular singular points which is sometimes taken as definition.

Proposition 3.3.3 *The singular point 0 is regular singular if and only if there is a fundamental solution $\mathscr{Y}(x)$ with moderate growth at 0.*

Moderate growth means that, $\mathscr{Y}(x)$ is defined on a neighborhood of 0 in the universal covering of $\mathbb{C}^*$ and that given any sector Δ with finite opening and small enough radius (for obvious reasons we assume that Δ contains no other singular point of the system) there exist constants $C > 0$ and λ such that

$$|\mathscr{Y}(x)| \leq C|x|^{\lambda} \quad \text{on } \Delta. \tag{3.7}$$

The constants C and λ depend on Δ. Note that when a fundamental solution has moderate growth all solutions have moderate growth (for, moderate growth is preserved by linear combinations).

Proof. Choose a direction θ. The main asymptotic existence theorem (cor. 3.4.2) applied to the series factor $\widetilde{F}(x)$ in $\widetilde{\mathscr{Y}}(x)$ provides a sector Δ bisected by θ and an analytic matrix $F_\theta(x)$ asymptotic to $\widetilde{F}(x)$ on Δ. Taking Δ smaller if necessary we can assume that $F_\theta(x)$ is bounded on Δ. Choosing a determination of the argument at θ allows to associate with the normal solution $\mathscr{Y}_0(x)$ and to $\widetilde{\mathscr{Y}}(x)$ analytic functions $\mathscr{Y}_{0,\theta}(x)$ and $\mathscr{Y}(x) = F_\theta(x)\mathscr{Y}_{0,\theta}(x)$ respectively, both defined on Δ. The necessary and sufficient condition results from the fact that a power and a logarithm have moderate growth whereas an exponential has not (at least, not for all θ). □

At an ordinary point one knows from the Cauchy-Lipschitz theorem that the system admits a fundamental solution consisting of convergent power series (in non negative integer powers of x), hence, of analytic functions in a neighborhood of 0. One proves (cf. Prop. 3.5.3, p. 93) that in the case of an apparent or a regular singular point, the series $\widetilde{F}(x)$ is convergent (cf. Prop. 3.5.3) and one obtains an actual fundamental solution by taking the sum of $\widetilde{F}(x)$ times a branched analytic version of $x^L e^{Q(1/x)}$ (cf. proof of proposition 3.3.3 above or definition of $\mathscr{Y}_{0,\underline{\alpha}}$ in Sect. 3.5.3.1, p. 98). In the case of an irregular singular point the series $\widetilde{F}(x)$ diverges in general. The theories of summation in chapters 5 and 7 will allow us to associate with $\widetilde{F}(x)$ well defined analytic functions (on various sectors with vertex 0) and thus to obtain in a natural manner analytic solutions in this case too.

The matrix $\widetilde{F}(x)$ satisfies the homological system

$$\frac{\mathrm{d}F}{\mathrm{d}x} = B(x)F - F B_0(x). \tag{3.8}$$

Compare formula (3.4). This is a linear differential system in the entries of F which admits the polynomials $q_\ell - q_j$ for $j,\ell = 1,\dots,J$ as determining polynomials. We split the matrix $\widetilde{F}(x)$ into column-blocks fitting the block-structure of Q (for $j = 1,\dots,J$, the matrix $\widetilde{F}_j(x)$ has n_j columns and $\sum n_j = n$):

$$\widetilde{F}(x) = \begin{bmatrix} \widetilde{F}_1(x) & \widetilde{F}_2(x) & \cdots & \widetilde{F}_J(x) \end{bmatrix}.$$

Thus, each block $\widetilde{F}_j$ comes in $\mathscr{Y}(x)$ with the exponential $e^{q_j(1/x)}$.

Below we refer to system (3.3) as "the system" and to system (3.8) as "the homological system". One can easily read formal invariants such as the levels of both systems, their Stokes and anti-Stokes directions and their Stokes arcs on the formal fundamental solution $\widetilde{\mathscr{Y}}(x)$. They are read as well on the normal solution $\mathscr{Y}_0(x)$ and they are defined as follows.

Definition 3.3.4 (Levels, Multi-Level)

(i) *The* levels $k_1, k_2, \ldots, k_\nu$ of the system *are the degrees of its (non-zero) determining polynomials* $q_1, q_2, \ldots, q_J$. *In this case, one says that the system has* multi-level $\underline{k} = (k_1, k_2, \ldots, k_\nu)$.

(ii) *The* levels of $\widetilde{F}_j(x)$ *are the degrees* $k_{j_1}, k_{j_2}, \ldots, k_{j_\nu}$ *of the polynomials* $q_\ell - q_j$ *for all* ℓ. *In this case, one says that* $F_j(x)$ *has* multi-level $(k_{j_1}, k_{j_2}, \ldots, k_{j_\nu})$.

(iii) The levels of the homological system *are the degrees of its determining polynomials* $q_\ell - q_j$ *for all* ℓ, j; *hence, the levels of all* $\widetilde{F}_j(x)$.

Observe that 0 is not a level since the polynomials q_j contain no constant term and $q_\ell \neq q_j$ for all $\ell \neq j$. Ordering the levels in increasing order we can then assume that multi-levels $\underline{k} = (k_1, k_2, \ldots, k_\nu)$ satisfy $0 < k_1 < k_2 < \cdots < k_\nu$ (cf. Def. 7.2.9, p. 205).

Definition 3.3.5 (Anti-Stokes and Stokes Directions)

(i) *The* anti-Stokes directions of a polynomial $q(1/x) = -a/x^k\,(1 + o(1/x))$ *with* $a \neq 0$ *are the directions in which the exponential* $e^{q(1/x)}$ *has maximal decay, that is, the directions* $\theta = \arg(a)/k \mod 2\pi/k$ *for which* $-a/x^k$ *is real negative.*

(ii) *The* Stokes directions of a polynomial $q(1/x) = -a/x^k\,(1 + o(1/x))$ *are the oscillatory directions of the exponential* $e^{q(1/x)}$, *that is, the directions in which* $-a/x^k$ *is purely imaginary.*

(iii) *The* anti-Stokes, resp. Stokes directions of the system *are the anti-Stokes,* resp. *the Stokes directions of its non-zero determining polynomials* $q_1, q_2, \ldots, q_J$.

(iv) *The* anti-Stokes, resp. Stokes directions, of $\widetilde{F}_j(x)$ *are the anti-Stokes,* resp. *the Stokes directions, of the polynomials* $q_\ell - q_j$ *for all* $\ell \neq j$.

(v) *The* anti-Stokes, resp. Stokes directions, of the homological system *are the anti-Stokes,* resp. *the Stokes directions, of the polynomials* $q_\ell - q_j$ *for all* ℓ, j *with* $\ell \neq j$; *hence, of all* $\widetilde{F}_j(x)$.

(vi) *The* Stokes values of $\widetilde{F}_j(x)$ *are the dominant coefficients* $a_{\ell,j}$ *in the polynomials*

$$(q_\ell - q_j)(1/x) = -a_{\ell,j}/x^{k_{\ell,j}}\,(1 + o(x)), \quad a_{\ell,j} \neq 0. \tag{3.9}$$

Notice, in the right-hand side of (3.9), the minus sign which we would not introduce if we worked at infinity.

The denominations "Stokes" versus "anti-Stokes" direction are not universal: they are sometimes interchanged.

Definition 3.3.6 (Stokes Arcs)

(i) *Let q be a polynomial of integer degree k in the variable $1/x$.*

The Stokes arcs of q *are the k closed arcs of S^1 of length π/k on which* $e^{q(1/x)}$ *is bounded; in particular,* $e^{q(1/x)}$ *is flat in all directions of the interior of the Stokes arcs of q. One also says the* Stokes arcs of $e^{q(1/x)}$ *or* associated with $q(1/x)$ *or* with $e^{q(1/x)}$.

(ii) *Let $q \in \mathbb{C}[1/x^{1/p}]$ be a polynomial with fractional degree k/p in $1/x$.*

The Stokes arcs *of q are defined similarly as above on the p-sheet covering of S^1 defined by $x = t^p$ or on the universal cover of S^1. They can be projected as proper sub-arcs of S^1 only when the degree of q is $> 1/2$ (see Exa. 3.3.7, p. 78). Otherwise, they live only on the p-sheet or the universal cover of S^1.*

(iii) *The* Stokes arcs of the system *are the Stokes arcs of its non-zero determining polynomials $q_1, q_2, \ldots, q_J$.*

(iv) *The* Stokes arcs of $\widetilde{F}_j(x)$ *are the Stokes arcs, for all $\ell \neq j$, of the non-zero polynomials $q_\ell - q_j$.*

(v) *The* Stokes arcs of the homological system *are the Stokes arcs of the polynomials $q_\ell - q_j$ for all ℓ, j with $\ell \neq j$; hence, of all $\widetilde{F}_j(x)$.*

Items (iii) and (v) in definitions 3.3.5 and 3.3.6 are coherent since the determining polynomials of the homological system are the polynomials $q_\ell - q_j$.

The Stokes arcs of a given polynomial $q(1/x)$ are regularly distributed on S^1 or on its universal cover. In the case when the system has a unique level, that is, when the non-zero determining polynomials have same degree the Stokes arcs of the system have all same size; they can be attached to different determining polynomials and they can overlap with each other but they cannot be properly included in each other. This is no more true when the system has several levels.

The Stokes arcs of a system are bisected by anti-Stokes directions and they have Stokes directions as boundary directions.

The definition stated for a linear differential system applies to a linear differential equation by looking at its companion system or at any meromorphically equivalent system.

Example 3.3.7

Suppose a determining polynomial of system (3.3), p. 71 be given by

$$q(1/x) = -1/x^{2/3}.$$

Then, the polynomials $jq(1/x)$ and $j^2q(1/x)$ (where $j^3 = 1$) are also determining polynomials of system (3.3). A fundamental solution of the system in the variable $t = x^{1/3}$ contains the three exponentials e^{-1/t^2}, e^{-j/t^2} and e^{-j^2/t^2} to which correspond the six Stokes arcs defined modulo 2π by $-\pi/4 \leq \arg(t) \leq +\pi/4 \bmod 2\pi/3$ and $3\pi/4 \leq \arg(t) \leq 5\pi/4 \bmod 2\pi/3$, that is, the six Stokes arcs $-\pi/4 \leq \arg(t) \leq +\pi/4 \bmod \pi/3$. By projection of on the circle S^1 in the variable x they provide the two Stokes arcs defined by $|\arg(x)| \leq +3\pi/4 \bmod \pi$, each one associated with the three polynomials.

One more formal invariant is the true Poincaré rank of the system which is however not so easy to read on the system. When the matrix B has a pole of order $r+1>0$ at 0 or no pole (set then $r+1=0$) the system is said to have *(Poincaré) rank* r. The rank is not a formal invariant of the system since it is not left invariant by gauge transformations. However, the possible ranks obtained when applying a meromorphic gauge transformation admit a lower bound $r_0 \geq -1$.

Definition 3.3.8 (True Poincaré Rank) *The* true Poincaré rank r_0 *of a linear differential system is the smallest Poincaré rank found in the formal class of the system.*

From its definition the true Poincaré rank is a formal invariant of the system. One can check that it is related to the maximal degree k/p of the determining polynomials by the formula $r_0 = k+p-1$.

Definition 3.3.9 (Generic Arc)

▷ *An open arc I is said to be* generic for a polynomial $q(1/x)$ *if it contains no Stokes arc of $q(1/x)$.*

▷ *It is said to be* generic for a differential equation or system *if it is generic for all its determining polynomials (it contains no Stokes arc of the equation or the system).*

▷ *A closed (or semi-closed) arc is said to be generic when it is contained in a generic open arc.*

Compare definition 7.3.4, p. 210.

It is worth to notice that it is always possible to permute the columns of a formal fundamental solution by writing it

$$\mathscr{Y}(x) = \widetilde{F}(x)\, P\, x^{P^{-1}LP}\, \mathrm{e}^{P^{-1}Q(1/x)P}$$

with P the chosen permutation. It is also always possible to normalize a given eigenvalue of L, say λ_1, and a given determining polynomial, say q_1, to zero by the change of variable $Y \leftarrow x^{-\lambda_1}\,\mathrm{e}^{-q_1}Y$ in the initial system (and at the same time, in its normal form). The Stokes arcs and the levels of $\widetilde{F}_1(x)$ are then the Stokes arcs and the degrees of the determining polynomials q_ℓ themselves.

For more definitions attached to the normal form such as the *formal monodromy* or the *exponential torus* we refer to [MS16, Sect. 2.2.3.1 and 2.2.3.2] in volume I.

3.3.3 The Equation Case

The meromorphic and the formal equivalence of linear differential operators of order n were given in definition 3.2.8 with a characterization in proposition 3.2.9, p. 70.

Like for systems the formal class of an equation is made explicit from a formal fundamental solution which can be read as the first row of a formal fundamental solution of its companion system. Each such solution takes the form

$$\phi(x)\,x^{\lambda}\,\mathrm{e}^{q(1/x)}$$

where the factors $\phi(x)$ are polynomials in $\ln(x)$ with formal series coefficients. The levels, the Stokes arcs and the anti-Stokes directions are defined similarly as for systems. The invariants are all the determining polynomials $q(1/x)$ with multiplicities, the corresponding exponents λ and the degrees in $\ln(x)$ of each associated $\phi(x)$.

The formal invariants are much easier to determine for an equation than for a system. Below we sketch a procedure one can follow to compute the formal invariants of an equation.

3.3.3.1 Newton Polygons

Newton polygons are a very convenient tool to identify the formal invariants of a linear differential equation $Dy = 0$ at a singular point. By means of a change of variable any singular point can be sent to the origin 0. However, we state the definitions both at 0 and at infinity.

Suppose the linear differential operator

$$D = b_n \frac{\mathrm{d}^n}{\mathrm{d}x^n} + b_{n-1} \frac{\mathrm{d}^{n-1}}{\mathrm{d}x^{n-1}} + \cdots + b_0$$

has meromorphic coefficients b_j expanded either in the variable x (for a study at $x = 0$) or in the variable $1/x$ (for a study at $x = \infty$). Temporarily, we do not need that the coefficients be convergent.

The valuation of a power series $b(x) = \sum_{m \geq m_0} \beta_m x^m$ at the origin is denoted by $v_0(b)$ and defined as the smallest degree with respect to x of the non-zero monomials $\beta_m x^m$ of b; thus, $v_0(b) = m_0$ when $\beta_{m_0} \neq 0$. The valuation of a power series $b(1/x) = \sum_{m \geq m_1} \beta_m / x^m$ at infinity is denoted by $v_\infty(b)$ and defined as the highest degree with respect to x of a non-zero monomial β_m / x^m of b; thus, $v_\infty(b) = -m_1$ when $\beta_{m_1} \neq 0$. When b is a polynomial in x, then $v_\infty(b)$ is the degree of b with respect to x.

Definition 3.3.10 (Newton Polygons)
(i) Newton polygon at 0. — *Suppose the coefficients b_j of D are formal or convergent meromorphic power series in x. With the operator D one associates in $\mathbb{R}^+ \times \mathbb{R}$ the set $\mathscr{P}_D$ of* marked points

$$\boxed{\mathscr{P}_D = \big\{(j, v_0(b_j) - j)\,;\, 0 \leq j \leq n\big\}}$$

The Newton polygon $\mathscr{N}_0(D)$ of D at 0 is the upper envelop in $\mathbb{R}^+ \times \mathbb{R}$ of the various attaching lines of $\mathscr{P}_D$ with non-negative slopes.

(ii) Newton polygon at infinity. — *Suppose the coefficients b_j of D are formal or convergent meromorphic power series in $1/x$. With the operator D one associates in $\mathbb{R}^+ \times \mathbb{R}$ the set $\mathscr{P}_D$ of* marked points

$$\boxed{\mathscr{P}_D = \left\{(j, v_\infty(b_j) - j)\,; 0 \leq j \leq n\right\}}$$

The Newton polygon $\mathscr{N}_\infty(D)$ of D at 0 is the lower envelop in $\mathbb{R}^+ \times \mathbb{R}$ of the various attaching lines of $\mathscr{P}_D$ with non-positive slopes.

Equivalently, we can say that the Newton polygon at 0 is the intersection of the closed upper half-planes limited by the various attaching lines of $\mathscr{P}_D$ with non-negative slopes while the Newton polygon at infinity is the intersection of the closed lower half-planes limited by the various attaching lines of $\mathscr{P}_D$ with non-positive slopes.

One obtains the same Newton polygon when one enlarges the set of marked points to any points $(j, m-j)$ corresponding to a non-zero monomial $x^m \frac{\mathrm{d}^j}{\mathrm{d}x^j}$ in D or to the horizontal segments issuing from the points of $\mathscr{P}_D$ backwards to the vertical axis.

Example 3.3.11

Consider the operator $D = x^m\, \mathrm{d}^j/\mathrm{d}x^j$. Since x^m is both a meromorphic series in x and in $1/x$ it makes sense to determine both its Newton polygon at 0 and at infinity. There corresponds to D the unique marked point $(j, m-j)$ and the corresponding Newton polygons are as shown on Fig. 3.1.

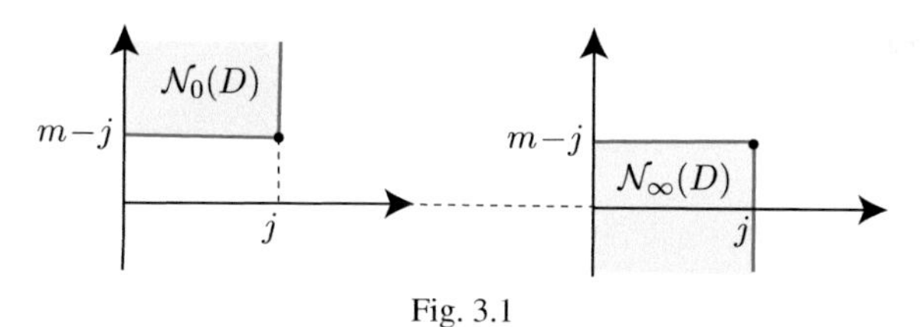

Fig. 3.1

When D has polynomial coefficients one can define its Newton polygons both at 0 and at infinity.

Definition 3.3.12 (Full Newton Polygon)

Suppose D has polynomial coefficients.

The full Newton polygon *$\mathscr{N}(D)$ is the intersection $\mathscr{N}(D) = \mathscr{N}_0(D) \cap \mathscr{N}_\infty(D)$ of the Newton polygons of D at* 0 *and at infinity.*

For simplicity and when there is no ambiguity, we denote by $\mathscr{N}(D)$ anyone of these Newton polygons.

Example 3.3.13

Here below are the full Newton polygons of the Euler operator $\mathscr{E} = x^2 \frac{\mathrm{d}}{\mathrm{d}x} + 1$, its homogeneous variant $\mathscr{E}_0 = x^3 \frac{\mathrm{d}^2}{\mathrm{d}x^2} + (x^2 + x)\frac{\mathrm{d}}{\mathrm{d}x} - 1$ and the hypergeometric operator

$$D_{3,1} = z\big(z\frac{d}{dz} + 4\big) - z\frac{d}{dz}\big(z\frac{d}{dz} + 1\big)\big(z\frac{d}{dz} - 1\big).$$

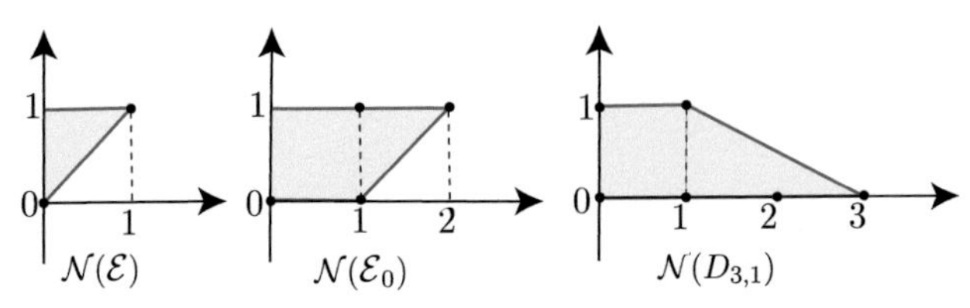

Fig. 3.2

From now on, unless otherwise specified, we work at the origin 0, that is, we suppose that D has formal or convergent meromorphic coefficients at 0.

Proposition 3.3.14 (Levels of D; Regular Singular Points)
Suppose 0 is a singular point of D, i.e., *at least one of the coefficients b_j/b_n has a pole at 0.*

(i) *The levels of D at* 0 *are the positive slopes of $\mathscr{N}_0(D)$.*

(ii) *The point* 0 *is regular singular for D if and only if the Newton polygon $\mathscr{N}_0(D)$ has no non-zero slope.*

Proposition 3.3.15 *Newton polygons satisfy the following properties.*

(i) *Let $D_m = x^m D$, $m \in \mathbb{Z}$. The Newton polygon of D_m is the Newton polygon of D translated vertically by m.*

(ii) *Let D_1 and D_2 be two linear differential operators meromorphic at* 0*. Then,*

$$\mathscr{N}_0(D_1 D_2) = \mathscr{N}_0(D_1) + \mathscr{N}_0(D_2).$$

Proof. Assertion (i) is elementary.
For a proof of (ii) we refer, for instance, to [DMR07, Lem. 1.4.1, p. 99]. □

As a consequence of (i), we may define the Newton polygon of an equation $Dy = 0$ as being the Newton polygon of D up to vertical translation.

On the set $\mathbb{C}[[x]][1/x, \mathrm{d}/\mathrm{d}x]$ of linear differential operators at 0 it is convenient to introduce a *weight* (or 0-weight) w by setting

$$w\Big(x^k \frac{\mathrm{d}^j}{\mathrm{d}x^j}\Big) = k - j \text{ and } w\Big(\sum x^k \frac{\mathrm{d}^j}{\mathrm{d}x^j}\Big) = \min_{k,j} w\Big(x^k \frac{\mathrm{d}^j}{\mathrm{d}x^j}\Big).$$

In particular, $w(x) = 1, w\big(\frac{\mathrm{d}}{\mathrm{d}x}\big) = -1$ and, to an operator D with weight $w(D) = w$, the product $x^{-w}D$ has weight 0. At our convenience, given a differential equation $Dy = 0$, we can then assume that D has weight 0.

Lemma 3.3.16 *Given $j \in \mathbb{N}$ and $k \in \mathbb{Z}$, one has*

$$\left(x^{k+1}\frac{\mathrm{d}}{\mathrm{d}x}\right)^j = x^{jk+j}\frac{\mathrm{d}^j}{\mathrm{d}x^j} + \sum_{\substack{1\le j''<j\\ j'-j''=jk}} c_{j',j''}x^{j'}\frac{\mathrm{d}^{j''}}{\mathrm{d}x^{j''}} \quad (c_{j',j''}\in\mathbb{C}).$$

Observe that all monomials in the right-hand side have weight jk and then, the whole expression in the left-hand side has weight jk. The marked point associated with $x^{jk+j}\,\mathrm{d}^j/\mathrm{d}x^j$ is $A=(j,jk)$. The marked points associated with the monomials in the sum are (j',jk) with $1\le j'\le j-1$, hence points lying on the horizontal segment between A and the vertical coordinate axis.

Proof. The formula in lemma 3.3.16 is trivially true for $j=1$. By the Leibniz rule we obtain the commutation law $\frac{\mathrm{d}}{\mathrm{d}x}x^{k+1} = x^{k+1}\frac{\mathrm{d}}{\mathrm{d}x} + (k+1)x^k$, from which it follows that

$$\left(x^{k+1}\frac{\mathrm{d}}{\mathrm{d}x}\right)^2 = x^{2k+2}\frac{\mathrm{d}^2}{\mathrm{d}x^2} + (k+1)x^{2k+1}\frac{\mathrm{d}}{\mathrm{d}x}.$$

Hence the formula for $j=2$. The general case is similarly obtained by recurrence. □

Proposition 3.3.17 *Given a differential operator D in the variable x denote by D_z the operator deduced from D by the change of variable $x=1/z$.*

The Newton polygons $\mathcal{N}_0(D)$ and $\mathcal{N}_\infty(D_z)$ are symmetric to each other with respect to the horizontal coordinate axis.

Proof. One has $\mathrm{d}/\mathrm{d}z = -x^2\,\mathrm{d}/\mathrm{d}x$. From lemma 3.3.16 we know that we can expand D in powers of the derivation $\delta = x^2\,\mathrm{d}/\mathrm{d}x$ with weight $w(\delta)=+1$:

$$D = c_n(x)\delta^n + c_{n-1}(x)\delta^{n-1} + \cdots + c_0(x)$$

and the set $\mathscr{P}_D$ of marked points is then given by $(j, v_0(c_j)+j)$ for $0\le j\le n$. Now, the operator D_z reads

$$D_z = (-1)^n c_n(1/z)\frac{d^n}{dz^n} + (-1)^{n-1}c_{n-1}(1/z)\frac{d^{n-1}}{dz^{n-1}} + \cdots + c_0(1/z)$$

and the associated marked points are $(j, v_\infty(c_j(1/z)) - j) = (j, -v_0(c_j) - j)$. □

3.3.3.2 Newton Polygon and Borel Transform

We consider here the classical Borel transform $\mathscr{B}$ (or 1-Borel transform) at 0 as defined in section 5.3.1, p. 145 below and we denote by ξ the variable in the Borel plane. We suppose that D has polynomial coefficients in x and $1/x$. As previously, we can expand D in powers of $\delta = x^2\frac{\mathrm{d}}{\mathrm{d}x}$:

$$D = c_n(x)\delta^n + c_{n-1}(x)\delta^{n-1} + \cdots + c_0(x).$$

We assume that the coefficients c_j are polynomials in $1/x$. If this were not the case, we would replace D by $x^{-N}D$ with N the degree of the c_j's with respect to x.

Let $\Delta = \mathscr{B}(D)$ denote the operator deduced from D by Borel transform. Since $\mathscr{B}(\delta) = \xi$ and $\mathscr{B}(1/x) = \mathrm{d}/\mathrm{d}\xi$ (cf. Sect. 5.3.1, p. 145) the operator Δ reads

$$\Delta = c_n\Big(\frac{\mathrm{d}}{\mathrm{d}\xi}\Big)\xi^n + c_{n-1}\Big(\frac{\mathrm{d}}{\mathrm{d}\xi}\Big)\xi^{n-1} + \cdots + c_0\Big(\frac{\mathrm{d}}{\mathrm{d}\xi}\Big)$$

and is then a linear differential operator with polynomial coefficients. The fact that D had polynomial coefficients in $1/x$ is a key point here. In the general case, due to the fact that $\mathscr{B}(fg) = \mathscr{B}(f) * \mathscr{B}(g)$, the Borel transform of a linear differential operator is a convolution operator. The proposition below is a corollary of [Mal91b, Thm. (1.4)].

Proposition 3.3.18 *With normalization as above, the following two properties are equivalent:*

(i) *the levels of D at* 0 *are* ≤ 1;

(ii) *the levels of Δ at infinity are* ≤ 1.

Proof. Let $v = \min_j v_0(c_j) \leq 0$ be the minimal valuation of the coefficients of D at 0. This implies that all marked points associated with D at 0 are on the line issuing from $(0,v)$ with slope 1 (Recall that δ has weight 1) or over and that at least one of them belongs to the line. As a consequence, all levels of D are ≤ 1 if and only if the point $(n, v+n)$ of the line is a marked point, i.e., if and only if $v_0(c_n) = v$.

To the other side, Δ has degree n and order $-v$. Similarly at 0, its Newton polygon at infinity has no slope > 1 if and only if the monomial $\xi^n \frac{\mathrm{d}^{-v}}{\mathrm{d}\xi^{-v}}$ does exist in Δ. And indeed, this is precisely what the condition $v_0(c_n) = v$ says. □

3.3.3.3 Calculating the Formal Invariants

We briefly sketch here how to calculate the formal invariants of a linear differential equation $Dy = 0$ with (formal) meromorphic coefficients at 0. Recall that the formal invariants at 0 of the equation are the determining polynomials $q(1/x)$ with multiplicities, the exponents of formal monodromy λ and how many logarithms are associated with.

▷ *Indicial equation.* — Suppose the Newton polygon $\mathscr{N}_0(D)$ has a horizontal side and consider the operator restricted to the marked points lying on that horizontal side. Up to a power of x to the left that operator reads

$$D_0 = \gamma_r x^r \frac{\mathrm{d}^r}{\mathrm{d}x^r} + \gamma_{r-1} x^{r-1} \frac{\mathrm{d}^{r-1}}{\mathrm{d}x^{r-1}} + \cdots + \gamma_1 x \frac{\mathrm{d}}{\mathrm{d}x} + \gamma_0, \quad \gamma_r, \ldots, \gamma_0 \in \mathbb{C}.$$

The indicial equation is the equation in the variable λ obtained by writing that x^λ satisfies the equation $D_0 y = 0$.

Denote $[\lambda]_r = \lambda(\lambda-1)\dots(\lambda-r+1)$.
Then, the indicial equation reads as follows:

$$\gamma_r [\lambda]_r + \gamma_{r-1} [\lambda]_{r-1} + \ldots + \gamma_1 [\lambda]_1 + \gamma_0 = 0.$$

Its roots λ_j (with multiplicities) are the exponents of factors x^λ associated with no exponential.

▷ *k-characteristic equation.* — The dominant coefficients of the exponentials of degree k are found on the side with slope k of the Newton polygon $\mathcal{N}_0(D)$.

Consider the differential operator restricted to the marked points lying on the side with slope k. This operator reads $x^{k'} D_k \frac{\mathrm{d}^{s''}}{\mathrm{d}x^{s''}}$ with

$$D_k = c_m x^{m(k+1)} \frac{\mathrm{d}^m}{\mathrm{d}x^m} + c_{m-1} x^{(m-1)(k+1)} \frac{\mathrm{d}^{m-1}}{\mathrm{d}x^{m-1}} + \cdots + c_1 x^{k+1} \frac{\mathrm{d}}{\mathrm{d}x} + c_0.$$

The k-characteristic equation is the equation

$$c_m X^m + c_{m-1} X^{m-1} + \cdots + c_1 X + c_0 = 0. \tag{3.10}$$

Its roots are all products ak for all exponentials $\mathrm{e}^{-a/x^k+\cdots}$ counted with multiplicity in a formal fundamental solution.

Indeed, one can check that the differential operator

$$\Delta_k = c_m \left(x^{(k+1)} \frac{\mathrm{d}}{\mathrm{d}x}\right)^m + c_{m-1} \left(x^{(k+1)} \frac{\mathrm{d}}{\mathrm{d}x}\right)^{m-1} + \cdots + \left(c_1 x^{k+1} \frac{\mathrm{d}}{\mathrm{d}x}\right)^1 + c_0$$

coincide with D_k on the side of slope k of their Newton polygons. Writing that the exponential $y = \mathrm{e}^{-a/x^k}$ satisfies $\Delta_k y = 0$ we obtain the relation

$$c_m k^m a^m + c_{m-1} k^{m-1} a^{m-1} + \cdots + c_1 k a + c_0 = 0$$

hence, Eq. 3.10 to determine the products ak.

▷ *Iterated characteristic equations.* — Once one has determined the dominant coefficient a in the exponentials the next coefficients including the factor x^λ attached to each exponential can be determined as follows. Select one root a and consider the differential operator $D_1 = D^{(-a/x^k)}$ deduced from D by the change of variable $y = \mathrm{e}^{-a/x^k} Y$ (and simplifying by the factor e^{-a/x^k}). The Newton polygon $\mathcal{N}_0(D_1)$ may have no slope $k' < k$ and no horizontal side; in that case e^{-a/x^k} is the exponential we look for and it comes factored with no x^λ. It may have no slope $k' < k$ but a horizontal side; in that case, there exist terms of the form $x^\lambda \mathrm{e}^{-a/x^k}$ for all root λ of the indicial equation of D_1. The Newton polygon $\mathcal{N}_0(D_1)$ may also have non-zero slopes $k' < k$; in that case, one has to solve the k'-characteristic equations to determine the next term in the exponential $\mathrm{e}^{-a/x^k+\cdots}$ and so on ... until all exponentials and associated x^λ are found.

▷ *Frobenius method.* — When the indicial equations have multiple roots modulo $\mathbb{Z}$ there might exist logarithmic terms. To determine which terms appear with logarithms there exist a classical algorithm called *Frobenius algorithm*. Although the procedure is easy and natural from a theoretical point of view (it might be long and laborious in practice) we do not develop it here and we refer to the classical literature, for instance, [CL55, Sect. 4.8], [Inc44, Sect. 16.1].

When one knows all normal solutions, to complete them by formal series to get solutions of the initial equation $Dy = 0$ one proceeds by like powers identification.

All these algorithms have been implemented in MAPLE packages such as *Isolde* or *gfun*.

The case of systems is much more difficult to treat practically. There exists however algorithms to determine formal fundamental solutions. One can always apply the cyclic vector algorithm (Sect. 3.1, p. 64) and proceed as before. However, this way, there appear, in general, huge coefficients making the calculation heavy. It is then, in general, recommended to operate directly on the system itself. One method, which relies on Moser's rank, was developed by M. Barkatou and his group (cf. [BCL03] for a sketched algorithm and references). A variant was developed by M. Miyake [Miy12].

3.4 The Main Asymptotic Existence Theorem

The main asymptotic existence theorem plays a fundamental role in the meromorphic study of differential equations. It is at the core of the Stokes phenomenon and the meromorphic classification of linear differential equations or systems.

Let us first consider a linear differential operator with analytic coefficients at 0

$$D = \sum_{j=1}^{n} b_j(x) \frac{\mathrm{d}^j}{\mathrm{d}x^j}.$$

The question here addressed is the following:

Is any formal solution of the equation $Dy = 0$
the asymptotic expansion of an asymptotic solution?

A positive answer is given by the main asymptotic existence theorem (M.A.E.T.) either in Poincaré asymptotics or in Gevrey asymptotics.

In the case of Poincaré asymptotics the theorem, precisely Cor. 3.4.2, p. 87, is mostly due to Hukuhara with a complete proof by Wasow [Was76]. An extension to Gevrey asymptotics is given by B. Malgrange in [Mal91a, Append. 1] and to non linear operators by J.-P. Ramis and Y. Sibuya in [RS89].

In its classical form the theorem roughly says that to a formal solution $\widetilde{f}$ of a differential equation (linear or non linear) there correspond actual solutions f asymptotic to $\widetilde{f}$ in any direction. Given a direction, it is possible to determine from the equation itself a minimal opening of the sector on which such an asymptotic solution exists. However, these asymptotic solutions are, in general, neither unique nor given by explicit formulæ. Let us first state the theorem in its stronger sheaf form.

Theorem 3.4.1 (Main Asymptotic Existence Theorem) *The operator D acts linearly and surjectively on the sheaf $\mathscr{A}^{<0}$ and on the sheaves $\mathscr{A}^{\leq -k}$ for all $k > 0$.*

In other words, the sequences

$$\mathscr{A}^{<0} \xrightarrow{D} \mathscr{A}^{<0} \longrightarrow 0 \quad \text{and} \quad \mathscr{A}^{\leq -k} \xrightarrow{D} \mathscr{A}^{\leq -k} \longrightarrow 0 \quad \text{for all } k > 0$$

are exact sequences of sheaves of $\mathbb{C}$-vector spaces. For the proof we refer to [Mal91a, Appendix 1; Thm. 1] where the theorem is stated and proved for all spaces $\mathscr{A}^{<k}$ and $\mathscr{A}^{\leq k}$ for all $k \in \mathbb{R}$ (see definitions in [Mal91a]).

The main asymptotic existence theorem is best known under the following form which results from the general theorem 3.4.1 above.

Corollary 3.4.2 *Let $\widetilde{f}(x) = \sum_{m\geq 0} a_m x^m$ be a power series solution of the differential equation $Dy = 0$.*

(i) *Given any direction $\theta \in S^1$, there exists a sector $\Delta_\theta = \Delta_{]\theta-\delta,\theta+\delta'[}(R)$ and a function $f \in \overline{\mathscr{A}}(\Delta_\theta)$ such that*

- ▷ *f is an analytic solution on Δ_θ: $Df(x) = 0$ for all $x \in \Delta_\theta$,*
- ▷ *f is asymptotic to $\widetilde{f}$ at 0 on Δ_θ: $T_{\Delta_\theta} f = \widetilde{f}$.*

(ii) *If the series $\widetilde{f}(x)$ is Gevrey of order s then Δ_θ and $f(x)$ can be chosen so that $f(x)$ be s-Gevrey asymptotic to $\widetilde{f}(x)$ on Δ_θ.*

(iii) *In both cases and in each direction θ, the sector Δ_θ can be chosen with opening at least π/k_v where k_v denotes the highest level of $\widetilde{f}(x)$* (cf. *Def. 3.3.4 (ii), p. 77).*

Proof. It suffices to prove the assertions in the case without ramification (i.e., when the degree of ramification p is equal to 1; otherwise one should work with the variable $t = x^{1/p}$).

(i) The Borel-Ritt theorem (cf. Thm. 1.3.1 (i), p. 22), provides for any sector Δ' containing the direction θ, a function $g \in \overline{\mathscr{A}}(\Delta')$ with asymptotic expansion $T_{\Delta'} g = \widetilde{f}$ on Δ'. Since $T_{\Delta'}$ is a morphism of differential algebras, $T_{\Delta'} Dg = DT_{\Delta'} g = D\widetilde{f} = 0$. Hence, the function Dg is flat: $Dg \in \overline{\mathscr{A}}^{<0}(\Delta')$. The main asymptotic existence theorem above (Cor. 3.4.2 applied to Dg in the θ direction) provides a sector $\Delta_\theta \subset \Delta'$ containing the direction θ and a function $h \in \overline{\mathscr{A}}^{<0}(\Delta_\theta)$ such that $Dh = Dg$. The function $f = g - h$ satisfies the required conditions on Δ_θ.

(ii) When the series $\widetilde{f}(x)$ is s-Gevrey the Borel-Ritt theorem with Gevrey conditions (cf. Thm. 1.3.1 (ii), p. 22) provides a function $g \in \overline{\mathscr{A}}_s(\Delta')$ over some sector Δ' containing the direction θ which is s-Gevrey asymptotic to $\widetilde{f}$ on Δ'. Its derivative Dg

is asymptotic to $D\widetilde{f}(x)=0$ and, from proposition 1.2.17, p. 20, we can assert that Dg is k-exponentially flat on $\boldsymbol{\Delta}'$. Hence, by the main asymptotic existence theorem, h belongs to $\overline{\mathscr{A}}^{\leq -k}$ and the conclusion follows as in the previous case.

(iii) Prove that $\boldsymbol{\Delta}_\theta$ can be chosen with opening at least π/k_ν. In the chosen case "without ramification" the levels are positive integers; in particular, $k_1 > 1/2$ so that the Stokes arcs of the equation are all proper sub-arcs of S^1. Let I be an open arc of size π/k_ν in S^1 containing the direction θ. Denote by $\mathscr{V}$ the sheaf of germs of solutions of the equation $Dy=0$ and denote by $\mathscr{V}^{<0}=\mathscr{V}\cap\mathscr{A}^{<0}$ the sub-sheaf of $\mathscr{V}$ of germs of flat solutions. From point (i) above we deduce that there exist sectors U_j making a covering $\mathscr{U}=(U_j)$ of I and solutions $f_j\in\Gamma(U_j\,;\mathscr{V})$ asymptotic to $\widetilde{f}(x)$ on U_j defining thus an element of $H^0\big(I\,;\mathscr{V}/\mathscr{V}^{<0}\big)$. We can assume that $\mathscr{U}$ is finite (apply point (i) at the boundary points of I and take the restriction to directions in I) and assume that the sectors U_j have same radius so that we can take $\boldsymbol{\Delta}_\theta=\cup_j U_j$. We have to prove that (f_j) has a representative in $H^0\big(I\,;\mathscr{V}\big)$. To the short exact sequence $0\longrightarrow\mathscr{V}^{<0}\longrightarrow\mathscr{V}\longrightarrow\mathscr{V}/\mathscr{V}^{<0}\longrightarrow 0$ there is the long exact sequence of cohomology

$$0\longrightarrow H^0\big(I\,;\mathscr{V}^{<0}\big)\longrightarrow H^0\big(I\,;\mathscr{V}\big)\longrightarrow H^0\big(I\,;\mathscr{V}/\mathscr{V}^{<0}\big)\longrightarrow H^1\big(I\,;\mathscr{V}^{<0}\big)\longrightarrow\cdots$$

and it suffices to prove that $H^1\big(I\,;\mathscr{V}^{<0}\big)=0$. This results from the lemma below after noticing that an open arc of size π/k_ν is too small to contain any Stokes arc; hence it is generic (cf. Def. 3.3.9, p. 79). To treat the Gevrey case it suffices to replace $\mathscr{V}$ by $\mathscr{V}_{1/k_\nu}$ and $\mathscr{V}^{<0}$ by $\mathscr{V}^{\leq -k_\nu}$ □

Lemma 3.4.3 *Let I be an arc of S^1 or of its universal covering.*

If I is generic for the differential equation $Dy=0$ (cf. *Def. 3.3.9, p. 79) then*

$$H^1(I\,;\mathscr{V}^{<0})=0.$$

Proof. Again, by means of a change of variable $x=t^r$ for $r\in\mathbb{N}$ large enough we can assume that the operator D belongs to the case without ramification and it follows that I is a proper sub-arc of S^1. We can also assume that I is open.

Let D' denote a normal form of D (cf. Thm. 3.3.1, p. 72) and $\mathscr{V}'$, resp. $\mathscr{V}'^{<0}$, the sheaf of germs of solutions, resp. of flat solutions, of the normal equation $D'y=0$. It results from the main asymptotic existence theorem that the sheaves $\mathscr{V}^{<0}$ and $\mathscr{V}'^{<0}$ are isomorphic over I and it suffices to prove the lemma with $\mathscr{V}'^{<0}$ instead of $\mathscr{V}^{<0}$. The space $\mathscr{V}'_\theta$ of germs of solutions of $D'y=0$ in a θ direction is spanned by n solutions (n the order of D) of the form $g_\ell(x)=h_\ell(x)\,\mathrm{e}^{q_\ell(1/x)}$ where the q_ℓ's are the determining polynomials of the equation and the h_ℓ's are germs of functions with moderate growth which can be continued to the whole Riemann surface of logarithm. The germ $g_\ell(x)=h_\ell(x)\,\mathrm{e}^{q_\ell(1/x)}$ of $\mathscr{V}'$ at θ belongs to $\mathscr{V}'^{<0}_\theta$ when $\mathrm{e}^{q_\ell(1/x)}$ is flat in the θ direction. Hence, it lives over the interior J of a Stokes arc $\overline{J}$ associated with q_ℓ (cf. Def. 3.3.6 (i), p. 78). The sub-sheaf it generates is a sheaf of $\mathbb{C}$-vector spaces of dimension 1 over J continued by 0 outside of J: this is a piecewise constant sheaf of the form $j_!\mathbb{C}_J$ for $j: J \hookrightarrow S^1$ the inclusion map of J into S^1 (cf. Def. 2.1.36, p. 48). The sheaf $\mathscr{V}'^{<0}$ itself is a direct sum of such sheaves and because of the linearity of cocycles we can assume that $\mathscr{V}'^{<0}$ is generated by just one g_ℓ and we denote it by $\mathscr{V}'^{<0}(g_\ell)$.

The situation is then as follows. Any 1-cocycle $\dot{\varphi}(x)$ of $H^1\big(I\,;\mathscr{V}'^{<0}(g_\ell)\big)$ is represented by a 1-cochain $\dot{f}(x)$ on a good (meaning with no 3-by-3 intersection) covering $\mathscr{I} = (I_p)$ having all its 2-by-2 intersections in $I \cap J$. Denote $I = (\alpha, \beta)$ and suppose that $I \cap J = (\gamma, \beta)$ with $\alpha \leq \gamma < \beta$ (the interior J of the Stokes arc $\overline{J}$ covers I up to its right-hand side; the symmetric case "up to its left-hand side" is similar). Suppose a component $\dot{f}_p(x) \in \Gamma\big(I_p \cap I_{p+1}\,; \mathscr{V}^{<0}(g_\ell)\big)$ of $\dot{f}(x)$ is not zero: there exists then a constant $c \in \mathbb{C}^*$ such that $\dot{f}_p(x) = c g_\ell(x)$ on $I_p \cap I_{p+1}$ and the 0-cochain $u = (u_p)$ with $u_{p'} = 0$ for $p' \leq p$ and $u_{p''} = c g_\ell(x)$ for $p'' \geq p+1$ conjugates $\dot{f}(x)$ to a 1-cochain with a trivial component on $I_p \cap I_{p+1}$. Iterating the process one can thus annihilate all components of $\dot{f}(x)$. Note however that, since the arc I is open, there might be infinitely many I_p's. This achieves the proof of the lemma. □

Since the proof of corollary 3.4.2, p. 87 relies on the Borel-Ritt theorem it does not provide uniqueness for the asymptotic solutions. And indeed, there exists in general many such solutions. We saw that the sectors Δ_θ in the main asymptotic existence theorem can always be chosen with opening at least π/k_v. However, particular sectors might have much larger opening: as proved above, any generic arc is convenient but also, the formal series could have good properties, say, be convergent for instance, so that sectors might be chosen with even larger opening.

Comments 3.4.4 (On the Examples of Section 1.1.2, p. 3)

▷ Example 1.1.4, p. 4. The Euler function is asymptotic to the Euler series on a sector of opening 3π and this sector is an asymptotic sector in any of its θ direction. The highest (and actually unique) level of the Euler equation is $k = 1$ and thus, the actual opening of 3π is larger than $\pi/k = \pi$. However, if we ask for a sector bisected by the θ direction the opening reduces to π in the $\theta = \pi$ direction.

▷ Example 1.1.6, p. 7. The hypergeometric function $g(z)$ is asymptotic to the hypergeometric series $\widetilde{g}(z)$ on a sector of opening 4π; however the minimal size given by the main asymptotic theorem is $\pi/k = 2\pi$ (the unique level of the hypergeometric equation $D_{3,1}y = 0$ is $k = 1/2$). The anti-Stokes (and singular) directions are the directions $\theta = 0 \bmod 2\pi$ since the exponentials of a formal fundamental solution are $\mathrm{e}^{\pm 2z^{1/2}}$. An asymptotic sector bisected by $\theta = 0$ has 2π as maximal opening.

▷ In the previous two examples there exists only one singular direction and the possible asymptotic sectors are much larger than the announced minimal value. Actually, considering two neighboring singular directions $\theta < \theta'$ an asymptotic sector always exists with opening $]\theta - \pi/(2k), \theta' + \pi/(2k)[$ taking k equal to the highest level of the equation. When the singular directions are irregularly distributed the asymptotic sectors are "irregularly" wide.

Let $\Delta Y \equiv \mathrm{d}Y/\mathrm{d}x - B(x)Y = 0$ be a linear differential system of order 1 and dimension n with a formal fundamental solution $\widetilde{F}(x)x^L\mathrm{e}^{Q(1/x)}$. Denote its normal form with the normal solution $\mathscr{Y}_0(x) = x^L\mathrm{e}^{Q(1/x)}$ by $\Delta_0 Y \equiv \mathrm{d}Y/\mathrm{d}x - B_0(x)Y = 0$.

The main asymptotic existence theorem 3.4.1 and its corollary 3.4.2, p. 87 are valid in the following form for systems.

Corollary 3.4.5 *The operator Δ acts surjectively in $\left(\mathscr{A}^{<0}\right)^n$ and in $\left(\mathscr{A}^{\leq -k}\right)^n$ for all $k>0$ and therefore, in any direction $\theta \in S^1$, it satisfies the following properties:*

(i) *There exists a sector $\Delta_\theta = \Delta_{]\theta-\omega,\theta+\omega'[}(R)$ and an invertible matrix function $F \in \mathrm{GL}\left(n,\overline{\mathscr{A}}(\Delta_\theta)\right)$ such that*

$$\begin{cases} \Delta\left(F(x)x^L\, \mathrm{e}^{Q(1/x)}\right) = 0 & \textit{for all } x \in \Delta_\theta,\\ T_{\Delta_\theta}F = \widetilde{F} & (F \textit{ is asymptotic to } \widetilde{F} \textit{ at } 0 \textit{ on } \Delta_\theta). \end{cases}$$

(ii) *If an entry of $\widetilde{F}$ is s-Gevrey then the corresponding entry of F can be chosen to be s-Gevrey asymptotic on a convenient sector Δ_θ.*

(iii) *The sector Δ_θ can be chosen open with opening at least π/k_ν for k_ν the highest level of the homological system associated with Δ and Δ_0* (cf. *Def. 3.3.4 (iii), p. 77)*.

Proof. This extension to differential systems follows from the fact that each entry of $\widetilde{F}(x)$ satisfies itself a linear differential equation with meromorphic coefficients deduced from the homological system (3.8): $\mathrm{d}F/\mathrm{d}x = BF - FB_0$. □

3.5 Meromorphic Classification

We consider again the case of linear differential systems with meromorphic coefficients. Contrary to the formal invariants which depend algebraically on the coefficients of the system the meromorphic invariants are transcendental with respect to the system. Algebraic reductions are then hopeless to find meromorphic normal forms like we did to characterize the formal classes. Since the meromorphic classification refines the formal classification one can restrict the study to a formal class with a given normal form, say, $\mathrm{d}Y/\mathrm{d}x = B_0(x)Y$. Recall the notation $\widetilde{\mathbb{G}}(B_0)$ (cf. Sect. 3.3.1, p. 71) for the set of the formal gauge transformations of the normal form $\mathrm{d}Y/\mathrm{d}x = B_0(x)Y$, that is, the set of the gauge transformations $T \in \widetilde{\mathbb{G}} \equiv \mathrm{GL}(n,\mathbb{C}[[x]][1/x])$ such that the transformed matrices ${}^{T}B_0(x)$ are convergent (cf. formula (3.2), p. 64). By definition, all systems in the formal class are obtained from the chosen normal form by means of a formal gauge transformation in $\widetilde{\mathbb{G}}(B_0)$. However, given a system $\mathrm{d}Y/\mathrm{d}x = B(x)Y$ in this formal class there might exist several $\widetilde{F}(x) \in \widetilde{\mathbb{G}}(B_0)$ to change the normal form into this system as we shall see below.

The main part of the classification consists in classifying the formal gauge transformations of the given normal form[4], that is, the elements of $\widetilde{\mathbb{G}}(B_0)$ under the action

[4] D.G. Babbitt and V.S. Varadarajan [BV89] talk of *meromorphic pairs* $(B_0,\widetilde{F})$.

to the left of the group $\mathbb{G}$ of the convergent gauge transformations; the corresponding classifying set is then the quotient $\mathbb{G}\backslash\widetilde{\mathbb{G}}(B_0)$. This part is referred to as the analysis of the *Stokes phenomenon*. The name "Stokes phenomenon" is commonly used in a wide meaning in reference to the fact that two solutions with same asymptotic series might differ of exponentially small quantities (calling to mind 1-cocycles!); cf. Exa. 1.1.4, p. 4 or 1.4.3, p. 30. With a view to a classification theorem one has to normalize "the exponentially small quantities" to be taken into consideration. Basically, the classification is achieved with non abelian Čech cohomological arguments based on the use of the main asymptotic existence theorem. Once a meromorphic class is identified with a cohomology class it is further possible to select in each cohomology class a special 1-cocycle called *Stokes cocycle*. This cocycle has a finite number of components (equal to the number of anti-Stokes directions) appearing in the form of unipotent automorphisms of the normal form and called *Stokes automorphisms*. It provides a full free set of meromorphic invariants. This latter part of the proof is constructible. And so, in the meromorphic classification, the only transcendental argument comes from the use of the main asymptotic existence theorem to start the analysis. Choosing a normal solution, say, $\mathscr{Y}_0 = x^L \mathrm{e}^{Q(1/x)}$ the Stokes cocycle can be given representations by finitely many constant unipotent matrices called *Stokes matrices*. Be aware of the fact that the term Stokes matrix is sometimes used in a wider meaning of matrix connecting any two solutions with same asymptotic series. We restrict the expression "Stokes matrix" (to avoid confusion, one also says *Stokes-Ramis matrix*) to those matrices representing the components of Stokes cocycles. These are closely related to summation as we will see below (Thm. 3.5.14, p. 106).

From the definition, the meromorphic classes of systems appear as follows. Let us first characterize the formal gauge transformations connecting the normal form to a same system. Two formal gauge transformations $\widetilde{F}_1(x)$ and $\widetilde{F}_2(x)$ transform the normal form into a same system if and only if they satisfy the relation ${}^{\widetilde{F}_1}B_0 = {}^{\widetilde{F}_2}B_0$, equivalent to ${}^{\widetilde{F}_2^{-1}\widetilde{F}_1}B_0 = B_0$. This means that there exists a formal gauge transformation T which leaves invariant the normal form $\mathrm{d}Y/\mathrm{d}x = B_0(x)Y$ and such that $\widetilde{F}_1(x) = \widetilde{F}_2(x)T(x)$. Notice that $T(x)$ acts on $\widetilde{F}_2(x)$, *to the right*. The gauge transformations T for which ${}^{T}B_0 = B_0$ form a group.

Definition 3.5.1 *The group $\mathbb{G}_0(B_0) \subset \widetilde{\mathbb{G}}(B_0)$ of formal gauge transformations T of the normal form $Y' = B_0(x)Y$ satisfying ${}^{T}B_0 = B_0$ is called* the group of (formal) isotropies *or* group of invariance of the normal form.

The classifying set of the meromorphic classes of systems in the formal class of the normal form $\mathrm{d}Y/\mathrm{d}x = B_0(x)Y$ can thus be defined as being the double quotient

$$\boxed{\mathbb{G}\backslash\widetilde{\mathbb{G}}(B_0)/\mathbb{G}_0(B_0)}$$

quotient of the previous classifying set $\mathbb{G}\backslash\widetilde{\mathbb{G}}(B_0)$ by the group $\mathbb{G}_0(B_0)$ of invariance of $\mathrm{d}Y/\mathrm{d}x = B_0(x)Y$ to the right (recall that isotropies act on gauge transformations to the right, cf. supra).

The group $\mathbb{G}_0(B_0)$ is in general small, even trivial, and it is easily determined in each particular case: it consists in all matrices $T(x)$ such that there exists a matrix $C \in \mathrm{GL}(n,\mathbb{C})$ satisfying $T(x)\mathscr{Y}_0(x) = \mathscr{Y}_0(x)C$; this corresponds to constant block-diagonal matrices C commuting[5] with Q and such that $x^L C x^{-L}$ is meromorphic. In the case when all diagonal terms q_j in $Q(1/x)$ are distinct the group $\mathbb{G}_0(B_0)$ is the group of all invertible constant diagonal matrices; if, in addition, we ask for tangent-to-identity gauge transformations then the group reduces to the identity.

Examples 3.5.2

Denote by I_j the identity matrix of dimension j and by J_j the irreducible nilpotent upper Jordan block of dimension j.

▷ Suppose the normal solution has the form

$$\mathscr{Y}_0(x) = x^{\lambda_1 I_1 \oplus(\lambda_2 I_3 + J_3)}\, \mathrm{e}^{q_1 I_1 \oplus q_2 I_3}$$

where $0 < \Re(\lambda_1), \Re(\lambda_2) < 1$ and where $q_1 \neq q_2$ are polynomials in $1/x$. The invertible matrices C such that $\mathscr{Y}_0(x)C\mathscr{Y}_0(x)^{-1}$ is a meromorphic transformation are those which commute both to $\mathrm{e}^{q_1 I_1 \oplus q_2 I_3}$ (this is a general fact) and to $x^{\lambda_1 I_1 \oplus(\lambda_2 I_3 + J_3)}$. One can check that this means that the matrix C has the form $C = C_1 \oplus C_2$ where $C_1 = cI_1$ with $c \in \mathbb{C}^*$ and $C_2 = c_1 I_3 + c_2 J_3 + c_3 J_3^2$ with $c_1, c_2, c_3 \in \mathbb{C}$ and $c_1 \neq 0$. All such constant matrices C form a group isomorphic to $\mathbb{G}_0(B_0)$.

▷ Suppose the normal solution has the form $\mathscr{Y}_0(x) = \bigoplus x^{L_j}\,\mathrm{e}^{q_j I_{n_j}}$ with distinct q_j's and matrices $L_j = \mathrm{diag}(\lambda_{j,1},\dots,\lambda_{j,n_j})$ with integer entries $\lambda_{j,1},\dots,\lambda_{j,n_j} \in \mathbb{Z}$. Then, $C = \bigoplus C_j$ is any constant invertible block-diagonal matrix with C_j of dimension n_j and the elements of $\mathbb{G}_0(B_0)$ are the transformations of the form $T(x) = \bigoplus x^{L_j} C_j x^{-L_j}$. Their coefficients are polynomials in x and $1/x$.

The difficult part of the meromorphic classification is to describe the set $\mathbb{G}\backslash\widetilde{\mathbb{G}}(B_0)$, that is, the Stokes phenomenon of the system.

▷ *Local meromorphic classification at a regular singular point.*

The case when 0 is a regular singular point (cf. Def. 3.3.2 (a)) of the system is quite simple. A formal fundamental solution of the system $\mathrm{d}Y/\mathrm{d}x = B(x)Y$ exists then in the form $\widetilde{\mathscr{Y}}(x) = \widetilde{F}(x)x^L$ with $\widetilde{F}(x) \in GL\big(n, \mathbb{C}[[x]][1/x]\big)$ and L a constant $n \times n$-matrix (cf. Thm. 3.3.1, p. 72). It is well known that $\widetilde{F}(x)$ is actually convergent. A proof of this fact is as follows: one reduces the problem to the case of a Fuchsian singular point at 0 (pole of order 1) by factoring $\widetilde{F}(x)$ in the form $\widetilde{F}(x) = P(x)\big(I + \widetilde{H}(x)\big)$ where $P(x)$ is a "polar part" (a polynomial matrix in x and $1/x$) and $\widetilde{H}(x)$ is an invertible (a priori, formal) analytic matrix. Then one proceeds by majorant series to prove the convergence of $\widetilde{H}(x)$. For details we refer to [Was76, Thm. 5.6 and Sect. 17.1] or [CL55, Thm. 2.1].

Let us sketch another proof [MS16, Cor. 1.37] based on the characterization of regular singular points by the moderate growth of the solutions (cf. Prop. 3.3.3). The idea is to consider an analytic fundamental solution $Y(x)$ in a neighborhood of 0

---

[5] If $Q = \bigoplus q_j I_{n_j}$ then $C = \bigoplus C_j$ with matrices C_j of size n_j.

on the universal covering of $\mathbb{C}^* = \mathbb{C} \setminus 0$ and its monodromy $e^{2\pi i E}$ (by the Cauchy-Lipschitz theorem and Grönwall's lemma any analytic solution can be continued along any path which avoids the singular points). Then the matrix $M(x) = Y(x)x^{-E}$ is analytic with trivial monodromy. Having moderate growth (and being single-valued) there exists a constant $C > 0$ and an integer $N \in Z$ such that $|M(x)| \leq C\,|x|^N$ on a punctured neighborhood of 0 in $\mathbb{C}$. The function $M(x)x^{-N}$ being both analytic and bounded on a punctured neighborhood of 0 the removable singularity theorem [Rud87, Thm. 10.20] allows us to conclude that $M(x)x^{-N}$ can be analytically continued at 0. Thus $M(x)$ is (convergent) meromorphic at 0. A formal fundamental solution reads $M(x)x^E = \widetilde{F}(x)x^L K$ for a convenient constant invertible matrix K. This implies that $x^E K^{-1} x^{-L}$ contains neither logarithms nor non-integer powers of x and that $\widetilde{F}(x)$ is convergent.

We conclude that when 0 is a regular singular point the system has no Stokes phenomenon at 0 and we can state:

Proposition 3.5.3 *Suppose the normal form* $\mathrm{d}Y/\mathrm{d}x = B_0(x)Y$ *has a regular singular point at 0. Then,* $\widetilde{\mathbb{G}}(B_0) = \mathbb{G}$ *so that the quotient* $\mathbb{G} \setminus \widetilde{\mathbb{G}}(B_0)$ *and the classifying set* $\mathbb{G} \setminus \widetilde{\mathbb{G}}(B_0)/\mathbb{G}_0(B_0)$ *are trivial.*

▷ *Local meromorphic classification at an irregular singular point.*

Just the opposite, the case when 0 is an irregular singular point is quite involved. The description is performed in several steps that we develop below. We begin with an auxiliary result, the infinitesimal isomorphism theorem, variant of the Cauchy-Heine theorem, which plays a central role in Sibuya's proof to come later.

It is worth to notice that, in the case of *linear* differential systems, the analysis of the Stokes phenomenon, including the construction of the Stokes cocycle, can be performed without the help of any theory of summation. On the contrary, as far as we know, the analysis of the Stokes phenomenon in the case of *non linear* differential equations is always achieved by means of a theory of summation or, so to speak equivalently, the theory of resurgence.

3.5.1 The Infinitesimal Isomorphism

In this section, we consider the sheaves $\mathscr{A}$ and $\mathscr{A}^{<0}$ (cf. Sect. 2.1.5, p. 41) seen as sheaves of additive, hence *abelian, groups* and we consider *abelian Čech cohomology*.

The Borel-Ritt theorem provides the exact sequence

$$0 \to \mathscr{A}^{<0} \longrightarrow \mathscr{A} \xrightarrow{T} \mathbb{C}[[x]] \to 0$$

(where T is the Taylor map; cf. Cor. 2.1.30, p. 46) and the Cauchy-Heine theorem implies that the natural map $H^1(S^1, \mathscr{A}^{<0}) \to H^1(S^1, \mathscr{A})$ is the null map (cf. Cor. 2.2.15, p. 56). Hence, from the long exact sequence of cohomology associated with the short exact sequence above we obtain the exact sequence

$$0 \to H^0(S^1\,;\mathscr{A}^{<0}) \longrightarrow H^0(S^1\,;\mathscr{A}) \longrightarrow H^0(S^1\,;\mathbb{C}[[x]]) \xrightarrow{\ \delta\ } H^1(S^1\,;\mathscr{A}^{<0}) \to 0$$

that is, the short exact sequence

$$0 \to \mathbb{C}\{x\} \to \mathbb{C}[[x]] \xrightarrow{\ \delta\ } H^1(S^1\,;\mathscr{A}^{<0}) \to 0 \tag{3.11}$$

and we can state:

Theorem 3.5.4 (Infinitesimal Isomorphism Theorem)
The coboundary map δ induces a linear isomorphism

$$\mu: \quad \mathbb{C}[[x]]/\mathbb{C}\{x\} \longrightarrow H^1(S^1\,;\mathscr{A}^{<0}). \tag{3.12}$$

The inverse map of μ is the Cauchy-Heine map defined as follows: to a 1-cocycle $\varphi = (\varphi_{j,j+1}) \in \prod \Gamma(U_j \cap U_{j+1}\,;\mathscr{A}^{<0})$ of a good covering $\mathscr{U} = (U_j)_{j\in\mathbb{Z}/J\mathbb{Z}}$ of S^1 by sectors U_j and to rays γ_j from 0 to some point in $U_j \cap U_{j+1}$ one associates the series

$$\widetilde{f}(x) = \sum_{m\in\mathbb{N}} f_m x^m \quad \textit{where} \quad f_m = \sum_{j\in J} \frac{1}{2\pi \mathrm{i}} \int_{\gamma_j} \frac{\varphi_{j,j+1}(t)}{t^{m+1}}\,\mathrm{d}t \quad \textit{for all } m;$$

Changing the rays γ_j changes $\widetilde{f}(x)$ by the addition of a convergent series, thus defining an element of $\mathbb{C}[[x]]/\mathbb{C}\{x\}$.

The identification of the arcs of a covering of S^1 to sectors of $\mathbb{C}^*$ or $\widetilde{\mathbb{C}}$ was explained in remark 2.2.11.

Recall that a good covering $\mathscr{U} = (U_j)_{j\in\mathbb{Z}/J\mathbb{Z}}$ of S^1 is a finite covering with no intersection 3-by-3. Its elements U_j are sectors. At least when $J \geq 3$, they can be cyclically indexed so that their 2-by-2 intersections reduce to $U_j \cap U_{j+1}$'s for all j (cf. Def. 2.2.9, p. 53 and the specific notation for the case when J equals 1 or 2). Both spaces $\mathbb{C}[[x]]/\mathbb{C}\{x\}$ and $H^1(S^1\,;\mathscr{A}^{<0})$ naturally have a structure of $\mathbb{C}$-vector spaces.

Recall also the definition of the coboundary map δ:

Given $\widetilde{f}(x) \in \mathbb{C}[[x]]$ and a covering $\mathscr{U} = (U_i)_{i\in I}$ of S^1 the Borel-Ritt theorem provides a 0-cochain $(f_i) \in \prod \Gamma(U_i\,;\mathscr{A})$ of $\mathscr{U}$ with values in $\mathscr{A}$ such that $Tf_i = \widetilde{f}$ for all i. Its coboundary determines a 1-cocycle $(\varphi_{i,j}) = (-f_i + f_j)$ with values in $\mathscr{A}^{<0}$. The functions $f_i(x)$ given by the Borel-Ritt theorem are not unique; however, the cohomology class of the 1-cocycle $(\varphi_{i,j})$ does not depend on these different choices. Indeed, let $(g_i(x))$ be other realizations of $\widetilde{f}(x)$ on $\mathscr{U}$ and set $\psi_{i,j} = -g_i + g_j$ for all i,j. Then, we can write

$$\psi_{i,j}(x) - \varphi_{i,j}(x) = -\big(g_i(x) - f_i(x)\big) + \big(g_j(x) - f_j(x)\big)$$

The right-hand-side is a coboundary with values in $\mathscr{A}^{<0}$ since $Tf_i = Tg_i$ for all i. The two 1-cocycles $(\varphi_{i,j})$ and $(\psi_{i,j})$ are then cohomologous in $H^1(S^1\,;\mathscr{A}^{<0})$. Moreover, the 1-coclycle ϕ induced in $H^1(S^1\,;\mathscr{A}^{<0})$ by $(\varphi_{i,j})$ does not depend on the choice of the covering $\mathscr{U}$. Indeed, let the covering $\mathscr{V} = (V_\alpha)_{\alpha\in\Omega}$ be finer than $\mathscr{U}$ and let $\sigma : \Omega \to I$ be a simplicial map from $\mathscr{V}$ to $\mathscr{U}$; hence, $V_\alpha \subseteq U_{\sigma(\alpha)}$ for all α. Associated with $\mathscr{V}$ choose the 0-cochain $(h_\alpha) \in \prod \Gamma(V_\alpha\,;\mathscr{A})$ defined by $h_\alpha = f_{\sigma(\alpha)}{}_{|V_\alpha}$. Denote by $[\varphi_{i,j}]$ the 1-cohomology class of $(\varphi_{i,j})$ in $H^1(\mathscr{U}\,;\mathscr{A}^{<0})$ and by $[\eta_{\alpha,\beta}]$ the 1-cohomology class of $(-h_\alpha + h_\beta)$ in $H^1(\mathscr{V}\,;\mathscr{A}^{<0})$. Then, one has $\mathfrak{S}^1(\mathscr{V},\mathscr{U})([\varphi_{i,j}]) = [\eta_{\alpha,\beta}]$. Hence the result.

Adding a convergent series to $\widetilde{f}(x)$ does not change the 1-cocycle $(\varphi_{i,j})$ above. Hence, δ can be factored through the quotient $\mathbb{C}[[x]]/\mathbb{C}\{x\}$ defining thus the map μ; its linearity is clear.

The inverse of μ in terms of Cauchy-Heine integrals is straightforward from the Cauchy-Heine theorem 1.4.2, p. 27 taking into account the additivity of 1-cocycles over good coverings. The fact that changing one or several γ_j changes $\widetilde{f}(x)$ by adding a convergent series has already been mentioned in section 1.4, p. 26. □

3.5.2 *Malgrange-Sibuya Theorems*

From now, we consider sheaves of *non-abelian groups* and *non-abelian Čech cohomology*. More precisely, we consider the following sheaves of multiplicative groups over S^1:

- ▷ $\Lambda = GL(n,\mathscr{A})$ the sheaf of germs of $n\times n$ invertible matrices with entries in $\mathscr{A}$;
- ▷ $\Lambda^{<0} = GL_I(n,\mathscr{A}^{<0})$ the subsheaf of germs of Λ asymptotic to the identity I;
- ▷ $\mathscr{G}L(n,\mathbb{C}[[x]])$ the constant sheaf whose stalks are all $n\times n$ invertible matrices with entries in $\mathbb{C}[[x]]$;
- ▷ $\Lambda(B_0)$ the sheaf of germs of gauge transformations of $\mathrm{d}Y/\mathrm{d}x = B_0(x)Y$;
- ▷ $\Lambda^{<0}(B_0)$ the subsheaf of germs of $\Lambda^{<0}$ leaving $\mathrm{d}Y/\mathrm{d}x = B_0(x)Y$ fixed.

Recall the notation

- ▷ $\mathbb{G} \equiv \mathrm{GL}(n,\mathbb{C}\{x\}[1/x])$,
- ▷ $\widetilde{\mathbb{G}} \equiv \mathrm{GL}(n,\mathbb{C}[[x]][1/x])$,
- ▷ $\widetilde{\mathbb{G}}(B_0)$ for the set of the formal gauge transformations of the normal form $\mathrm{d}Y/\mathrm{d}x = B_0(x)Y$, that is, the set of gauge transformations $T\in\widetilde{\mathbb{G}}$ such that the transformed matrix ${}^{T}B_0(x)$ is convergent (cf. Sect. 3.3.1, p. 71 and formula (3.2), p. 64).

We assume that $\mathrm{d}Y/\mathrm{d}x = B_0(x)Y$ is a normal form with the fundamental solution $\mathscr{Y}_0 = x^L e^{Q(1/x)}$. Given a θ direction, a germ in $\Lambda_\theta(B_0)$ is a germ $f\in\Lambda_\theta$ such that there exists a system $\mathrm{d}Y/\mathrm{d}x = B(x)Y$ in the formal class of $\mathrm{d}Y/\mathrm{d}x = B_0(x)Y$ satisfying ${}^{f}B_0 = B$ in a neighborhood of the θ direction. The germ f belongs to $\Lambda_\theta^{<0}(B_0)$ if it is both flat (that is, asymptotic to the identity matrix $I = I_n$) and an isotropy of the normal form (that is, ${}^{f}B_0 = B_0$). The germs in $\Lambda^{<0}(B_0)$ are called *germs of flat isotropy of the normal form.*

For $n > 1$, these sheaves are sheaves of non abelian groups. They all have a neutral element: the neutral section, that is, the constant section equal to the identity matrix I; with no harm, we denote also the neutral section by I. The Taylor map T and the Borel-Ritt theorem can be extended to matrices yielding the short exact sequences of sheaves

$$I \longrightarrow \Lambda^{<0} \longrightarrow \Lambda \stackrel{T}{\longrightarrow} \mathscr{G}L(n,\mathbb{C}[[x]]) \longrightarrow I.$$

It follows that $\Lambda^{<0}$ is normal in Λ and that we have an exact sequence of cohomology

$$I \longrightarrow \mathbb{G} \longrightarrow \widetilde{\mathbb{G}} \stackrel{\delta}{\longrightarrow} H^1\big(S^1\,;\Lambda^{<0}\big) \longrightarrow H^1\big(S^1\,;\Lambda\big) \qquad (3.13)$$

after identification of $H^0\big(S^1\,;\Lambda\big)$ and $H^0\big(S^1\,;\mathscr{G}L(n,\mathbb{C}[[x]])\big)$ to $\mathbb{G}$ and $\widetilde{\mathbb{G}}$ respectively.

The definition of the coboundary map δ is similar to the one in the abelian case. Because of its importance we make it explicit again. Let the series $\widetilde{F}(x)$ belong to $\widetilde{\mathbb{G}}$ and let $\mathscr{U} = (U_j)_{j\in\mathbb{Z}/J\mathbb{Z}}$ be a good covering of S^1, that is, a covering without intersection 3-by-3. From the Borel-Ritt theorem there exists a 0-cochain (F_j) in $H^0\big(\mathscr{U}\,;\Lambda\big)$ satisfying $TF_j = \widetilde{F}$ for all $j \in J$. There corresponds the 1-cocycle $(\varphi_{i,j} = F_i^{-1}F_j)$ in $H^1\big(\mathscr{U}\,;\Lambda^{<0}\big)$ (there is no cocycle condition to be checked since the covering has no intersection 3-by-3). Given a different choice (F_j') for the 0-cochain there corresponds a 1-cocycle $(\varphi_{i,j}')$ cohomologous to $(\varphi_{i,j})$ in $H^1\big(\mathscr{U}\,;\Lambda^{<0}\big)$. Indeed, from the condition $TF_j = TF_j'$ we deduce that the functions $f_j = F_j^{-1}F_j'$ belong to $\Gamma(U_j\,;\Lambda^{<0})$ for all j and that they satisfy the relations $\varphi_{i,j}' = f_i^{-1}\varphi_{i,j}f_j$ for all i,j. Thus, we define an application from $\widetilde{\mathbb{G}}$ to the cohomology set $H^1\big(\mathscr{U}\,;\Lambda^{<0}\big)$ and finally to $H^1\big(S^1\,;\Lambda^{<0}\big)$ by taking the direct limit (see the previous section).

The coboundary map δ is constant on analytic classes $\{f\widetilde{F}\,;f\in\mathbb{G}\}$:

Indeed, to any representative $f\widetilde{F}$ of a given class there corresponds a same 1-cocycle $(fF_i)^{-1}(fF_j) = F_i^{-1}F_j = \varphi_{i,j}$ whatever the choice of the convergent series f. It can then be factored through the quotient $\mathbb{G}\backslash\widetilde{\mathbb{G}}$ inducing a map

$$\mathrm{Exp}_\mu\,:\ \mathbb{G}\backslash\widetilde{\mathbb{G}} \longrightarrow H^1\big(S^1\,;\Lambda^{<0}\big).$$

We can now state the isomorphism theorem.

Theorem 3.5.5 (Malgrange-Sibuya Isomorphism Theorem)

The map $\mathrm{Exp}_\mu\,:\ \mathbb{G}\backslash\widetilde{\mathbb{G}} \longrightarrow H^1\big(S^1\,;\Lambda^{<0}\big)$ *is bijective.*

Recall that, a priori, $H^1\big(S^1\,;\Lambda^{<0}\big)$ is a set, not a group.

From the exactness of the cohomology sequence 3.13, p. 96 one has just to prove that the natural map $H^1\big(S^1\,;\Lambda^{<0}\big) \to H^1\big(S^1\,;\Lambda\big)$ is the trivial map sending any element in $H^1(S^1\,;\Lambda^{<0})$ to the trivial cohomology class I in $H^1(S^1\,;\Lambda)$. The result is similar to Cartan's lemma on holomorphic matrices but the fact that we have here an extra asymptotic condition.

A first proof was given by Y. Sibuya in 1976 in an article in Japanese summarized in [Sib77] and developed in [Sib90]. The method consists in step-by-step approximations obtained by successive applications of the infinitesimal isomorphism theorem; a limit process gives the result thanks to careful estimates. A second proof is mentioned by B. Malgrange in [Mal79]. It relies on exact sequences of complexes of Dolbeault type: B. Malgrange considers sheaves of $\mathscr{C}^\infty$ germs of functions in two real variables satisfying a $\overline{\partial}$ condition. A third proof was given by J. Martinet and J.-P. Ramis in [MarR82] following an idea of B. Malgrange in [Mal82] which is based on a theorem of Newlander-Nirenberg about the integration of quasi-complex structures. See also [BJL79b].

For seek of completeness, Sibuya's proof is included at the end of the section.

The classifying set $\mathbb{G}\backslash\widetilde{\mathbb{G}}(B_0)$ of the formal gauge transformations of a given normal form $\mathrm{d}Y/\mathrm{d}x = B_0Y$ modulo the convergent ones can be identified to a subset of $\mathbb{G}\backslash\widetilde{\mathbb{G}}$ and one can adapt the construction of Exp_μ to this subset. The Malgrange-Sibuya isomorphism theorem becomes the Malgrange-Sibuya classification theorem below.

Theorem 3.5.6 (Malgrange-Sibuya Classification theorem)

The map Exp_μ *induces a bijection*

$$\exp_\mu:\ \mathbb{G}\backslash\widetilde{\mathbb{G}}(B_0) \longrightarrow H^1\big(S^1;\Lambda^{<0}(B_0)\big)$$

thus providing a one-to-one correspondence between the meromorphic classes of formal gauge transformations of the normal form $\mathrm{d}Y/\mathrm{d}x = B_0(x)Y$ *and the cohomology classes in* $H^1\big(S^1;\Lambda^{<0}(B_0)\big)$.

Proof. In restriction to $\mathbb{G}\backslash\widetilde{\mathbb{G}}(B_0)$ the map Exp_μ has values in $H^1\big(S^1;\Lambda^{<0}(B_0)\big)$. Indeed, let $\widetilde{F}(x)$ be in $\widetilde{\mathbb{G}}(B_0)$ and denote $B = {}^{\widetilde{F}}B_0$. The matrix $\widetilde{F}(x)$ is solution of the linear differential system $\mathrm{d}F/\mathrm{d}x = BF - FB_0$. To define the coboundary map δ one can then use the main asymptotic existence theorem instead of the Borel-Ritt theorem and the result follows.

The map $\exp_\mu$ is injective, for, there is the exact sequence of cohomology

$$I \longrightarrow \mathbb{G} \longrightarrow \widetilde{\mathbb{G}}(B_0) \xrightarrow{\ \delta_0\ } H^1\big(S^1\,;\Lambda^{<0}(B_0)\big) \longrightarrow H^1\big(S^1\,;\Lambda(B_0)\big) \qquad (3.14)$$

associated with the short exact sequence

$$I \longrightarrow \Lambda^{<0}(B_0) \longrightarrow \Lambda(B_0) \xrightarrow{\ T\ } \widetilde{\mathbb{G}}(B_0) \longrightarrow I.$$

To prove that $\exp_\mu$ is surjective let φ belong to $H^1(S^1; \Lambda^{<0}(B_0))$ and $(\varphi_{j,j+1})$ be a 1-cocycle representing φ over a good covering $\mathscr{U} = (U_i)$ of S^1. The 1-cocycle has also values in $\Lambda^{<0}$ and its cohomology class in $H^1(S^1; \Lambda^{<0})$ is associated with a formal series $\widetilde{F}(x)$ by the Malgrange-Sibuya isomorphism theorem. In particular, (since the natural map $H^1(S^1; \Lambda^{<0}) \longrightarrow H^1(S^1; \Lambda)$ is trivial) there exists a good covering $\mathscr{V} = (V_j)_{j\in\mathbb{Z}/J\mathbb{Z}}$ finer than $\mathscr{U}$ and sections $F_j(x) \in \Gamma(V_j; \Lambda)$ such that the 1-cocycle $(F_j^{-1}F_{j+1})$ is equal to the image $(\psi_{j,j+1}) = \mathfrak{S}(\mathscr{V}, \mathscr{U})(\varphi_{j,j+1})$ of $(\varphi_{j,j+1})$ over $\mathscr{V}$. This implies $F_{j+1} = F_j\,\psi_{j,j+1}$ on $V_{j,j+1} = V_j \cap V_{j+1}$ and then ${}^{F_{j+1}}B_0 = {}^{F_j\psi_{j,j+1}}B_0 = {}^{F_j}B_0$. Thus, the matrices ${}^{F_j}B_0$ glue together into an analytic matrix $B(x)$ around 0 with asymptotic expansion ${}^{\widetilde{F}}B_0$. The matrix $B(x)$ is then meromorphic at 0, hence, $\widetilde{F}(x)$ belongs to $\widetilde{\mathbb{G}}(B_0)$; the 1-cocycle φ and therefore, the 1-cocycle $(\psi_{j,j+1})$ having values in $\Lambda^{<0}(B_0)$ we deduce that φ and $(\psi_{j,j+1})$ both represent $\exp_\mu(\widetilde{F})$.

This achieves the proof of the theorem. □

3.5.3 *The Stokes Cocycle Theorem*

It might be unclear that working with 1-cohomology classes is quite often more convenient than working with formal series modulo the convergent ones; however this is the case and especially when the series are determined inductively term after term since a finite number of terms is always equivalent to I modulo convergent series. Nevertheless, the Malgrange-Sibuya theorem remains a theoretical theorem. Actually, it can be given a more practical form, without having to consider equivalence classes. And indeed, although we were unable to select a normal form to characterize the meromorphic classes of systems we are able to select in each cohomology class of $H^1(S^1; \Lambda^{<0}(B_0))$ a special 1-cocycle, called *Stokes cocycle* to characterize the class. The aim of this section is to describe this new characterization. To this end we need first to introduce some definitions.

3.5.3.1 The Stokes Groups and the Stokes Automorphisms

Let $\mathfrak{A}$ denote the set of anti-Stokes directions of the homological system linking the normal form $\mathrm{d}Y/\mathrm{d}x = B_0(x)Y$ to any system in its formal class. This means all directions of maximal decay for at least one (non-zero) polynomial $q_\ell - q_j$ (cf. Def. 3.3.5 (v), p. 77).

Definition 3.5.7

▷ *The* Stokes group $\mathrm{Sto}_\alpha(B_0)$ associated with a normal form $\mathrm{d}Y/\mathrm{d}x = B_0(x)Y$ in an $\alpha \in S^1$ direction *is the subgroup of the stalk* $\Lambda_\alpha^{<0}(B_0)$ *of the sheaf* $\Lambda^{<0}(B_0)$, *at α, consisting of all germs of flat isotropies of* $\mathrm{d}Y/\mathrm{d}x = B_0(x)Y$ *having maximal decay in the α direction.*

▷ *The elements of* $\mathrm{Sto}_\alpha(B_0)$ *are said to be the* Stokes automorphisms *of the normal form* $\mathrm{d}Y/\mathrm{d}x = B_0(x)Y$ *in the α direction.*

Let us describe the groups $\mathrm{Sto}_\alpha(B_0)$ in more details.

▷ When $\alpha \notin \mathfrak{A}$ the group $\mathrm{Sto}_\alpha(B_0)$ is trivial: no flat isotropy has maximal decay but the identity. Recall that the term "flat" is taken here in the multiplicative sense of "asymptotic to identity".

▷ When $\alpha \in \mathfrak{A}$ the group $\mathrm{Sto}_\alpha(B_0)$ can be given a linear representation as follows: give first an analytic meaning to the (formal) normal solution

$$\mathscr{Y}_0(x) = x^L \mathrm{e}^{Q(1/x)} \quad \text{with} \quad Q(1/x) = \bigoplus_{j=1}^{J} q_j(1/x) I_j \quad \text{and distinct} \quad q_j\text{'s}.$$

To this end we choose a determination $\underline{\alpha}$ of α and we give an analytic meaning to the powers of x and the logarithms in a neighborhood of α by choosing this determination. Then, we fix an initial value for the exponentials[6], say, $\mathrm{e}^0 = 1$ and we denote by $\mathscr{Y}_{0,\underline{\alpha}}(x)$ the function thus defined. Since there is no ambiguity, we denote again by x^L and e^Q the analytic functions thus obtained near the $\underline{\alpha}$ direction. An element $\varphi_\alpha(x)$ of $\mathrm{Sto}_\alpha(B_0)$ being a germ of flat isotropy of the normal form $\mathrm{d}Y/\mathrm{d}x = B_0(x)Y$ there exists a unique constant invertible matrix $I_n + C_{\underline{\alpha}}$ such that

$$\varphi_\alpha(x)\,\mathscr{Y}_{0,\underline{\alpha}}(x) = \mathscr{Y}_{0,\underline{\alpha}}(x)\,(I_n + C_{\underline{\alpha}}). \tag{3.15}$$

This implies that $\varphi_\alpha(x) = x^L \mathrm{e}^{Q(1/x)}\,(I_n + C_{\underline{\alpha}})\,\mathrm{e}^{-Q(1/x)} x^{-L}$ with the given choice of exponentials and arguments near α.

Denote by $C_{\underline{\alpha}} = [C_{\underline{\alpha}}^{(\ell,j)}]$ the decomposition of $C_{\underline{\alpha}}$ into blocks fitting the structure of Q. Then, the germ $\varphi_\alpha(x)$ reads in the form

$$\varphi_\alpha(x) = x^L\left(I_n + \left[C_{\underline{\alpha}}^{(\ell,j)}\,\mathrm{e}^{(q_\ell - q_j)(1/x)}\right]\right)x^{-L}. \tag{3.16}$$

Moreover, $\varphi_\alpha(x)$ is flat, meaning asymptotic to I_n, in the α direction if and only if $C_{\underline{\alpha}}^{(\ell,j)} = 0$ for all (ℓ, j) such that the exponential $\mathrm{e}^{q_\ell(1/x) - q_j(1/x)}$ has not maximal decay for $\arg(x) = \underline{\alpha}$. Recall that an exponential $\mathrm{e}^{q(1/x)} = \mathrm{e}^{-a/x^k(1+o(1/|x^k|))}$ has maximal decay in an $\arg(x) = \underline{\alpha}$ direction if and only if $-a\,\mathrm{e}^{-ik\underline{\alpha}}$ is real negative (k might be fractional); the exponential is then said to be *led* by $\underline{\alpha}$ or $\underline{\alpha}$ to be a *leading direction* of the exponential. In particular, for $j = \ell$, the exponential $\mathrm{e}^{q_j - q_\ell}$ does not have maximal decay and the corresponding diagonal block $C_{\underline{\alpha}}^{(j,j)}$ is zero; if $\mathrm{e}^{q_\ell - q_j}$ has maximal decay in the $\underline{\alpha}$ direction, hence in particular, if the block $C_{\underline{\alpha}}^{(\ell,j)}$ is not

[6] Recall that a formal exponential $\mathrm{e}^{q(x)}$ is a symbol which formally satisfies the differential equation $y' - q'(x)y = 0$; hence it can equally been identified with any of the functions $c\exp(q(x))$ with $c \in \mathbb{C}^*$.

equal to zero, then $e^{q_j - q_\ell}$ has not maximal decay, and the symmetric block $C_{\underline{\alpha}}^{(j,\ell)}$ is necessarily zero. This implies that the matrix $I_n + C_{\underline{\alpha}}$ is unipotent. Reciprocally, any constant unipotent matrix with the necessary blocks of zeros characterizes a unique element of $\mathrm{Sto}_\alpha(B_0)$. As a consequence, $\mathrm{Sto}_\alpha(B_0)$ naturally has the structure of a *unipotent Lie group*.

The following proposition holds true.

Proposition 3.5.8

▷ *The Stokes groups* $\mathrm{Sto}_\alpha(B_0)$ *are unipotent Lie groups.*

▷ *Their elements can be represented by matrices fitting the block structure of Q; their diagonal blocks are unit matrices, some blocks are necessarily zero whereas the possible non zero blocks may have arbitrary complex entries. These matrices are called* Stokes matrices.

▷ *In particular, the Stokes groups* $\mathrm{Sto}_\alpha(B_0)$ *have dimension* $\leq n(n-1)/2$ *(Recall that n is the dimension of* B_0*).*

▷ *The global dimension* $N = \dim\left(\prod_{\alpha \in \mathfrak{A}} \mathrm{Sto}_\alpha(B_0)\right)$ *is the irregularity of the systems in the formal class of* $\mathrm{d}Y/\mathrm{d}x = B_0(x)Y$.

Proof. The first three assertions result from the explanations above. The irregularity of an operator (replace the normal form $\mathrm{d}Y/\mathrm{d}x = B_0(x)Y$ or any system of its formal class by a one dimensional differential operator by means of the cyclic vector lemma 3.1.2, p. 64) is defined in the next chapter. The last assertion will result from theorem 4.2.3-3 and remark 4.2.4 in the next chapter. □

Let us comment the notation: the Stokes automorphisms are independent of any choice of the determination of the argument and are then denoted by φ_α, indexed by α and not $\underline{\alpha}$, whereas their representations depend on the choice of a determination $\underline{\alpha}$ of α, and this explains the notation $I + C_{\underline{\alpha}}$ indexed by $\underline{\alpha}$ for the Stokes matrices. The Stokes matrices depend also on the choice of initial constants for the exponentials but, for simplicity and since we always choose $e^0 = 1$, we do not mention this dependence in the notation.

Remark 3.5.9 The name *Stokes automorphism* is used with different, though closely related, meanings. We introduced it in the form of a special gauge transform, namely as a maximally flat isotropy of the normal form in an anti-Stokes direction (Def. 3.5.7). It is best known as a linear automorphism in the space of analytic solutions in the anti-Stokes direction. This is how we interpreted it in order to make explicit its matrix representations in Eq. 3.15. For further use, in differential Galois theory or in resurgence theory for instance, one needs to extend the Stokes automorphisms into automorphisms of differential algebras, called again Stokes automorphisms. For this point of view we refer for instance to the first volume of this book [MS16, Sect. 2.2.3.3] and to [LR11, Sect. 4.4.2].

3.5.3.2 The Stokes Cocycle Theorem

Consider a covering $\mathscr{U} = (U_\alpha)_{\alpha\in\mathfrak{A}}$ of S^1 of the following form. $\mathscr{U}$ is a good covering containing as many sectors U_α as anti-Stokes directions α. Given α denote by α^+ the next anti-Stokes direction according to a chosen orientation[7] of S^1. Set $\dot{U}_\alpha = U_\alpha \cap U_{\alpha^+}$ and assume that $\dot{U}_\alpha$ contains the unique anti-Stokes direction α. A 1-cocycle of $H^1\big(\mathscr{U}\,;\Lambda^{<0}(B_0)\big)$ is by definition a family of functions $\phi_\alpha \in \Gamma(\dot{U}_\alpha\,;\Lambda^{<0}(B_0))$. It is determined as well by the germs φ_α induced at α by the ϕ_α's. Moreover, given another covering $\mathscr{U}'$ with same characteristics and a 1-cocycle (ϕ'_α) of $H^1\big(\mathscr{U}'\,;\Lambda^{<0}(B_0)\big)$ inducing the same germs at α (i.e., $\phi_\alpha = \phi'_\alpha$ in restriction to $\dot{U}_\alpha \cap \dot{U}'_\alpha$), there corresponds the same element in $H^1\big(S^1\,;\Lambda^{<0}(B_0)\big)$. We can then identify a family of germs in $\Lambda^{<0}(B_0)$ to a class of cohomologous 1-cocycles in $H^1\big(S^1\,;\Lambda^{<0}(B_0)\big)$ and set the following definition.

Definition 3.5.10 (Stokes Cocycle)

A Stokes cocycle *(for the normal form $Y' = B_0(x)Y$) is a family of Stokes automorphisms given in each anti-Stokes direction:*

$$\varphi = (\varphi_\alpha)_{\alpha\in\mathfrak{A}} \in \prod_{\alpha\in\mathfrak{A}} \mathrm{Sto}_\alpha(B_0)$$

The Stokes cocycle theorem below says, in particular, that the Stokes cocycle is a special 1-cocycle which exists and is uniquely determined in each cohomology class of $H^1\big(S^1\,;\Lambda^{<0}(B_0)\big)$. One can thus improve the Malgrange-Sibuya classification theorem by replacing classes of cohomologous 1-cocycles by just one special 1-cocycle: the Stokes cocycle. Let us develop somehow these results.

We denote by h the map that identifies a Stokes cocycle to a cohomology class of 1-cocycles in $H^1\big(S^1\,;\Lambda^{<0}(B_0)\big)$ as explained above.

Theorem 3.5.11 (Stokes Cocycle Theorem [Lod94, Lod03])

The map

$$h:\ \prod_{\alpha\in\mathfrak{A}} \mathrm{Sto}_\alpha(B_0) \longrightarrow H^1\big(S^1\,;\Lambda^{<0}(B_0)\big)$$

is bijective.

Its inverse map

$$c:\ H^1\big(S^1\,;\Lambda^{<0}(B_0)\big) \longrightarrow \prod_{\alpha\in\mathfrak{A}} \mathrm{Sto}_\alpha(B_0)$$

is constructive and it is natural in the sense that it commutes with isomorphisms.

---

[7] Any orientation is equally possible. We usually orient S^1 counterclockwise about infinity and clockwise about 0 so that all formulæ be compatible when moving from 0 to infinity or vice-versa by the change of variable $x = 1/z$.

The following corollary is straightforward from the Malgrange-Sibuya classification theorem.

Corollary 3.5.12 (Classification by Means of Stokes Cocycles)
There is a bijection

$$\mathbb{G}\backslash\widetilde{\mathbb{G}}(B_0) \xrightarrow{\sim} \prod_{\alpha\in\mathfrak{A}} \mathrm{Sto}_\alpha(B_0)$$

between the classifying set $\mathbb{G}\backslash\widetilde{\mathbb{G}}(B_0)$ *of the meromorphic classes of formal gauge transformations of* $\mathrm{d}Y/\mathrm{d}x = B_0(x)Y$ *and the Stokes group* $\prod_{\alpha\in\mathfrak{A}} \mathrm{Sto}_\alpha(B_0)$*, product of the Stokes groups in each anti-Stokes direction.*
The classifying set $\mathbb{G}\backslash\widetilde{\mathbb{G}}(B_0)$ *inherits thus the structure of a unipotent Lie group.*

In practice, given a normal form $\mathrm{d}Y/\mathrm{d}x = B_0(x)Y$ and a gauge transformation $\widetilde{F}(x) \in \widetilde{\mathbb{G}}(B_0)$ there are N (the irregularity; cf. Def. 4.1.1, p. 122) meromorphic invariants to characterize the meromorphic class of $\widetilde{F}(x)$: they are the non-trivial entries (the *Stokes multipliers*) of the Stokes matrices $I_n + C_{\underline{\alpha}}$ relative to the choice of analytic normal solutions $\mathscr{Y}_{0,\underline{\alpha}}(x)$ in each α direction (cf. Sect. 3.5.3.1, p. 98). These numerical invariants are relative to the normal solutions $\mathscr{Y}_{0,\underline{\alpha}}(x)$; they are not intrinsic.

The link between the Stokes cocycle theorem and the k- or multi-sums of $\widetilde{F}(x)$ is given in the next section. In [Lod94] the Stokes cocycle theorem is applied to the factorization of $\widetilde{F}(x)$ by k_j-summable factors for $j = 1,2,\ldots,\nu$ (where $k_1,k_2,\ldots,k_\nu$ are the levels of any systems in the formal class of $\mathrm{d}Y/\mathrm{d}x = B_0(x)Y$); it is also used to provide a tannakian proof of Ramis's density theorem in differential Galois theory.

The cocycle theorem was first given an abstract proof in [BV89, Thm. 1.1.2]; however, the proof does not provide the constructibility of the map c. A direct constructive proof is given in [Lod94]. For details we refer to this latter reference and to [Lod03] where the injectivity of the map h is further developed. Here, we limit ourselves to sketch some main points.

It suffices to consider the case of integer levels. Otherwise, we can go to the case of integer levels by a change of variable $x = t^p$. A descent process to turn back to the initial variable is made possible by the fact that the matrices $\widetilde{F}(x)$ are power in x with no root of x (cf. [Lod94, Sect. II.4]). From now we suppose that all levels are integers.

▷ *Proof of theorem 3.5.11 in the case with a single level.*

This case is quite simple (at least, the combinatorial description of the Stokes arcs is quite simple). The proof contains ideas used in the general case.

Assume that the homological system $\mathrm{d}Y/\mathrm{d}x = B_0(x)Y - YB_0(x)$ associated with the normal form $\mathrm{d}Y/\mathrm{d}x = B_0(x)Y$ has the single level $k \in \mathbb{N}^*$ (cf. Def. 3.3.4 (iii), p. 77). This means that all non-zero polynomials $q_\ell - q_j$ have degree k and the same property holds for the homological system $\mathrm{d}Y/\mathrm{d}x = BY - YB_0$ where the system $\mathrm{d}Y/\mathrm{d}x = B(x)Y$ is any system in the formal formal class of $\mathrm{d}Y/\mathrm{d}x = B_0(x)Y$.

In this case, a Stokes arc $\overline{I}_\alpha$ is a closed arc bisected by an anti-Stokes direction $\alpha \in \mathfrak{A}$ with length π/k; in particular, all Stokes arcs have the same length. A given Stokes arc may be attached to several polynomials $q_j - q_\ell$ but two distinct Stokes arcs are led by distinct anti-Stokes directions. Thus, the Stokes arcs can be ordered cyclically like are ordered the anti-Stokes directions on S^1. Given a direction α we denote by α^- and α^+ the anti-Stokes directions respectively preceding and following α. We say that two Stokes arcs are *contiguous* if they are led by two contiguous anti-Stokes directions α and α^+; they overlap by at least one point and they might overlap by a proper sub-arc.

Suppose first that there are more than two anti-Stokes directions so that two contiguous directions are not opposite and the union of two contiguous Stokes arcs is not S^1. We define the covering $\mathscr{U} = (U_\alpha)$ of S^1 as follows: the open arc U_α is the interior of the closed arc $\overline{I}_{\alpha^-} \cup \overline{I}_\alpha$ (recall that two contiguous Stokes arcs overlap by at least one point). We obtain so an open covering of S^1 by arcs U_α with the following property: $\dot{U}_\alpha = U_\alpha \cap U_{\alpha^+}$ is the interior of the Stokes arc $\overline{I}_\alpha$.

The limit case with only two (necessarily opposite) anti-Stokes directions α and $\alpha^+ = \alpha^-$ and two Stokes arcs (necessarily two opposite half-circles corresponding to level $k = 1$) requires a special convention. In this case the covering $\mathscr{U} = (U_\alpha, U_{\alpha^+} = U_{\alpha^-})$ is defined as follows: let β and β^+ be the two common boundary points of the two Stokes arcs and set $U_\alpha = S^1 \setminus \{\beta\}$ and $U_{\alpha^+} = S^1 \setminus \{\beta^+\}$. Thus, $U_\alpha \cap U_{\alpha^+}$ has two connected components $\dot{U}_\alpha$ and $\dot{U}_{\alpha^+} = \dot{U}_{\alpha^-}$. To fit the general notation we should have $U_\alpha \cap U_{\alpha^+} = \dot{U}_\alpha$ and $U_{\alpha^-} \cap U_\alpha = \dot{U}_{\alpha^-}$. However $U_{\alpha^+} = U_{\alpha^-}$ and $U_\alpha \cap U_{\alpha^+} = U_{\alpha^-} \cap U_\alpha = \dot{U}_\alpha \cup \dot{U}_{\alpha^-}$. From now, to avoid this notational specificity, we place ourselves in the general case, leaving to the reader the translation of the proof to this limit case. In particular, we assume that intersections of two contiguous arcs U_α reduce always to only one arc.

In general, the covering $\mathscr{U}$ is not a good covering in the sense of definition 2.2.9, p. 53: there might exist 3-by-3, 4-by-4, etc... intersections. But we can use it in the same way by considering only partial 1-cocycles $(\dot{\varphi}_\alpha)_{\alpha\in\mathfrak{A}}$ on the intersections $\dot{U}_\alpha$, not on all 2-by-2 intersections. We observe that intersections of non contiguous U_α are contained in at least one $\dot{U}_\alpha$ and the components on 2-by-2 intersections of non contiguous U_α, are uniquely determined from the partial 1-cocycle by the cocycle condition (cf. Def. 2.3.2, p. 58). It results that partial 1-cochains generate automatically 1-cocycles without having to check any cocycle condition. Observe also that a section of $\Lambda^{<0}(B_0)$ over a Stokes arc $\dot{U}_\alpha$ induces at α a Stokes germ led by α, that is, an element of $\mathrm{Sto}_\alpha(B_0)$; and reciprocally any Stokes germ led by α can be continued as a section of $\Lambda^{<0}(B_0)$ all over $\dot{U}_\alpha$. On the other hand, there is no 0-cochain of $\mathscr{U}$ with values in $\Lambda^{<0}(B_0)$ since $\Lambda^{<0}(B_0)$ admits no non trivial section over arcs larger than π/k (no exponential of order k is flat on arcs larger than π/k). There is then no 1-cohomology relation and we can identify the set of partial 1-cochains of $\mathscr{U}$ with values in $\Lambda^{<0}(B_0)$ to the set of cohomology classes in $H^1\big(\mathscr{U}\,;\Lambda^{<0}(B_0)\big)$. We have thus proved the existence of a bijection

$$h^* : \prod_{\alpha\in\mathfrak{A}} \mathrm{Sto}_\alpha(B_0) \longrightarrow H^1\big(\mathscr{U}\,;\Lambda^{<0}(B_0)\big).$$

By composition with the natural inclusion

$$i^* : H^1\big(\mathscr{U}\,;\Lambda^{<0}(B_0)\big) \hookrightarrow H^1\big(S^1\,;\Lambda^{<0}(B_0)\big)$$

we obtain an injective map

$$h = i^* \circ h^* : \prod_{\alpha\in\mathfrak{A}} \mathrm{Sto}_\alpha(B_0) \longrightarrow H^1\big(S^1\,;\Lambda^{<0}(B_0)\big).$$

We are left to prove that i^* is surjective. To this end, consider a good covering $\mathscr{V} = (V_j) \preceq \mathscr{U}$ of S^1 finer than $\mathscr{U}$ and $\varphi = (\varphi_{i,i+1})$ a 1-cocycle of $H^1\big(\mathscr{V}\,;\Lambda^{<0}(B_0)\big)$. We must prove that there is a refinement $\mathscr{V}'$ of both $\mathscr{U}$ and $\mathscr{V}$ and a 1-cocycle ψ of $H^1\big(\mathscr{U}\,;\Lambda^{<0}(B_0)\big)$ (i.e., a Stokes cocycle) such that φ and ψ induce cohomologous 1-cocycles on $\mathscr{V}'$. Set $\dot{V}_j = V_j \cap V_{j+1}$. Since the covering $\mathscr{V}$ is good the $\dot{V}_j$'s are 2-by-2 disjoint. One obtains refinements of $\mathscr{V}$ when shrinking an arc $\dot{V}_j$ or when inserting one more disjoint $\dot{V}$. Let $\mathscr{V}'$ be deduced from $\mathscr{V}$ by adding one $\dot{V}$ (this is done by splitting one arc V_j into two arcs V'_j, V''_j overlapping over $\dot{V}$). Define the simplicial map $\sigma_{\mathscr{V}',\mathscr{V}}$ from $\mathscr{V}'$ to $\mathscr{V}$ by setting $\sigma_{\mathscr{V}',\mathscr{V}}(V'_j) = \sigma_{\mathscr{V}',\mathscr{V}}(V''_j) = V_j$ and otherwise $\sigma_{\mathscr{V}',\mathscr{V}}(V_\ell) = V_\ell$. The image of φ by $\sigma^*_{\mathscr{V}',\mathscr{V}}$ is the "same" cocycle completed with a trivial component over $\dot{V}$. After refinement we can then assume that $\mathscr{V}$ has the following property: for each anti-Stokes direction α there is an open arc $\dot{V}_\alpha$ of the 1-nerve of $\mathscr{V}$ which contains α, no other anti-Stokes direction and which is contained in the Stokes arc $\overline{I}_\alpha$. Denote by $\dot{\varphi}_j$ the component of φ on $\dot{V}_j$. We have to prove that φ is cohomologous to a 1-cocycle $\psi = (\dot{\psi}_j)$ of the following form: for all anti-Stokes directions α, the component $\dot{\psi}_\alpha$ on $\dot{V}_\alpha$ "is" a Stokes germ led by α; otherwise, $\dot{\psi}_j$ is equal to the identity.

To prove this result we set first the following lemma.

Lemma 3.5.13 *Let W be an arc of S^1. Suppose that W is contained in $r \geq 1$ Stokes arcs $\dot{U}_{\alpha_1}, \dot{U}_{\alpha_2}, \ldots, \dot{U}_{\alpha_r}$ (assume $\alpha_1 < \alpha_2 < \cdots < \alpha_r$ on S^1 oriented clockwise).*

Then, given a section $\psi \in \Gamma\big(W\,;\Lambda^{<0}(B_0)\big)$, there exist sections

$$f_1 \in \Gamma(\dot{U}_{\alpha_1}\,;\Lambda^{<0}(B_0)),\ f_2 \in \Gamma(\dot{U}_{\alpha_2}\,;\Lambda^{<0}(B_0)), \ldots \text{ and } f_r \in \Gamma(\dot{U}_{\alpha_r}\,;\Lambda^{<0}(B_0))$$

such that $\psi = f_1 f_2 \ldots f_r$ on W.

Roughly speaking, ψ is a product of Stokes germs $f_1, f_2, \ldots, f_r$ respectively led by $\alpha_1, \alpha_2, \ldots, \alpha_r$. For an example, see Exe. 4, p. 238.

Proof. For seek of notational simplicity let us do the proof with $r = 2$. Choosing a normal solution $\mathscr{Y}_{0,\underline{\alpha}}(x)$ of the normal form $\mathrm{d}Y/\mathrm{d}x = B_0(x)Y$ the section ψ reads in the form given in formula (3.16):

$$\psi(x) = x^L\left(I_n + \left[C_{\underline{\alpha}}^{(\ell,j)}\, \mathrm{e}^{(q_\ell - q_j)(1/x)}\right]\right)x^{-L}.$$

The non zero terms in $C_{\underline{\alpha}}^{(\ell,j)}\,\mathrm{e}^{(q_\ell-q_j)(1/x)}$ are found among those led either by α_1 or by α_2. It is an exercise of linear algebra to factor the terms led by α_1 to the left and the terms led by α_2 to the right. One can, for instance, order the determining polynomials q_j so that the matrix $I_n + C_{\underline{\alpha}}$ be upper diagonal, then factor to the left or to the right one after another term in the successive sup-diagonals. □

Fix an anti-Stokes direction α and consider the arc $\dot{U}_\alpha$. We look at the $\dot{V}_j$'s included in $\dot{U}_\alpha$ and for simplicity, we suppose that there are only three such arcs $\dot{V}_{j^-}, \dot{V}_\alpha$ and $\dot{V}_{j^+}$. From lemma 3.5.13 the components of φ read in the form $f_{j^-}^- f_{j^-}^\alpha f_{j^-}^+$ on $\dot{V}_{j^-}$, $f_\alpha^- f_\alpha f_\alpha^+$ on $\dot{V}_\alpha$ and $f_{j^+}^- f_{j^+}^\alpha f_{j^+}^+$ on $\dot{V}_{j^+}$ where the exponents $-$, $+$ and α refer to sections led respectively by anti-Stokes directions to the left of α, to its right or to α itself. Each factor with a $\pm$ as exponent may split into a product of several Stokes germs that we keep together. Moving these factors to the right or to the left provides cohomologous 1-cocycles: thus, the 1-cocycle φ is cohomologous to the cocycle with new components $f_\alpha^- f_\alpha f_\alpha^+ f_{j^+}^-$ on $\dot{V}_\alpha$ and $f_{j^+}^\alpha f_{j^+}^+$ on $\dot{V}_{j^+}$ ($f_{j^+}^-$ lives all over the arc from $\dot{V}_\alpha$ to $\dot{V}_{j^+}$). Now, we can factor again $f_\alpha^- f_\alpha f_\alpha^+ f_{j^+}^-$ to put to the left all germs led by anti-Stokes directions to the left of α. Observe that doing so, new terms attached to α may appear due to relations of the form $\mathrm{e}^{q_i-q_j}\,\mathrm{e}^{q_j-q_\ell} = \mathrm{e}^{q_i-q_\ell}$ (the exponential to the right-hand side is led by an anti-Stokes direction located between those to the left-hand side). Then, we can move all factors f^- to the component over $\dot{V}_{j^-}$. We can symmetrically move all factors f^+ to the component over $\dot{V}_{j^+}$ and finally all factors attached to α to the component over $\dot{V}_\alpha$. Iterating the procedure to all anti-Stokes directions yields the result. □

Notice that the covering $\mathscr{U}$ is the *unique less fine covering* which is acyclic for the sheaf $\Lambda^{<0}(B_0)$, that is, for which $H^1\big(\mathscr{U}\,;\Lambda^{<0}(B_0)\big) \simeq H^1\big(S^1\,;\Lambda^{<0}(B_0)\big)$.

▷ *Case with several levels.*

In the case of a single level we were able to find a covering $\mathscr{U}$ less fine than any other covering $\mathscr{U}'$ for which $H^1\big(\mathscr{U}'\,;\Lambda^{<0}(B_0)\big)$ is isomorphic to $H^1\big(S^1\,;\Lambda^{<0}(B_0)\big)$. We had thus no cocycle condition and no cohomology condition to check. With several levels such a covering does not exist anymore. The idea is to proceed inductively on levels and, at each stage, to build coverings as similar as possible to $\mathscr{U}$. We thus introduce coverings called *cyclic coverings* because they comprise finitely many open arcs which we can again order cyclically. With cyclic coverings we can again limit ourselves to consider partial 1-cocycles defined by the components on intersections $\dot{U}_j$ of contiguous arcs. There exist 0-cochains to conjugate such 1-cocycles but they are free of terms of the highest level k_ν. This property and the fact that the sheaf $\Lambda^{<0}(B_0)$ splits into direct products

$$\Lambda^{<0}(B_0) \;\simeq\; \Lambda^{k_\nu}(B_0) \rtimes \Lambda^{<k_\nu}(B_0) \;\simeq\; \Lambda^{<k_\nu}(B_0) \ltimes \Lambda^{k_\nu}(B_0)$$

allows to treat the case of factors of the highest level k_ν. The fact that $\Lambda^{k_\nu}(B_0)$ is normal in $\Lambda^{<0}(B_0)$ results from the fact that the product of a flat exponential of order k_ν by a flat exponential of order $\le k_\nu$ is a flat exponential of order k_ν. We can then iterate the process to $\Lambda^{<k_\nu}(B_0)$, split it, factor and so on... until the smallest level. For details we refer to [Lod94] and to [Lod03] (where injectivity of h is given in more details).

3.5.3.3 Stokes Cocycle and Summation

Stokes cocycles and Stokes matrices are connected as follows to the theory of summation developed in Chapters 5 and 7. Suppose we are given a formal fundamental solution $\widetilde{\mathscr{Y}}(x) = \widetilde{F}(x)\widetilde{\mathscr{Y}_0}(x)$ at 0 and an anti-Stokes direction $\alpha \in \mathfrak{A}$ and denote by $F_\alpha^-(x)$ and $F_\alpha^+(x)$ the sums (k-sums or multisums) of $\widetilde{F}(x)$ respectively to the left and to the right of the α direction. As before, we choose a determination of the argument and we denote by $\underline{\alpha}$ the α direction with this choice of determination.

Theorem 3.5.14 *The Stokes cocycle $(\varphi_\alpha)_{\alpha\in\mathfrak{A}}$, that is, each Stokes automorphism φ_α, satisfies*

$$\boxed{\varphi_\alpha = F_\alpha^-(x)^{-1}F_\alpha^+(x)}$$

For each $\alpha \in \mathfrak{A}$, the Stokes matrix $I_n + C_{\underline{\alpha}}$ is the matrix, in the basis $F_\alpha^-(x)\,\mathscr{Y}_{0,\underline{\alpha}}(x)$, of the Stokes automorphism φ_α seen as a linear automorphism of the space of solutions spanned by $F_\alpha^-(x)\,\mathscr{Y}_{0,\underline{\alpha}}(x)$ in a sector that contains the α direction. It is characterized by the relation

$$\boxed{F_\alpha^+(x)\,\mathscr{Y}_{0,\underline{\alpha}}(x) = F_\alpha^-(x)\,\mathscr{Y}_{0,\underline{\alpha}}(x)\,(I_n + C_{\underline{\alpha}})}$$

The latter relation results from the former one and relation (3.15) p. 99. Different interpretations of the Stokes automorphisms have been mentioned in remark 3.5.9, p. 100. For more details we refer to [Lod94, Sect. III. 2]. See also exercise 8, p. 240.

Formerly, one used to call Stokes matrices all matrices $I_n + C$ satisfying a condition of the type

$$F_j(x)\,\mathscr{Y}_{0,\underline{\alpha}}(x) = F_\ell(x)\,\mathscr{Y}_{0,\underline{\alpha}}(x)\,(I_n + C)$$

linking two overlapping asymptotic solutions, i.e., any matrix representing a germ of flat isotropy $F_\ell(x)^{-1}F_j(x) = \mathscr{Y}_{0,\underline{\alpha}}(x)\,(I_n + C)\,\mathscr{Y}_{0,\underline{\alpha}}(x)^{-1}$, not necessarily a Stokes germ. This appeared to be not restrictive enough to easily characterize the meromorphic classes of systems or to exhibit good Galoisian properties: an example of a non-Galoisian "Stokes matrix" in this wide meaning is given in [Lod94, Sect. III.3.3.2]. We always use the expression Stokes matrix in the restrictive meaning of associated to a Stokes cocycle. As already said, these Stokes matrices are sometimes called Stokes-Ramis matrices.

3.5.4 *Sibuya's Proof of the Malgrange-Sibuya Isomorphism Theorem*

The Malgrange-Sibuya isomorphism theorem 3.5.5, p. 96 says that the set of invertible matrices of formal modulo convergent series is isomorphic to the non-abelian Čech 1-cohomology set $H^1(S^1\,;\Lambda^{<0})$. The aim of this section is to develop Sibuya's proof [Sib77, Sib90].

We already know that the map

$$\mathrm{Exp}_\mu : \mathbb{G}\backslash\widetilde{\mathbb{G}} \longrightarrow H^1(S^1;\Lambda^{<0})$$

is injective. To prove that it is surjective we have to prove that the natural map

$$u:\ H^1(S^1;\Lambda^{<0}) \longrightarrow H^1(S^1;\Lambda)$$

is the null map: given a good covering $\mathscr{U} = (U_i)$ of S^1 and a 1-cocycle φ with values in $\Lambda^{<0}$ (denote by $\dot{\varphi}_i = \varphi_{i,i+1}$its components) we must prove that there is a covering $\mathscr{V} = (V_j)$ finer than $\mathscr{U}$ such that the 1-cocycle $\psi = \sigma^*(\varphi)$ defined by $\dot{\psi}_j = \varphi_{\sigma(j),\sigma(j+1)}|_{\dot{V}_j}$ on $\dot{V}_j = V_j \cap V_{j+1}$ for all j is trivial in $H^1(\mathscr{V};\Lambda)$. This means that there exist a 0-cochain $(f_j) \in \prod \Gamma(V_j;\Lambda)$ such that $\dot{\psi}_j = f_j^{-1} f_{j+1}$ on $\dot{V}_j$ for all j.

▷ Prove first that *it suffices to consider the case when ψ (hence, a fortiori φ) is elementary* (i.e., with at most one non trivial component).

Let $\psi = (\dot{\psi}_1,\ldots,\dot{\psi}_p)$ be a 1-cocycle of $\mathscr{V}$ with values in $\Lambda^{<0}$ having $p>1$ (non trivial) components and assume that any 1-cocycle with at most $p-1$ non trivial components is trivial in $H^1(S^1;\Lambda)$. Consider the 1-cocycle $\widetilde{\Phi} = (\dot{\psi}_1,\ldots,\dot{\psi}_{p-1},I_n)$ extracted from ψ and an elementary 1-cocycle of the form $\Phi = (I_n,\ldots,I_n,\phi)$ to be made explicit later. By recurrence assumption, there exist 0-cochains $(g_1,\ldots,g_p)$ and $(h_1,\ldots,h_p)$ with values in $\Lambda^{<0}$ such that

$$\begin{cases} g_1^{-1}g_2 &= I_n\\ g_2^{-1}g_3 &= I_n\\ \cdots & \\ g_{p-1}^{-1}g_p &= I_n\\ g_p^{-1}g_1 &= \phi \end{cases} \qquad \text{and} \qquad \begin{cases} h_1^{-1}h_2 &= \dot{\psi}_1\\ h_2^{-1}h_3 &= \dot{\psi}_2\\ \cdots & \\ h_{p-1}^{-1}h_p &= \dot{\psi}_{p-1}\\ h_p^{-1}h_1 &= I_n \end{cases}$$

Choosing $\phi = h_p\,\dot{\psi}_p\,h_1^{-1}$ we obtain the cohomology relation

$$\begin{cases} h_1^{-1}g_1^{-1}g_2h_2 &= \dot{\psi}_1\\ h_2^{-1}g_2^{-1}g_3h_3 &= \dot{\psi}_2\\ \cdots & \\ h_{p-1}^{-1}g_{p-1}^{-1}g_ph_p &= \dot{\psi}_{p-1}\\ h_p^{-1}g_p^{-1}g_1h_1 &= \dot{\psi}_p \end{cases}$$

The 1-cocycle ψ is then trivial in $H^1(S^1;\Lambda)$ and the proof of this point is achieved.

▷ *From now we suppose that φ is elementary and defined as follows.*

Let $\mathscr{U} = (U_1, U_2)$ be a good covering of S^1 consisting of two sectors U_1 and U_2 overlapping on the two sectors $\dot{U}_1$ and $\dot{U}_2$. By means of a rotation we can assume that $\dot{U}_1$ is bisected by the $\theta = 0$ direction and we denote by R the common radius of $\dot{U}_1$ and $\dot{U}_2$, by $2\Omega_1$ and $2\Omega_2$ the openings of $\dot{U}_1$ and $\dot{U}_2$ respectively. We set $\varphi = (\varphi_1, \varphi_2)$ with

$$\begin{cases} \varphi_1 = I_n + \phi \in \Gamma(\dot{U}_1\,;\Lambda^{<0}) \\ \varphi_2 = I_n \in \Gamma(\dot{U}_2\,;\Lambda^{<0}). \end{cases}$$

We look for a covering $\mathscr{V} = (V_1, V_2)$ finer than $\mathscr{U}$ and sections $f_j = I_n + F_j$ in $\Gamma(V_j\,;\Lambda)$ for $j = 1,2$ such that

$$\begin{cases} I_n + \phi = (I_n + F_1)^{-1}(I_n + F_2) & \text{on } \dot{V}_1 \\ F_1 = F_2 & \text{on } \dot{V}_2 \end{cases} \tag{3.17}$$

The condition "$\mathscr{V} = (V_1, V_2)$ finer than $\mathscr{U}$" means $V_1 \subset U_1$ and $V_2 \subset U_2$ and therefore $\dot{V}_1 \subset \dot{U}_1$ and $\dot{V}_2 \subset \dot{U}_2$. We can again choose $\dot{V}_1$ bisected by $\theta = 0$, a common radius $r < R$ for $\dot{V}_1$ and $\dot{V}_2$ and openings $2\omega_1$ of $\dot{V}_1$ and $2\omega_2$ of $\dot{V}_2$ such that $\omega_1 = \Omega_1 - \delta$, $\delta > 0$ and $\omega_2 = \Omega_2$. Convenient values for r and δ will be made explicit below.

The proof proceeds by successive approximations given by iterated application of the Cauchy-Heine theorem starting with φ in order to determine the series F_1 and F_2 at the limit.

▷ *The sequence of coverings $\mathscr{V}_m$.*

We set $\mathscr{V}_0 = \mathscr{U}$ and, for $m \geq 1$, we define the covering $\mathscr{V}_m = (V_{1,m}, V_{2,m})$ by the following conditions:

$$\begin{cases} \dot{V}_{1,m} \text{ is bisected by } \theta = 0 \text{ for all } m. \\ \dot{V}_{1,0} = \dot{U}_1;\ \text{hence } \dot{V}_{1,0} \text{ has opening } 2\omega_0 = 2\Omega_1 \text{ and radius } r_0 = R. \\ \text{for } m \geq 1,\ \dot{V}_{1,m} \text{ has opening } 2\omega_m \text{ such that } \omega_m = \omega_{m-1} - \delta_m \text{ with } \delta_m = \delta/2^m, \\ \text{and } \dot{V}_{1,m} \text{ has radius } r_m \text{ satisfying } r_m = r_{m-1} - R\delta/2^{m+1}. \end{cases}$$

Choosing $\delta \leq 1$ implies $r_m > 0$ for all m and $r = \lim_{m \to +\infty} r_m = R(1 - \delta/2)$; in particular $R/2 \leq r < R$. The sectors $\dot{V}_{1,m}$ are proper sub-sectors of each other:

$$\dot{V}_{1,m} \Subset \dot{V}_{1,m-1}.$$

To the other side, the sector $\dot{V}_{2,m}$ is chosen to be equal to $\dot{U}_2$ truncated at radius r_m. The covering $\mathscr{V}$ is the limit of the coverings $\mathscr{V}_m$ in the sense that $V_1 = \cap_m V_{1,m}$ and $V_2 = \cap_m V_{2,m}$.

With a view to applying the Cauchy-Heine theorem we define the paths $\gamma_{1,m}$ and $\gamma_{2,m}$ as showed on Fig. 3.3. These paths are drawn in $\dot{V}_{1,m-1}$ on each side of $\dot{V}_{1,m}$. They begin at 0 with rays making an angle $\beta_m = \delta_m/2 = \delta/2^{m+1}$ with $\dot{V}_{1,m}$ and they continue around $\dot{V}_{1,m}$ along a circle centered at 0 to the point $a_m = (r_{m-1}+r_m)/2$ on the positive real axis.

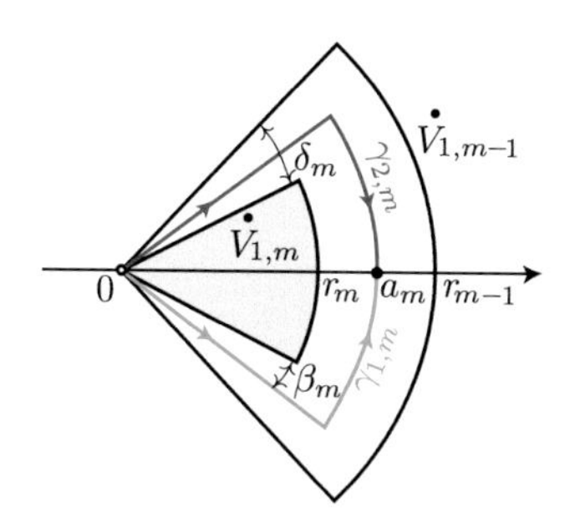

Fig. 3.3

▷ *The sequence of 0-cochains* $(I_n + F_{1,m}, I_n + F_{2,m})$.

The functions $F_{1,m}$ and $F_{2,m}$ are respectively sections of $\Gamma(V_{1,m}\,;\Lambda)$ and $\Gamma(V_{2,m}\,;\Lambda)$ obtained by the Cauchy-Heine theorem applied to a same flat function over the paths $\gamma_{1,m}$ and $\gamma_{2,m}$.

Set $\phi^0 = \phi$.

Since the two functions F_1 and F_2 must satisfy the condition $F_1(0) = F_2(0) = 0$ we ask all $F_{j,m}$ for $j = 1,2$ and $m \geq 1$ to satisfy the same condition. To this end, we apply the Cauchy-Heine theorem to $\phi^0(x)/x$ instead of applying it to $\phi^0(x)$. Thus, we define

$$F_{j,1}(x) = \frac{x}{2\pi \mathrm{i}} \int_{\gamma_{j,1}} \frac{\phi^0(\xi)}{\xi\,(\xi - x)}\,\mathrm{d}\xi \quad \text{on } V_{j,1}.$$

The Cauchy-Heine theorem asserts that $-F_{1,1}(x) + F_{2,1}(x) = \begin{cases} \phi^0(x) & \text{on } \dot{V}_{1,1} \\ 0 & \text{on } \dot{V}_{2,1}. \end{cases}$

On $\dot{V}_{2,1}$, the couple $(F_{1,1}, F_{2,1})$ satisfies the second condition (3.17), p. 108.

On $\dot{V}_{1,1}$ we would like to have $\big(I_n + F_{2,1}(x)\big) = \big(I_n + F_{1,1}(x)\big)\big(I_n + \phi^0\big)$, which means $-F_{1,1}(x) + F_{2,1}(x) = \phi^0(x) + O\big(x\phi^0(x)\big)$. Henceforth, with this choice of $F_{1,1}$ and $F_{2,1}$, the first condition (3.17) is satisfied by $(F_{1,1}, F_{2,1})$ only "at order one".

Let $\phi^1(x)$ be defined on $\dot{V}_{1,1}$ by the relation

$$\big(I_n + \phi^1(x)\big)\,\big(I_n + F_{2,1}(x)\big) = \big(I_n + F_{1,1}(x)\big)\,\big(I_n + \phi^0(x)\big)$$

(ϕ^1 measures the gap between the 1-cocycles $\big(I_n + F_{1,1}(x)\big)^{-1}\big(I_n + F_{2,1}(x)\big)$ and $\big(I_n + \phi^0(x)\big)$. We can now define the functions $F_{j,2}(x)$ from $\phi^1(x)$ on $V_{j,2}$ like $F_{j,1}(x)$

from $\phi^0(x)$ on $V_{j,1}$. Observe that $\phi^1 = F_{1,1}\,\phi^0\,(I_n+F_{2,1})^{-1} = O(x\phi^0)$. Iterating the process we set, for all $m \geq 1$,

$$F_{j,m}(x) = \frac{x}{2\pi\mathrm{i}} \int_{\gamma_{j,m}} \frac{\phi^{m-1}(\xi)}{\xi\,(\xi-x)}\,\mathrm{d}\xi \quad \text{on } V_{j,m} \quad \text{for } j=1,2, \tag{3.18}$$

from which we deduce $-F_{1,m}(x)+F_{2,m}(x) = \begin{cases} \phi^{m-1}(x) & \text{on } \dot{V}_{1,m} \\ 0 & \text{on } \dot{V}_{2,m}. \end{cases}$

Let $\phi^m(x)$ be defined on $\dot{V}_{1,m}$ by the relation

$$\big(I_n+\phi^m(x)\big)\,\big(I_n+F_{2,m}(x)\big) = \big(I_n+F_{1,m}\big)\,\big(I_n+\phi^{m-1}(x)\big) \tag{3.19}$$

equivalent to

$$\phi^m(x) = F_{1,m}(x)\,\phi^{m-1}(x)\,\big(I_n+F_{2,m}(x)\big)^{-1} \text{ on } \dot{V}_{1,m}. \tag{3.20}$$

We have now $\phi^m = O(x^m\phi^0)$. The relations (3.19), p. 110 imply for all m

$$I_m+\phi^0 = (I_n+F_{1,1})^{-1}(I_n+F_{1,2})^{-1}\ldots(I_n+F_{1,m})^{-1}(I_n+\phi^m)\,(I_n+F_{2,m})\ldots(I_n+F_{2,1}).$$

To achieve the proof it suffices to prove the following three points:

(a) $I_n+F_j(x) = \lim_{m\to+\infty}(I_n+F_{j,m}(x))(I_n+F_{j,m-1}(x))\ldots(I_n+F_{j,1}(x))$ exist on V_j for $j=1,2$;

(b) $\lim_{m\to+\infty}\phi^m(x) = 0$ on $\dot{V}_1$;

(c) The functions I_n+F_j for $j=1,2$ belong to $\Gamma\big(n\,;\overline{\mathscr{A}}(V_j)\big)$.

▷ *Proof of points* (a) *and* (b).
In $GL(n,\mathbb{C})$ we choose a norm of algebra $|\cdot|$ satisfying the condition $|AB| \leq |A|\,|B|$ say, $\big|[a_{j,\ell}]\big| = n\max(|a_{j,\ell}|)$. We prove first, by recurrence on m, the estimates

$$\begin{cases} \sup\limits_{x\in\dot{V}_{1,m}} \left|\dfrac{\phi^m(x)}{x}\right| \leq \dfrac{1}{2^{2m}} & (3.21) \\[2ex] \sup\limits_{x\in V_{1,m}} \big|F_{j,m}(x)\big| \leq \dfrac{1}{2^m} & (3.22) \end{cases}$$

From relations (3.18) and (3.20), p. 110 we deduce, for all $x \in V_{j,m}$,

$$\big|F_{j,m}(x)\big| \leq \frac{|x|}{2\pi} L \sup_{\xi\in\gamma_{j,m}} \left|\frac{\phi^{m-1}(\xi)}{\xi}\right| \frac{1}{|x|\sin\beta_m}$$

where $2L$ denotes an upper bound for the length of the boundary of the sector $\dot{U}_1$ and then of all sectors $\dot{V}_{j,m}$. With the condition $\delta \leq 1$ all β_m satisfy $\beta_m < 2\sin(\beta_m)$

and then,

$$\sup_{x\in V_{j,m}} \left|F_{j,m}(x)\right| \leq \frac{L2^{m+1}}{\pi\delta} \sup_{\xi\in\dot{V}_{1,m-1}} \left|\frac{\phi^{m-1}(\xi)}{\xi}\right| \tag{3.23}$$

and

$$\sup_{x\in\dot{V}_{1,m}} \left|\frac{\phi^m(x)}{x}\right| \leq \sup_{x\in V_{1,m}} \left|F_{1,m}(x)\right| \sup_{x\in\dot{V}_{1,m-1}} \left|\frac{\phi^{m-1}(x)}{x}\right| \sum_{\ell\geq 0} \sup_{x\in V_{2,m}} \left|F_{2,m}(x)\right|^{\ell}. \tag{3.24}$$

Assume that R is chosen small enough for the estimate $\sup_{\xi\in\dot{V}_{1,0}} \left|\phi^0(\xi)/\xi\right| \leq 1$ to hold, that is, estimate (3.21), p. 110 for $m=0$.

Recall that the covering $\mathscr{U}=(U_1,U_2)$ is a covering of S^1 and the sectors U_1,U_2 chosen to represent arcs of S^1 can be chosen with an arbitrary small radius R. For later use, we assume that R is also small enough for the estimate $L2^6 \leq \pi\delta$ to hold.

Assume now that estimate (3.21) holds for $m-1$ and prove (3.21) and (3.22) for m. The recurrence hypothesis and the estimates above imply

$$\sup_{x\in V_{j,m}} \left|F_{j,m}(x)\right| \leq \frac{L2^{m+1}}{\pi\delta}\frac{1}{2^{2(m-1)}} = \frac{L2^3}{\pi\delta}\frac{1}{2^m} \leq \frac{1}{2^m}$$

In particular, $\sup_{x\in V_{j,m}} \left|F_{j,m}(x)\right| \leq 1/2$ and we obtain

$$\sup_{x\in\dot{V}_{1,m}} \left|\frac{\phi^m(x)}{x}\right| \leq \frac{L2^{m+1}}{\pi\delta}\left(\frac{1}{2^{2m-2}}\right)^2 2 = \frac{L2^6}{\pi\delta}\frac{1}{2^m}\frac{1}{2^{2m}} \leq \frac{1}{2^{2m}}.$$

This achieves the proof of estimates (3.21) and (3.22).

It follows that, for $j=1,2$, the series $\sum_m F_{j,m}$ converge normally on $V_j=\cap V_{j,m}$, hence also the infinite products

$$\prod_{+\infty\geq m\geq 1} (I_n+F_{j,m}) = \cdots(I_n+F_{j,m})\cdots(I_n+F_{j,2})(I_n+F_{j,1}).$$

Denote by I_n+F_j their limit. The sequence $(\phi^m)_{m\geq 1}$ converges normally to 0 on $\dot{V}_1=\cap\dot{V}_{1,m}$.

▷ *Proof of point* (c). As uniform limit of holomorphic functions the two functions I_n+F_j are holomorphic on V_j. There is to prove that they admit a same asymptotic expansion $I_n+\widetilde{F}$ satisfying $\widetilde{F}(0)=0$.

Study first the asymptotic expansion of $I_n+F_{j,m}$. We proceed like for the Euler function (Exa. 1.1, p. 4) writing

$$F_{j,m}(x) = \sum_{\ell=1}^{N-1} a_{\ell,m}x^\ell + \frac{x^N}{2\pi\mathrm{i}}\int_{\gamma_{j,m}} \frac{\phi^{m-1}(\xi)}{(\xi-x)\xi^N}\,\mathrm{d}\xi \quad \text{with } a_{\ell,m} = \frac{1}{2\pi\mathrm{i}}\int_{\gamma_{j,m}} \frac{\phi^{m-1}}{\xi^{\ell+1}}\,\mathrm{d}\xi.$$

By Cauchy's theorem the coefficients $a_{\ell,m}$ do not depend on j. As above we deduce from relation (3.20), p. 110 the estimate

$$\sup_{x\in\dot{V}_{1,m}}\left|\frac{\phi^m(x)}{x^{N+1}}\right|\le\frac{1}{2^m}\sup_{\xi\in\dot{V}_{1,m-1}}\left|\frac{\phi^{m-1}(\xi)}{\xi^{N+1}}\right|.$$

Hence, by recurrence, the estimate

$$\sup_{x\in\dot{V}_{1,m}}\left|\frac{\phi^m(x)}{x^{N+1}}\right|\le\frac{1}{2^m}\frac{1}{2^{m-1}}\cdots\frac{1}{2^2}\frac{1}{2}\sup_{x\in\dot{V}_{1,m}}\left|\frac{\phi^0(x)}{x^{N+1}}\right|\le\frac{C'_{N+1}}{2^{2m}}$$

where the constant $C'_{N+1}>0$ do not depend on m.

It follows that there exists a constant $A_\ell>0$, independent of m, such that

$$|a_{\ell,m}|\le\frac{A_\ell}{2^{2m}} \tag{3.25}$$

Moreover, from

$$\begin{aligned}|\xi-x| &\ge \min\big(|\xi|\sin(\beta_m),(r_{m-1}-r_m)/2m\big)\\ &\ge \min(|\xi|,R)\,\delta/2^{m+2}\end{aligned}$$

for all $x\in\dot{V}_{1,m}$ and $\xi\in\gamma_{j,m}$ we deduce the asymptotic conditions

$$\left|F_{j,m}(x)-\sum_{\ell=1}^{N-1}a_{\ell,m}x^\ell\right|\le\frac{C_N}{2^m}|x|^N\quad\text{for }x\in V_{j,m} \tag{3.26}$$

for all m and N, choosing $C_N=\max(C'_{N+1},C'_N/R)$.

Now, consider the infinite products $I_n+F_j(x)=\prod_{+\infty\ge m\ge1}\big(I_n+F_{j,m}(x)\big)$. We can substitute the asymptotic series $\sum_{\ell\ge1}a_{\ell,m}x^\ell$ for $F_{j,m}(x)$ to obtain a power series in x. Indeed, the infinite product $\prod_{+\infty\ge m\ge1}(1+\frac{w}{2^{2m}})$ converges normally on compact sets and defines thus an analytic function of u. It results that the infinite product $\prod_{+\infty\ge m\ge1}\big(1+\frac{1}{2^{2m}}\sum_{\ell\ge1}A_\ell x^\ell\big)$ defines a formal power series in x and, by an argument of majorant series, that the same property holds true for the infinite product $\prod_{+\infty\ge m\ge1}\big(1+\sum_{\ell\ge1}a_\ell x^\ell\big)$.

Prove that this latter formal series is the asymptotic expansion of $I_n+F_j(x)$ on V_j for all j.

Fix N and, given M, consider the difference

$$D_{j,M}(x)=\prod_{M\ge m\ge1}(I_n+F_{j,m})-\prod_{M\ge m\ge1}\big(I_n+\sum_{\ell<N}a_{\ell,m}x^\ell\big)$$

$$
\begin{aligned}
&= (I_n+F_{j,M})\dots(I_n+F_{j,2})(I_n+F_{j,1})\\
&\qquad -(I_n+F_{j,M})\dots(I_n+F_{j,2})\big(I_n+\sum_{\ell<N}a_{\ell,1}x^\ell\big)\\
&\quad + \quad \cdots\\
&\quad +(I_n+F_{j,M})\big(I_n+\sum_{\ell<N}a_{\ell,M-1}x^\ell\big)\dots\big(I_n+\sum_{\ell<N}a_{\ell,1}x^\ell\big)\\
&\qquad -\big(I_n+\sum_{\ell<N}a_{\ell,M}x^\ell\big)\big(I_n+\sum_{\ell<N}a_{\ell,M-1}x^\ell\big)\dots\big(I_n+\sum_{\ell<N}a_{\ell,1}x^\ell\big).
\end{aligned}
$$

This implies

$$
\begin{aligned}
&|D_{j,M}(x)|\\
&\quad\le \prod_{M\ge m\ge 1}(1+|F_{j,m}|)\prod_{M\ge m\ge 1}\big(1+\sum_{1\le\ell<N}|a_{\ell,m}x^\ell|\big)\sum_{m=1}^{M}\big|F_{j,m}(x)-\sum_{\ell<N}a_{\ell,m}x^\ell\big|\\
&\quad\le \prod_{+\infty\ge m\ge 1}\big(1+\frac{1}{2^m}\big)\prod_{+\infty\ge m\ge 1}\big(1+\frac{B_N}{2^{2m}}\big)\big(\sum_{m\ge 1}\frac{1}{2^m}\big)C_N|x|^N<+\infty
\end{aligned}
$$

where $B_N=\sup_{x\in V_1\cup V_2}\sum_{1\le\ell<N}A_\ell|x|^\ell$. There exists thus a constant K'_N such that, for all M and all $x\in V_j$ the inequality $|D_{j,M}(x)|\le K'_N|x|^N$ holds.

Taking the limit as M tends to infinity we obtain

$$
\Big|\prod_{+\infty\ge m\ge 1}(I_n+F_{j,m})-\prod_{+\infty\ge m\ge 1}\big(I_n+\sum_{1\le\ell\le N}a_{\ell,m}x^\ell\big)\Big|\le K'_N|x|^N \quad \text{on } V_j.
$$

Due to estimate (3.25) and to the convergence, for w on compact sets, of the infinite product $\prod_{+\infty\ge m\ge 1}\big(1+w/2^{2m}\big)$ the function

$$
\psi_N(x)=\prod_{+\infty\ge m\ge 1}\big(I_n+\sum_{1\le\ell\le N}a_{\ell,m}x^\ell\big)
$$

is analytic on $\mathbb{C}$. If $T_{N-1}(x)$ denotes its Taylor polynomial of order $N-1$ at 0 then, there exists a constant $K''_N>0$ such that $\big|\psi_N(x)-T_{N-1}(x)\big|\le K''_N|x|^N$. Setting $K_N=\max(K'_N,K''_N)$ we have thus for $j=1,2$

$$
\Big|\prod_{+\infty\ge m\ge 1}\big(I_n+F_{j,m}(x)\big)-T_{N-1}(x)\Big|\le K_N|x|^N \quad \text{on } V_j.
$$

This proves that the matrices $I_n+F_j(x)=\prod_{+\infty\ge m\ge 1}\big(I_n+F_{j,m}(x)\big)$ admit a same asymptotic expansion on V_j with constant term I_n. This achieves the proof of the theorem. □

3.6 Infinitesimal Neighborhoods at an Irregular Singular Point

While algebraic functions have moderate growth the form of formal solutions given above and the main asymptotic existence theorem show that solutions of linear differential equations at an irregular singular point may exhibit exponential growth or decay. Infinitesimal neighborhoods of algebraic geometry are then insufficient to discriminate between the various solutions. Below, we define infinitesimal neighborhoods for the irregular singular points of linear differential equations as suggested by P. Deligne [DMR07] in a letter to J.-P. Ramis dated $7/01/1986$. This point of view is developed in [LP97] with an application to index theorems.

3.6.1 *Infinitesimal Neighborhoods Associated with the Exponential Order*

We begin with a concept related only to the exponential order of growth or decay of the singularity under consideration. This concept will show up to be slightly too poor for a good characterization of k-summable series but it is a necessary step, at least for clarity.

▷ *Base space X.* — From this viewpoint the infinitesimal neighborhood X of 0 in $\mathbb{C}$ is defined as a full copy of $\mathbb{C}$ compactified by the adjunction of a circle at infinity and endowed with a structural sheaf $\mathscr{F}$ defined as below. For obvious reasons we represent the infinitesimal neighborhood of 0 as a compact disc in place of the origin 0 in $\mathbb{C}$. The "outside world" $\mathbb{C}^* = \mathbb{C}\setminus\{0\}$ is not affected by the construction and remains being endowed with the sheaf of germs of analytic functions.

▷ *Sheaf* $\mathscr{A}^{\leq k}$, $k>0$. — The definition is analogous to the one of k-exponentially flat functions (cf. definition in Sect. 2.1.5, p. 41), replacing decay by growth: A function f is said to have *exponential growth of order k on a sector* $\boldsymbol{\delta}$ if, for any proper subsector $\boldsymbol{\delta}' \Subset \boldsymbol{\delta}$, there exist constants K and $A>0$ such that the following estimate holds for all $x \in \boldsymbol{\delta}'$:

$$|f(x)| \leq K \exp\left(A/|x|^k\right).$$

The set of all functions with exponential growth of order k on $\boldsymbol{\delta}$ is denoted by $\overline{\mathscr{A}}^{\leq k}(\boldsymbol{\delta})$ and one defines a sheaf $\mathscr{A}^{\leq k}$ over S^1 of germs with exponential growth of order k (or, with k-exponential growth) in a similar way as for $\mathscr{A}^{\leq -k}$ (cf. Sect. 2.1.5, p. 41).

▷ *Presheaf* $\overline{\mathscr{F}}$. — With a view to define the sheaf $\mathscr{F}$ it suffices to define the presheaf $\overline{\mathscr{F}}$ on a basis of open sets of X. We consider the following open sets (cf. Fig. 3.4, p. 115):

$$\begin{cases} \text{the discs } D(0,k) \text{ for all } k>0, \\ \text{the (truncated) sectors } I\times]k',k''[=\{x=r\mathrm{e}^{i\theta}\,;\,\theta\in I \text{ and } 0<k'<r<k''\}, \\ \text{the (truncated) sectors } I\times]k,\infty]=\{x=r\mathrm{e}^{i\theta}\,;\,\theta\in I \text{ and } 0<k<r\leq\infty\}. \end{cases}$$

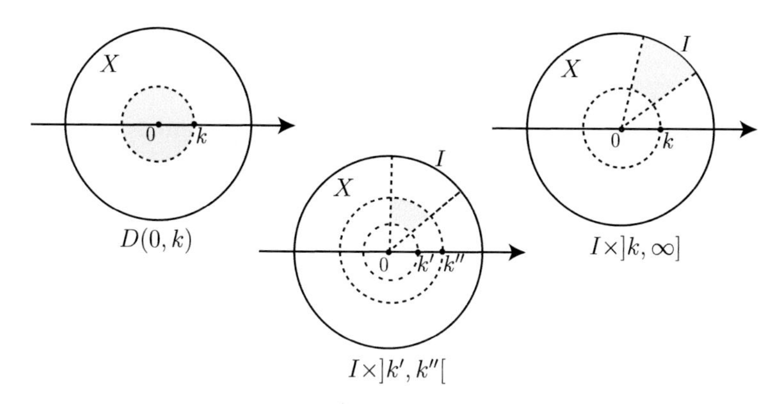

Fig. 3.4. Basis of open sets in X

and we set

$$\begin{cases} \overline{\mathscr{F}}\big(D(0,k)\big)=\mathbb{C}[[x]]_s \quad (\text{recall } s=1/k), \\ \overline{\mathscr{F}}\big(I\times]k',k''[\big)=H^0\big(I\,;\mathscr{A}^{\leq k'}/\mathscr{A}^{\leq -k''}\big), \\ \overline{\mathscr{F}}\big(I\times]k,\infty]\big)=H^0\big(I\,;\mathscr{A}^{\leq k}\big). \end{cases}$$

The restriction map between (truncated) sectors is the canonical restriction of functions and quotients. The restriction map from a disc $D(0,k)$ to a sector is made possible by the isomorphism between $\mathbb{C}[[x]]_s$ and $H^0\big(S^1;\mathscr{A}_s/\mathscr{A}^{\leq -k}\big)$, consequence of the Borel-Ritt theorem (cf. Seq. (2.2), p. 46 and Cor. 2.2.8, p. 53).

Notation 3.6.1 From now, a point $x=k\,\mathrm{e}^{i\theta}$ with $k>0$ is also denoted by its polar coordinates (θ,k).

▷ *Sheaf* $\mathscr{F}$. — The sheaf $\mathscr{F}$ over X is the sheaf associated with the presheaf $\overline{\mathscr{F}}$.
It is a sheaf of $\mathbb{C}$-algebras. The stalk $\mathscr{F}_0$ of $\mathscr{F}$ at 0 consists of all Gevrey series. If useful, it could be extended to any series of $\mathbb{C}[[x]]$, the germs of non-Gevrey series having the point $\{0\}$ as support. To define the stalk at the other points (θ,k) we introduce the sheaves

$$\mathscr{A}^{\leq k_-}=\varinjlim_{\varepsilon\to 0+}\mathscr{A}^{\leq k-\varepsilon} \quad \text{and} \quad \mathscr{A}^{\leq -k_+}=\varinjlim_{\varepsilon\to 0+}\mathscr{A}^{\leq -(k+\varepsilon)}.$$

A germ f at θ belongs to $\mathscr{A}^{\leq k_-}$ if there exist a sector Δ in $\mathbb{C}^*$ containing the θ direction, an $\varepsilon>0$, and constants $K,C>0$ such that

$$|f(x)|\leq K\exp\big(C/|x|^{k-\varepsilon}\big) \quad \text{for all } x\in\Delta.$$

A germ f is in $\mathscr{A}^{\leq -k_+}$ if under the same conditions it satisfies

$$|f(x)| \leq K \exp\big(-C/|x|^{k+\varepsilon}\big) \quad \text{for all } x \in \boldsymbol{\Delta}.$$

The stalk of $\mathscr{F}$ at (θ, k) is given by

$$\mathscr{F}_{(\theta,k)} = \mathscr{A}_\theta^{\leq k_-} / \mathscr{A}_\theta^{\leq -k_+}.$$

Example 3.6.2 (Definition Domain and Support of Exponentials)

An exponential $\exp\big(-a/x^k + q(1/x)\big)$ where q is a polynomial of degree less than k can be seen as a section of the complement of the closed disc $\overline{D}(0,k)$ since it has exponential growth of order less than k' for all $k' > k$.

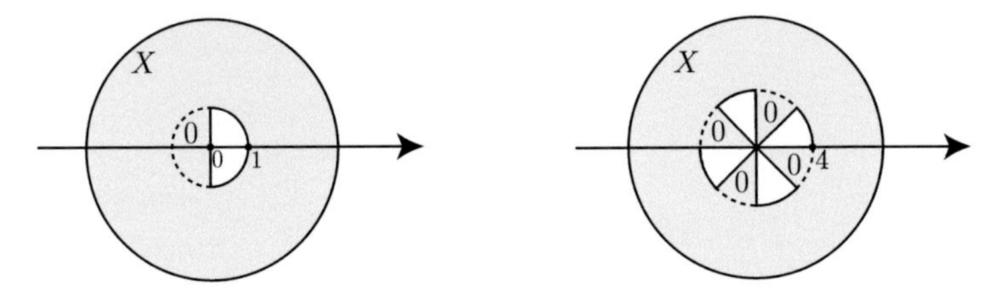

Fig. 3.5. Definition domain of $\exp(1/x)$ — Definition domain of $\exp\big(\mathrm{i}/x^4 + q(1/x)\big)$

On another hand, the exponential is flat on the k open sectors

$$\big|\arg(x) - \alpha/k\big| < \pi/(2k) \bmod 2\pi/k$$

where α denotes the argument of a. The exponential can then be continued on all of these open sectors and is equal to 0 inside $D(0,k)$. It cannot be continued any further. Its support is the complement of the open disc $D(0,k)$ in its definition domain. The arcs on the circle of radius k limiting the sectors where the exponential is equal to zero are of length π/k. By analogy with the big points of algebraic geometry, their closure is called *k-big points associated with the exponential* $\exp\big(-a/x^k + q(1/x)\big)$ (one also says associated with the polynomial $-a/x^k + q(1/x)$). The open shadowed domains in figure 3.5 are two examples of the definition domain of an exponential. The big points are the closed arcs drawn in dotted lines.

This example shows that the sheaf $\mathscr{F}$ is in no way a coherent sheaf, hence its surname of "wild analytic" sheaf.

The following properties of the sheaf $\mathscr{F}$ are elementary and their proof is left to the reader.

Proposition 3.6.3 *The sheaf $\mathscr{F}$ satisfies the following properties:*

1. *The* restriction $\mathscr{F}_{|S^1 \times \{k\}}$ *of $\mathscr{F}$ to the circle centered at 0 with radius k in X is isomorphic to the quotient sheaf* $\mathscr{A}^{\leq k_-}/\mathscr{A}^{\leq -k_+}$ *over* S^1.

2. Sections over an open disc:

$$H^0\big(D(0,k);\mathscr{F}\big) = \varprojlim_{\varepsilon \to 0+} \mathbb{C}[[x]]_{s+\varepsilon} = \bigcap_{\varepsilon > 0} \mathbb{C}[[x]]_{s+\varepsilon} =: \mathbb{C}[[x]]_{s+} \supsetneq \mathbb{C}[[x]]_s.$$

3. Sections over a closed disc:

$$H^0(\overline{D}(0,k);\mathscr{F}) = \varinjlim_{\varepsilon\to 0+} \mathbb{C}[[x]]_{s-\varepsilon} = \bigcup_{\varepsilon>0} \mathbb{C}[[x]]_{s-\varepsilon} =: \mathbb{C}[[x]]_{s-} \subsetneq \mathbb{C}[[x]]_s.$$

3.6.2 *Infinitesimal Neighborhoods Associated with the Exponential Order and Type*

As it follows from proposition 3.6.3, p. 116 the Gevrey space $\mathbb{C}[[x]]_s$ does not appear as a space of sections of $\mathscr{F}$ over some disc or any other domain. To fill in that gap we enrich the sheaf $\mathscr{F}$ by taking into account both the exponential order and the exponential type.

For a given $k>0$ we define extensions $X^k, \mathscr{F}^k$ of the space X and the sheaf $\mathscr{F}$ as follows.

▷ *Base space* X^k. — While, building X, we replaced the origin 0 by a copy of $\mathbb{C}$ compactified with a circle at infinity; now, we replace the circle $S^1\times\{k\}$ of radius k in X by a copy Y^k of $\mathbb{C}^* = S^1 \times]0,\infty[$ compactified by the two circles $S^1\times\{0\}$ and $S^1\times\{\infty\}$. Precisely, we glue the lower boundary $S^1\times\{0\}$ of Y^k to the boundary of the disc $D(0,k)$ and the upper boundary $S^1\times\{\infty\}$ of Y^k to the lower boundary of the complement $X\setminus D(0,k)$ (See Fig. 3.6).

As topological spaces, X and X^k are both isomorphic to $\overline{\mathbb{C}}$, that is, $\mathbb{C}$ compactified by a circle. A basis of open sets in X^k is given by open discs centered at 0 and truncated sectors like in X. We denote by $(\theta,\{k,\rho\})$ the polar coordinates of the points of $\overline{Y}^k$.

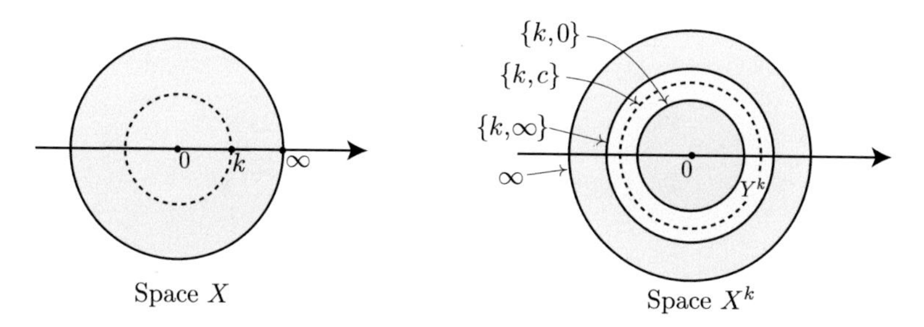

Fig. 3.6

▷ *Presheaf* $\overline{\mathscr{F}}^k$. — Given $c>0$ we take into account the type c of exponentials of order k by introducing the subsheaf $\mathscr{A}^{\leq k,c^-}$ of $\mathscr{A}^{\leq k}$ and the subsheaf $\mathscr{A}^{\leq -k,c^+}$ of $\mathscr{A}^{\leq -k}$ over S^1 defined as follows.

A germ of $\mathscr{A}^{\leq k,c^-}$ at θ is a germ $f\in\mathscr{A}^{\leq k}_\theta$ which satisfies the condition: there exist an open sector Δ containing the θ direction, an $\varepsilon>0$ and a constant $K>0$

such that

$$\left|f(x)\right| \le K \exp\left((c-\varepsilon)/|x|^k\right) \quad \text{for all } x \in \mathcal{A}.$$

A germ of $\mathscr{A}^{\le -k,c^+}$ at θ is a germ $f \in \mathscr{A}_\theta^{\le -k}$ which satisfies the condition: there exist an open sector $\mathcal{A}$ containing the θ direction, an $\varepsilon > 0$ and a constant $K > 0$ such that

$$\left|f(x)\right| \le K \exp\left(-(c+\varepsilon)/|x|^k\right) \quad \text{for all } x \in \mathcal{A}.$$

The space $\mathbb{C}[[x]]_{s,C}$ of series with fixed Gevrey order s and type C is the subspace of $\mathbb{C}[[x]]$ consisting of the series $\sum_{n\ge 0} a_n x^n$ whose coefficients satisfy an estimate of the form

$$|a_N| \le K(n!)^s C^{ns} \text{ for all } n \ge 0 \text{ and a convenient } K > 0.$$

Beside the spaces $\mathbb{C}[[x]]_{s,C}$ it is useful to introduce the spaces

$$\mathbb{C}[[x]]_{s,C^+} = \bigcap_{\varepsilon>0} \mathbb{C}[[x]]_{s,C+\varepsilon}.$$

Thus, a series $\sum_{n\ge 0} a_n x^n$ belongs to $\mathbb{C}[[x]]_{s,C^+}$ if for all $\varepsilon > 0$ there exists $K > 0$ such that

$$|a_N| \le K(n!)^s (C+\varepsilon)^{ns} \quad \text{for all } n \ge 0.$$

With a view to define the sheaf $\mathscr{F}^k$ it suffices to define $\overline{\mathscr{F}}^k$ on a basis of open sets by setting:

$$\begin{cases} \text{inside } X \setminus \overline{Y}^k, \text{ no change: } \overline{\mathscr{F}}^k = \overline{\mathscr{F}}, \\ \text{inside } Y^k: \quad \overline{\mathscr{F}}^k(I \times]\{k,c'\},\{k,c''\}[) = H^0\left(I; \mathscr{A}^{\le k,c'^-} / \mathscr{A}^{\le -k,c''^+}\right), \\ \text{across } \partial Y^k: \quad \overline{\mathscr{F}}^k\left(D(0,\{k,c\})\right) = \mathbb{C}[[x]]_{s,(1/c)^+}, \text{ for } 0 < c < +\infty \\ \qquad \overline{\mathscr{F}}^k\left(I \times]k',\{k,c\}[\right) = H^0\left(I; \mathscr{A}^{\le k'} / \mathscr{A}^{\le -k,c^+}\right), \\ \qquad \overline{\mathscr{F}}^k\left(I \times]\{k,c\},k''[\right) = H^0\left(I; \mathscr{A}^{\le k,c^-} / \mathscr{A}^{\le -k''}\right). \end{cases}$$

As for the presheaf $\overline{\mathscr{F}}$, the application of restriction in $\overline{\mathscr{F}}^k$ is defined on sectors by the natural restriction of functions and quotient. The restriction to an intersection of a sector and a disc is made consistent by the exact sequence

$$0 \to \mathscr{A}^{\le -k,c^-} \longrightarrow \mathscr{A}_{s,(1/c)^+} \xrightarrow{T} \mathbb{C}[[x]]_{s,(1/c)^+} \to 0,$$

analog to the Borel-Ritt exact sequence (2.2), p. 46 [LP97, Sect. 1]. In this sequence the notation $\mathscr{A}_{s,C^+}$ stands for the following sheaf. A germ of $\mathscr{A}_{s,C^+}$ at θ is a germ $f \in \mathscr{A}_\theta$ which satisfies the condition: there exist an open sector $\mathcal{A}$ containing the θ direction and a series $\sum_{n\ge 0} a_n x^n$ such that for all $\varepsilon > 0$ there is a constant $K > 0$ such that

$$\left|f(x) - \sum_{n=0}^{N-1} a_n x^n\right| \leq K\,(N!)^s\,|x|^N (C+\varepsilon)^{Ns} \quad \text{on } \Delta \text{ for all } N \in \mathbb{N}.$$

▷ *Sheaf* $\mathscr{F}^k$. — The sheaf $\mathscr{F}^k$ is the sheaf over X^k associated with the presheaf $\overline{\mathscr{F}}^k$. It is a sheaf of $\mathbb{C}\{x\}$-modules and no longer a sheaf of $\mathbb{C}$-algebras since the product of two functions of $\mathscr{A}^{\leq k,c^-}$ belongs to $\mathscr{A}^{\leq k,(2c)^-}$ and not to $\mathscr{A}^{\leq k,c^-}$ in general. The stalk at a point $(\theta,\{k,c\})$ of $\overline{Y}^k$ is given by

$$\mathscr{F}^k_{(\theta,\{k,c\})} = \begin{cases} \mathscr{A}_\theta^{\leq k^-}/\mathscr{A}_\theta^{\leq -k} & \text{if } c = 0, \\ \mathscr{A}_\theta^{\leq k,c^-}/\mathscr{A}_\theta^{\leq -k,c^+} & \text{if } 0 < c < +\infty, \\ \mathscr{A}_\theta^{\leq k}/\mathscr{A}_\theta^{\leq -k^+} & \text{if } c = +\infty. \end{cases}$$

Example 3.6.4 (The Definition Domain and Support of Exponentials)

An exponential $\exp\big(-a/x^k + q(1/x)\big)$ where $a = A\,\mathrm{e}^{i\alpha}$, $A > 0$ and q is a polynomial of degree less than k is well defined in $\overline{Y}^k$ for all $(\theta,\{k,\rho\})$ such that $-A\cos(\alpha - k\theta) < \rho$.
When $\cos(\alpha - k\theta) \leq 0$ this means all points out of the "arch" $\rho = -A\cos(\alpha - k\theta)$.
When $\cos(\alpha - k\theta) > 0$ this leads to no constraint on ρ; moreover, the exponential is equal to 0 inside the arch $\rho = A\cos(\alpha - k\theta)$.

The *k-big points associated with the exponential* are now the closures of the arches defined by $\rho < A\cos(\alpha - k\theta)$ and $\cos(\alpha - k\theta) > 0$ where the exponential vanishes in $\overline{Y}^k$.

In the example drawn in figure 3.7 the definition domain of the exponential is the shadowed part of the infinitesimal neighborhood of 0. We indicated by "0" the open regions where the exponential vanishes. The big points are the closure of the arches (colored in orange in case of a color copy) in which a small 0 is indicated.

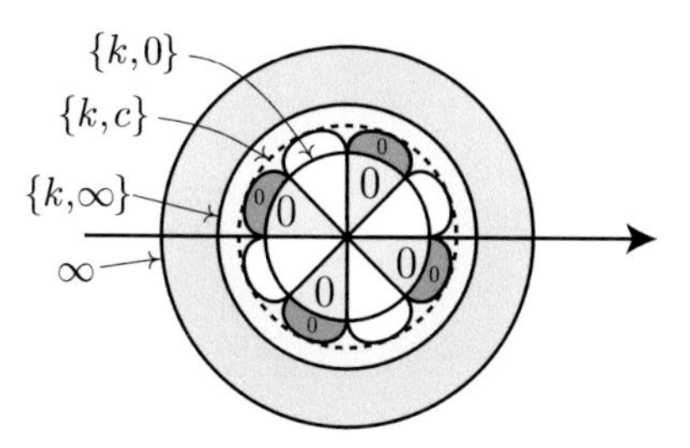

Fig. 3.7. Definition domain of $\exp\big(\frac{i}{x^4} + q(\frac{1}{x})\big)$ (here, $k = 4$ and $c = 1$)

We can now see the space $\mathbb{C}[[x]]_s$ as a space of sections of the sheaf $\mathscr{F}^k$.

Proposition 3.6.5 *Let* $\overline{D}(0,\{k,0\})$ *be the closure in* X^k *of the disc* $D(0,k)$ *centered at* 0 *with radius k. Then,*

$$H^0\big(\overline{D}(0,\{k,0\});\mathscr{F}^k\big) = \mathbb{C}[[x]]_s.$$

Proof. The equality follows from the fact that $\overline{D}(0,\{k,0\}) = \bigcap_{c>0} D(0,\{k,c\})$ and that $H^0\big(D(0,\{k,c\});\mathscr{F}^k\big) = \mathbb{C}[[x]]_{s,(1/c)^+}$. □

3.6.3 More Infinitesimal Neighborhoods

The previous construction can be repeated twice at levels $k = k_1$ and $k = k_2 > k_1$ or finitely many times at levels $k_1 < k_2 < \cdots < k_\nu$. One obtains thus base spaces X^{k_1,k_2} or $X^{k_1,k_2,\cdots,k_\nu}$ and sheaves $\mathscr{F}^{k_1,k_2}$ or $\mathscr{F}^{k_1,k_2,\cdots,k_\nu}$ in a trivial way. Such spaces are useful for handling multisummable series.

Chapter 4
Irregularity and Gevrey Index Theorems for Linear Differential Operators

Abstract In this chapter, the results of the preceding sections are applied to prove index theorems for linear differential operators in the spaces $\mathbb{C}[[x]]_s$ of s-Gevrey series as well as in the space $\mathbb{C}[[x]]_\infty = \mathbb{C}[[x]]$ of formal series and the space $\mathbb{C}[[x]]_0 = \mathbb{C}\{x\}$ of convergent series, following a method by Deligne and Malgrange. The existence and value of the irregularity follow. An application to the Maillet-Ramis theorem which makes explicit the Gevrey order of solutions of linear ordinary differential equations is included. We also sketch a method based on wild analytic continuation, that is, continuation in the infinitesimal neighborhood.

4.1 Introduction

A linear map $D : E \longrightarrow E$ is said to have an index in E if it has finite dimensional kernel $\ker(D,E)$ and cokernel $\operatorname{coker}(D,E)$. If so, the index is defined as being the number

$$\chi(D,E) = \dim \ker(D,E) - \dim \operatorname{coker}(D,E).$$

An index is the Euler characteristic of the complex

$$\cdots \to 0 \longrightarrow 0 \longrightarrow E \xrightarrow{\;D\;} E \longrightarrow 0 \longrightarrow 0 \to \cdots$$

where D is placed in degree 0 or even. It meets then all algebraic properties of Euler characteristics. In case $\operatorname{coker}(D,E) = 0$ the index $\chi(D,E)$ gives the number of solutions of the equation $Dy = 0$ in E. More generally, one says that a linear morphism $L : E \to E'$ between two vector spaces E and E' has an index if its kernel and its cokernel have finite dimension, the index being again the difference of these dimensions.

From now on, we suppose that D is a linear differential operator

$$D = b_n(x)\frac{\mathrm{d}^n}{\mathrm{d}x^n} + b_{n-1}(x)\frac{\mathrm{d}^{n-1}}{\mathrm{d}x^{n-1}} + \cdots + b_1(x)\frac{\mathrm{d}}{\mathrm{d}x} + b_0(x)$$

M. Loday-Richaud, *Divergent Series, Summability and Resurgence II*, Lecture Notes in Mathematics, DOI 10.1007/978-3-319-29075-1_4

where the coefficients $b_p(x)$ are convergent series at $0 \in \mathbb{C}$. The operator D is a linear operator in any of the spaces $\mathbb{C}[[x]]_s$ for $0 \leq s \leq +\infty$ and in any of the quotients $\mathbb{C}[[x]]_s/\mathbb{C}\{x\}$.

The irregularity of D was introduced by B. Malgrange in [Mal74] as follows.

Definition 4.1.1 (Irregularity) *The irregularity of D at 0 is the index of D seen as a linear operator in the quotient $\mathbb{C}[[x]]/\mathbb{C}\{x\}$.*

It was proved in [Mal74] that D has an index both in $\mathbb{C}[[x]]$ and in $\mathbb{C}\{x\}$, the irregularity being then equal to $\chi(D,\mathbb{C}[[x]]) - \chi(D,\mathbb{C}\{x\})$. It was also proved the relation $\operatorname{coker}(D,\mathbb{C}[[x]]/\mathbb{C}\{x\}) = 0$ which shows that the irregularity is the maximal number of divergent series solutions of the equations $Dy = g(x) \in \mathbb{C}\{x\}$ linearly independent modulo convergent ones. These indices were computed in terms of the coefficients $b_p(x)$ of D. The calculation of $\chi(D,\mathbb{C}[[x]])$ is elementary calculus (cf. Prop. 4.2.5 (i), p. 127). The calculation of $\chi(D,\mathbb{C}\{x\})$ follows from an adequate application of Ascoli's theorem.

By a similar analytical method, based on direct or projective limits of Banach spaces and compact perturbations of operators, J.-P. Ramis [Ram84] extended these indices to a large family of Gevrey series spaces: the Gevrey spaces $\mathbb{C}[[x]]_s$ as introduced above but also the Gevrey-Beurling spaces

$$\mathbb{C}[[x]]_{(s)} = \varprojlim_{C>0} \mathbb{C}[[x]]_{s,C} = \bigcap_{C>0} \mathbb{C}[[x]]_{s,C}$$

where

$$\mathbb{C}[[x]]_{s,C} = \Big\{ \sum_{n\geq 0} a_n x^n \in \mathbb{C}[[x]]\,;\, \exists K > 0 \text{ such that } |a_n| \leq K(n!)^s C^{ns} \quad \text{for all } n \Big\}$$

and the spaces $\mathbb{C}[[x]]_{s,C^+} = \bigcap_{\varepsilon>0} \mathbb{C}[[x]]_{s,C+\varepsilon}$ and $\mathbb{C}[[x]]_{s,C^-} = \bigcup_{\varepsilon>0} \mathbb{C}[[x]]_{s,C-\varepsilon}$.

B. Malgrange [Mal74] and J.-P. Ramis [Ram84] computed also the indices of D acting on the fraction fields of formal, convergent or Gevrey series. They proved that these indices differ by $-\chi(D,\mathbb{C}[[x]])$ from those of D acting in $\mathbb{C}[[x]]$, $\mathbb{C}\{x\}$ or $\mathbb{C}[[x]]_s$ for $0 < s < +\infty$. In particular, they are all computed in terms of the Newton polygon of D up to vertical translations and appear thus as formal meromorphic invariants of the equation. These indices are extended to systems by means of a cyclic vector.

A differential operator has no index in the spaces $\mathbb{C}[[x]]_{s,C}$ themselves in general. A counter-example (cf. [LP97, p. 1420]) is given by the Euler operator in $\mathbb{C}[[x]]_{1,1}$ as we prove below.

Proposition 4.1.2 *The Euler operator*

$$\mathscr{E} = x^2 \frac{\mathrm{d}}{\mathrm{d}x} - 1 : \mathbb{C}[[x]]_{1,1} \longrightarrow \mathbb{C}[[x]]_{1,1}$$

has no index when acting in $\mathbb{C}[[x]]_{1,1}$ for, $\operatorname{coker}(\mathscr{E},\mathbb{C}[[x]]_{1,1})$ *has infinite dimension.*

Proof. Check first that $\mathcal{E}$ acts in $\mathbb{C}[[x]]_{1,1}$.

Suppose $\sum_{n\geq 0} a_n x^n$ satisfy $|a_n| \leq K n!$ for all n.

Then,

$$\mathcal{E}\Big(\sum_{n\geq 0} a_n x^n\Big) = \sum_{n\geq 0} b_n x^n$$

where $b_n = (n-1)a_{n-1} - a_n$ satisfy $|b_n| \leq 2Kn!$ for all n and the series $\sum_{n\geq 0} b_n x^n$ belongs to $\mathbb{C}[[x]]_{1,1}$.

Consider now the family of series of $\mathbb{C}[[x]]_{1,1}$

$$g_\alpha(x) = \sum_{n>0} (n-1)!\, n^\alpha x^n, \quad 0 < \alpha < 1.$$

The unique series $\sum_{n\geq 0} c_n x^n$ solution of $\mathcal{E}(y) = g_\alpha(x)$ is given by $c_0 = 0$ and for $n > 0$ by $c_n = -(n-1)!\big(1^\alpha + 2^\alpha + \cdots + n^\alpha\big)$. The coefficients c_n have an asymptotic behavior of the form

$$c_n = \frac{1}{\alpha+1}(n-1)!\, n^{1+\alpha}\big(1 + O(1/n)\big) = \frac{1}{\alpha+1} n!\, n^\alpha\big(1 + O(1/n)\big)$$

with $\alpha > 0$ [Die80, p.119, Exer. 27 or p. 305, Formula (7.5.1)]. It results that the series $\sum_{n\geq 0} c_n x^n$ does not belong to $\mathbb{C}[[x]]_{1,1}$ and $g_\alpha(x)$ does not belong to the range of $\mathcal{E}$. Any non trivial linear combination $\sum \lambda_j g_{\alpha_j}(x)$ has the same property.

To prove that $\mathrm{coker}(\mathcal{E}, \mathbb{C}[[x]]_{1,1})$ has infinite dimension it suffices to prove that the series g_α are linearly independent. To this end, suppose that the g_α's satisfy a linear relation of the form $a_1 g_{\alpha_1} + a_2 g_{\alpha_2} + \cdots + a_r g_{\alpha_r} = 0$. This means that $a_1 n^{\alpha_1} + a_2 n^{\alpha_2} + \cdots + a_r n^{\alpha_r} = 0$ for all $n > 0$. Choose $n_0 \neq 1$. Applying the relation for $n = n_0, n = n_0^2, \ldots, n = n_0^r$ provides the van der Monde system based on $(\lambda_1 = n_0^{\alpha_1}, \lambda_2 = n_0^{\alpha_2}, \lambda_3 = n_0^{\alpha_3}, \ldots, \lambda_r = n_0^{\alpha_r})$:

$$\begin{cases} \lambda_1 a_1 + \lambda_2 a_2 + \cdots + \lambda_r a_r &= 0 \\ \lambda_1^2 a_1 + \lambda_2^2 a_2 + \cdots + \lambda_r^2 a_r &= 0 \\ \cdots \\ \lambda_1^r a_1 + \lambda_2^r a_2 + \cdots + \lambda_r^r a_r &= 0 \end{cases}$$

to determine the coefficients $a_1, a_2, \ldots, a_r$ which are then all equal to 0. Hence the g_α's are linearly independent and $\mathrm{coker}(\mathcal{E}, \mathbb{C}[[x]]_{1,1})$ has infinite dimension. □

4.2 The Deligne-Malgrange Approach

The proofs given in this section are due to B. Malgrange [Mal74] and to P. Deligne [DMR07, p. 21, letter to B. Malgrange, 22 août 1977].

We introduce the following notation [1]:

▷ $\mathscr{V}$ is the sheaf over S^1 of germs of solutions of D;

▷ $\mathscr{V}^{\leq k}$ the subsheaf of the germs of $\mathscr{V}$ with exponential growth of order at most k;

▷ $\mathscr{V}^{<0}$ the sheaf over S^1 of flat germs of solutions of D;

▷ $\mathscr{V}^{\leq -k}$ the subsheaf of germs of $\mathscr{V}^{<0}$ with exponential decay of order at least k.

The sheaf $\mathscr{V}^{\leq k}$ is a subsheaf of $\mathscr{A}^{\leq k}$, the sheaf $\mathscr{V}^{\leq -k}$ a subsheaf of $\mathscr{A}^{\leq -k}$ and the sheaf $\mathscr{V}^{<0}$ a subsheaf of $\mathscr{A}^{<0}$. All these sheaves are sheaves of $\mathbb{C}$-vector spaces. The dimensions of the stalks of $\mathscr{V}^{<0}$ and $\mathscr{V}^{\leq -k}$ at $\theta \in S^1$ are denoted respectively by

$$N^{<0}(\theta) = \dim \mathscr{V}_\theta^{<0} \quad \text{and} \quad N^{\leq -k}(\theta) = \dim \mathscr{V}_\theta^{\leq -k}.$$

Example 4.2.1

Here below are drawn the unique Stokes arc associated with the unique determining polynomial of the Euler equation (cf. Exa. 1.1.4, p. 4) and the graph of the function $\theta \mapsto N^{\leq -1}(\theta)$. In this case, $N^{<0}(\theta) = N^{\leq -1}(\theta)$.

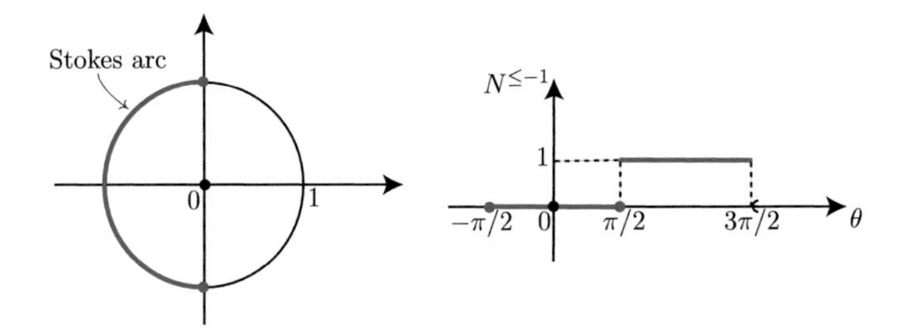

Fig. 4.1

Lemma 4.2.2 *The sheaves $\mathscr{V}^{<0}$ and $\mathscr{V}^{\leq -k}$ for all $k > 0$ are piecewise constant. The functions $\theta \mapsto N^{<0}(\theta)$ and $\theta \mapsto N^{\leq -k}(\theta)$ are lower semi-continuous with jumps occuring only when entering or exiting a Stokes arc of D, that is, when passing a Stokes direction.*

Proof. We proceed like in the proof of corollary 3.4.2, p. 87 of the main asymptotic existence theorem.

Let D' be a normal form of D. We denote by $\mathscr{V}', \mathscr{V}'^{\leq k}, \ldots$ the sheaves associated with D' like $\mathscr{V}, \mathscr{V}^{\leq k}, \ldots$ are associated with D. By the main asymptotic existence theorem (Thm. 3.4.1, p. 87) the sheaves $\mathscr{V}^{<0}$ and $\mathscr{V}^{\leq -k}$ for all $k > 0$ are isomorphic to $\mathscr{V}'^{<0}$ and $\mathscr{V}'^{\leq -k}$ respectively and it is sufficient to prove the lemma for D' instead of D. The space of solutions of $D'y = 0$ is spanned by functions of the form $h(x)\,\mathrm{e}^{q_j(1/x)}$ where $h(x)$ has moderate growth at $x = 0$ and is defined on the full germ of the universal cover of $\mathbb{C}^*$ at 0. Such functions belong to $\mathscr{V}^{<0}$ in a θ direction if and only if $\mathrm{e}^{q_j(1/x)}$ belongs to $\mathscr{A}_\theta^{<0}$. If $q_j(1/x)$ appears with multiplicity m_j in a

[1] The notations $\mathscr{A}^{\leq 0}$ and $\mathscr{V}^{\leq 0}$, in the continuation of the exponential case $\mathscr{A}^{\leq -k}$ and $\mathscr{V}^{\leq -k}$ for $k > 0$, are usually kept for germs with moderate growth.

formal fundamental solution of $Dy = 0$ then the solutions of the form $h(x)\,\mathrm{e}^{q_j(1/x)}$ generate a constant sheaf isomorphic to $\mathbb{C}^{m_j}$ over the interior of each Stokes arc generated by $\mathrm{e}^{q_j(1/x)}$ and 0 outside.

The same result is valid for $\mathscr{V}^{\leq -k}$ with respect to the exponentials $\mathrm{e}^{q_j(1/x)}$ of degree at least k. □

Theorem 4.2.3 (Deligne-Malgrange) *Any linear differential operator D with analytic coefficients satisfies the following properties.*

1. $\ker\big(D, \mathbb{C}[[x]]_s/\mathbb{C}\{x\}\big) \simeq \begin{cases} H^1\big(S^1;\mathscr{V}^{<0}\big) & \textit{for } s = +\infty, \\ H^1\big(S^1;\mathscr{V}^{\leq -k}\big) & \textit{for } 0 < s = 1/k < +\infty; \end{cases}$
2. $\operatorname{coker}\big(D, \mathbb{C}[[x]]_s/\mathbb{C}\{x\}\big) = 0$ *for* $0 < s \leq +\infty$;
3. $\dim H^1\big(S^1;\mathscr{V}^{<0}\big) = \frac{1}{2}\operatorname{var}\big(N^{<0}\big)$;
4. $\dim H^1\big(S^1;\mathscr{V}^{\leq -k}\big) = \frac{1}{2}\operatorname{var}\big(N^{\leq -k}\big)$ *for all* $k > 0$.

Proof. 1.–2. Consider first the case $s = +\infty$. The long exact sequence of cohomology associated with the short exact sequence $0 \to \mathscr{A}^{<0} \to \mathscr{A} \to \mathscr{A}/\mathscr{A}^{<0} \to 0$ reads

$$\begin{array}{ccccccc} 0 \to H^0\big(S^1;\mathscr{A}\big) & \longrightarrow & H^0\big(S^1;\mathscr{A}/\mathscr{A}^{<0}\big) & \longrightarrow & H^1\big(S^1;\mathscr{A}^{<0}\big) & \longrightarrow & H^1\big(S^1;\mathscr{A}\big) \\ \wr\wr & & \wr\wr & & \searrow & \swarrow & \\ \mathbb{C}\{x\} & & \mathbb{C}[[x]] & & & 0 & \end{array}$$

for, $H^0(S^1;\mathscr{A}^{<0}) = 0$, $H^0(S^1;\mathscr{A}/\mathscr{A}^{<0})$ is isomorphic to $\mathbb{C}[[x]]$ by the Borel-Ritt theorem (cf. Cor. 2.2.8, p. 53) and the map $H^1(S^1;\mathscr{A}^{<0}) \to H^1(S^1;\mathscr{A})$ factors through 0 by the Cauchy-Heine theorem (cf. Cor. 2.2.15, p. 56). Hence,

$$H^1(S^1;\mathscr{A}^{<0}) \simeq \mathbb{C}[[x]]/\mathbb{C}\{x\}.$$

The main asymptotic existence theorem in sheaf form (Thm. 3.4.1, p. 87) provides the short exact sequence $0 \to \mathscr{V}^{<0} \to \mathscr{A}^{<0} \xrightarrow{D} \mathscr{A}^{<0} \to 0$. The associated long exact sequence of cohomology reads

$$\begin{array}{ccccc} 0 \to H^1\big(S^1;\mathscr{V}^{<0}\big) & \longrightarrow & H^1\big(S^1;\mathscr{A}^{<0}\big) & \xrightarrow{D} & H^1\big(S^1;\mathscr{A}^{<0}\big) \to 0. \\ & & \wr\wr & & \wr\wr \\ & & \mathbb{C}[[x]]/\mathbb{C}\{x\} & & \mathbb{C}[[x]]/\mathbb{C}\{x\} \end{array}$$

Hence, $\ker(D, \mathbb{C}[[x]]/\mathbb{C}\{x\}) \simeq H^1\big(S^1;\mathscr{V}^{<0}\big)$ and $\operatorname{coker}(D, \mathbb{C}[[x]]/\mathbb{C}\{x\}) \simeq 0$.

The case when $s < +\infty$ is proved similarly from the short exact sequences

$$0 \to \mathscr{A}^{\leq -k} \to \mathscr{A}_s \xrightarrow{T} \mathscr{A}_s/\mathscr{A}^{\leq -k} \to 0 \ \text{ and } \ 0 \to \mathscr{V}^{\leq -k} \to \mathscr{A}^{\leq -k} \xrightarrow{D} \mathscr{A}^{\leq -k} \to 0$$

using the Gevrey parts of the Borel-Ritt and the Cauchy-Heine theorems.

3. Denote by α_ℓ for $\ell \in \mathbb{Z}/p\mathbb{Z}$, the boundary points of the Stokes arcs of D ordered cyclically on S^1 and by $i_\ell : \{\alpha_\ell\} \hookrightarrow S^1$ and $j_\ell :]\alpha_\ell, \alpha_{\ell+1}[\hookrightarrow S^1$ the canonical inclusions. Since S^1 is a real variety of dimension 1 the Euler characteristic[2] of the sheaf $\mathscr{V}^{<0}$ satisfy

$$\chi(\mathscr{V}^{<0}) = \dim H^0\big(S^1;\mathscr{V}^{<0}\big) - \dim H^1\big(S^1;\mathscr{V}^{<0}\big).$$

Then, $\chi(\mathscr{V}^{<0}) = -\dim H^1\big(S^1;\mathscr{V}^{<0}\big)$ since $\dim H^0\big(S^1;\mathscr{V}^{<0}\big) = 0$ (there exists no flat analytic function and, a fortiori, no flat solution all around 0 but the null function) and we are left to estimate the Euler characteristic of the sheaf $\mathscr{V}^{<0}$.

Consider the short exact sequence

$$0 \to \bigoplus_\ell j_{\ell!}\mathscr{V}^{<0} \longrightarrow \mathscr{V}^{<0} \longrightarrow \bigoplus_\ell i_{\ell *}\mathscr{V}^{<0} \to 0$$

The additivity of Euler characteristics allows us to write

$$\chi(\mathscr{V}^{<0}) = \sum_\ell \chi(j_{\ell!}\mathscr{V}^{<0}) + \sum_\ell \chi(i_{\ell *}\mathscr{V}^{<0}).$$

The space $H^0\big(S^1; j_{\ell!}\mathscr{V}^{<0}\big)$ is 0 since $]\alpha_\ell, \alpha_{\ell+1}[$ is not a closed subset of S^1 (a germ at a point of the boundary is the null germ by definition and generates null germs in the neighborhood, hence all over $]\alpha_\ell, \alpha_{\ell+1}[$). The sheaf $j_{\ell!}\mathscr{V}^{<0}$ is a constant sheaf in restriction to $]\alpha_\ell, \alpha_{\ell+1}[$ and 0 outside. Hence, the space $H^1\big(S^1; j_{\ell!}\mathscr{V}^{<0}\big) \simeq H^1\big(]\alpha_\ell, \alpha_{\ell+1}[; j_{\ell!}\mathscr{V}^{<0}\big)$ is isomorphic to the stalk of $\mathscr{V}^{<0}$ at any point α'_ℓ of $]\alpha_\ell, \alpha_{\ell+1}[$ and therefore, $\chi(j_{\ell!}\mathscr{V}^{<0}) = -\dim \mathscr{V}^{<0}_{\alpha'_\ell}$ for all ℓ.

The space $H^1\big(S^1; i_{\ell *}\mathscr{V}^{<0}\big)$ is 0 since the support of $i_{\ell *}\mathscr{V}^{<0}$ has dimension 0 and the space $H^0\big(S^1; i_{\ell *}\mathscr{V}^{<0}\big)$ is isomorphic to the stalk $\mathscr{V}^{<0}_{\alpha_\ell}$ of $\mathscr{V}^{<0}$ at α_ℓ. Thus, we obtain $\chi(i_{\ell *}\mathscr{V}^{<0}) = \dim \mathscr{V}^{<0}_{\alpha_\ell}$ for all ℓ.

The number $\dim \mathscr{V}^{<0}_{\alpha_\ell} - \dim \mathscr{V}^{<0}_{\alpha'_\ell}$ is both the variation of $N^{<0}$ at α_ℓ and at $\alpha_{\ell+1}$. Hence, the $\frac{1}{2}$ in the formula $\sum_\ell \chi(j_{\ell!}\mathscr{V}^{<0}) + \sum_\ell \chi(i_{\ell *}\mathscr{V}^{<0}) = \frac{1}{2}\operatorname{var}\big(N^{<0}\big)$. This ends the proof of point 3.

4. The extension of the previous proof to the sheaf $\mathscr{V}^{\leq -k}$ is straightforward. Denote then $N^{\leq -k}$ instead of $N^{<0}$. □

Remark 4.2.4 The half variation of $N^{<0}$ around S^1 is also the number of Stokes arcs of the determining polynomials q_j of D counted with multiplicity (cf. Def. 3.3.6, p. 78 and Exa. 3.3.7, p. 78), or equivalently, the sum of the (possibly fractional) degrees of all the exponentials of a formal fundamental solution of D. The half variation of $N^{\leq -k}$ around S^1 is the number of Stokes arcs of the determining polynomials q_j with degree at least k and counted with multiplicity, or equivalently, the sum of the degrees of all the exponentials of degree at least k.

[2] The number $\chi(\mathscr{F}) = \sum(-1)^j \dim H^j\big(X;\mathscr{F}\big)$ is, by definition, the Euler characteristic of a sheaf $\mathscr{F}$ over a space X.

Corollary 4.2.5 *Let $0 < k_1 < k_2 < \cdots < k_r < +\infty$ be the slopes of the Newton polygon of D and denote, as usually, $s_j = 1/k_j$ for $j = 1, \ldots, r$. The linear differential operator D has an index in all spaces $\mathbb{C}[[x]]_s$ for $0 \le s \le +\infty$ with values*

$$\text{(i)}\quad \chi\big(D,\mathbb{C}[[x]]\big) = -\ \textit{lower ordinate of the Newton polygon } \mathcal{N}(D);$$

$$\text{(ii)}\quad \chi(D,\mathbb{C}\{x\}) = \begin{cases} \chi\big(D,\mathbb{C}[[x]]\big) - \sharp\ \textit{Stokes arcs of any level}, \text{ i.e.,} \\ -\textit{lower ordinate of the vertical edge in } \mathcal{N}(D); \end{cases}$$

$$\text{(iii)}\quad \chi(D,\mathbb{C}[[x]]_s) = \begin{cases} \chi\big(D,\mathbb{C}[[x]]\big) - \sharp\ \textit{Stokes arcs of level } < k_j, \text{ i.e.,} \\ -\textit{lower ordinate of the edge of slope } k_j \textit{ in } \mathcal{N}(D) \\ \textit{when } s \textit{ satisfies } s_{j+1} < s \le s_j \end{cases}$$

where $\sharp$ stands for "the number of".

In particular, its irregularity $\mathrm{irr}_0(D)$ *satisfies*

$$\text{(iv)}\quad \mathrm{irr}_0(D) = \textit{height of the Newton polygon } \mathcal{N}(D) \textit{ of } D \ (\textit{off the vertical edge}).$$

Proof. Denote by $v(b_p)$ the valuation of the coefficient $b_p(x)$ of $\mathrm{d}^p/\mathrm{d}x^p$ in D and by $m = \inf\big(v(b_p) - p\big)$ the lower ordinate of the Newton polygon $\mathcal{N}(D)$ of D.

(i) Prove that D has an index in $\mathbb{C}[[x]]$, equal to $-m$.

From the definition of m the valuation of b_p satisfies $v(b_p) \ge p + m$ for all p, the equality being reached on a non-empty set $\mathcal{P}$ of indices p. Hence, the coefficient b_p reads $b_p(x) = \alpha_p x^{p+m} + A_p(x)$ with $v(A_p) > p + m$ for all p and the constant coefficient α_p is non-zero for $p \in \mathcal{P}$. For $r \ge -m$, we have

$$b_p(x)\frac{\mathrm{d}^p}{\mathrm{d}x^p}x^r = r(r-1)\cdots(r-p+1)\alpha_p x^{r+m} + \text{higher order terms.}$$

Hence,

$$Dx^r = \beta_{r+m}x^{r+m} + B_{r+m}(x)$$

where $v(B_{r+m}) > r + m$. The constant $\beta_{r+m} = \sum_{p\in\mathcal{P}} r(r-1)\cdots(r-p+1)\alpha_p$ being a polynomial with respect to r is non-zero for $r \ge r_0$ large enough. Denote by $\mathcal{M}$ the maximal ideal of $\mathbb{C}[[x]]$ (ideal generated by x). The previous calculation shows that, for $r \ge r_0$, the operator D induces a morphism $D : \mathcal{M}^r \to \mathcal{M}^{r+m}$.

Prove that this morphism is an isomorphism. Let $g(x) = g_{r+m}x^{r+m} + G_{r+m}(x)$ with $v(G_{r+m}) > r + m$ be given. A series $f(x) = f_r x^r + F_r(x)$ with $v(F_r) > r$ satisfies the equation $Df = g$ if and only if $f_r = g_{r+m}/\beta_{r+m}$ and $DF_r = G_{r+m} + C_{r+m}$ for an adequate formal series C_{r+m} with valuation $v(C_{r+m}) > r + m$. The same reasoning applied to this new equation proves that the next term in $f(x)$ is also uniquely determined and so on by recurrence. This achieves the proof of the fact that $D : \mathcal{M}^r \to \mathcal{M}^{r+m}$ is an isomorphism.

Now, consider the commutative diagram

$$\begin{array}{ccccccccc}
0 & \longrightarrow & \mathcal{M}^r & \longrightarrow & \mathbb{C}[[x]] & \longrightarrow & \mathbb{C}[[x]]/\mathcal{M}^r & \longrightarrow & 0 \\
 & & \downarrow \wr & & \downarrow D & & \downarrow & & \\
0 & \longrightarrow & \mathcal{M}^{r+m} & \longrightarrow & \mathbb{C}[[x]] & \longrightarrow & \mathbb{C}[[x]]/\mathcal{M}^{r+m} & \longrightarrow & 0
\end{array}$$

The left vertical morphism has an index equal to 0. The spaces $\mathbb{C}[[x]]/\mathcal{M}^r$ and $\mathbb{C}[[x]]/\mathcal{M}^{r+m}$ being of finite dimension equal to r and $r+m$ respectively the right vertical morphism has an index equal to $-m$. We can conclude that the morphism in the middle has also an index and, by additivity of Euler characteristics, this index satisfies $-0+\chi(D,\mathbb{C}[[x]])-(-m)=0$. Hence, the result.

(ii) To prove that D has an index in $\mathbb{C}\{x\}$ with the value given in the statement we consider the exact sequence

$$0 \to \mathbb{C}\{x\} \longrightarrow \mathbb{C}[[x]] \longrightarrow \mathbb{C}[[x]]/\mathbb{C}\{x\} \to 0.$$

Since D has an index both in $\mathbb{C}[[x]]$ and in $\mathbb{C}[[x]]/\mathbb{C}\{x\}$ (cf. Thm. 4.2.3, p. 125) it has also an index in $\mathbb{C}\{x\}$ and the three indices satisfy Euler's addition formula

$$\chi(D,\mathbb{C}\{x\})-\chi(D,\mathbb{C}[[x]])+\chi(D,\mathbb{C}[[x]]/\mathbb{C}\{x\})=0.$$

Hence, the result.

(iii) The same argument applied to the exact sequence

$$0 \to \mathbb{C}\{x\} \longrightarrow \mathbb{C}[[x]]_s \longrightarrow \mathbb{C}[[x]]_s/\mathbb{C}\{x\} \to 0$$

proves that D has an index in $\mathbb{C}[[x]]_s$ and again, Euler's addition formula provides the value given in the statement.

(iv) follows directly from (i) and (ii). □

Remarks 4.2.6 The following remarks are straightforward.

▷ The proof above shows that the cokernel of $D:\mathbb{C}[[x]]\to\mathbb{C}[[x]]$ can be generated by polynomials.

▷ The irregularity of D is zero if and only if the Newton polygon has only a horizontal edge. This corresponds, by definition, to a regular singular point at 0. Hence, the irregularity is zero if and only if 0 is a regular singular point.

▷ The function $s \mapsto \chi(D,\mathbb{C}[[x]]_s)$ is piecewise constant, increasing and left continuous.

▷ The indices $\chi(D,\mathbb{C}[[x]]_s), 0\le s\le+\infty$, are not themselves formal meromorphic invariants since, in a gauge transformation, the Newton polygon is translated vertically. However, the irregularity is a meromorphic invariant as well as any difference of two indices $\chi(D,\mathbb{C}[[x]]_s)$ for $0\le s\le+\infty$.

In Fig. 4.2, we denote by $-\mathcal{N}_0(D)$ the polygon symmetric of $\mathcal{N}_0(D)$ with respect to the horizontal coordinate axis. We thus obtain the value of the indices on the vertical axis, not their opposite value.

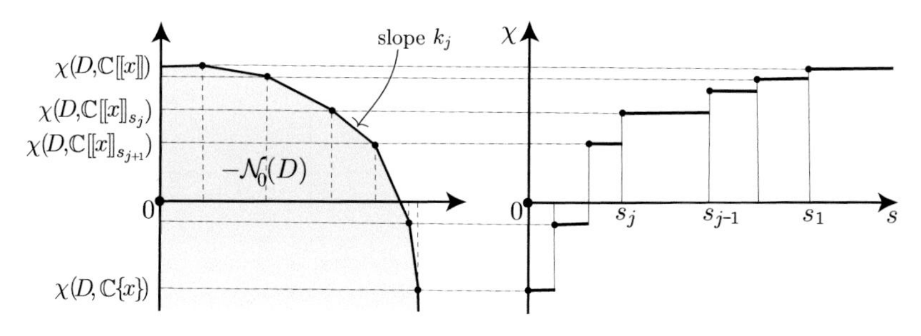

Fig. 4.2. Symmetric $-\mathcal{N}_0(D)$ of the Newton polygon $\mathcal{N}_0(D)$ and the curve $s \to \chi(D,\mathbb{C}[[x]]_s)$.

One consequence of the previous index theorem is the Maillet-Ramis theorem. Maillet's theorem asserts that series solutions of a linear or non linear differential equation are Gevrey of a certain order k. J.-P. Ramis made the theorem more precise in the linear case by proving that the possible k are the levels of the equation.

Theorem 4.2.7 (Maillet-Ramis)
A series $\widetilde{f}(x)$, solution to the differential equation $Dy = 0$, is either convergent or s-Gevrey for some $s > 0$. The best possible s reads $s = 1/k_i$ where k_i is one of the levels $k_1 < k_2 < \cdots < k_r$ of $\widetilde{f}(x)$.

A series solution $\widetilde{f}(x)$ is a formal solution of the form $\widetilde{f}(x)\,\mathrm{e}^0$ coming with a determining polynomial $q_j = 0$. The levels of $\widetilde{f}(x)$ are the degrees of the polynomials $q_\ell - q_j = q_\ell$ for all $\ell \neq j$ (cf. Def. 3.3.4 (ii), p. 77). They are then the non zero slopes of the Newton polygon $\mathcal{N}_0(D)$ of D at 0 (cf. Prop. 3.3.14 (i), p. 82).

Proof. Let $0 < s < +\infty$ and denote $k_0 = 0$, $k_\infty = +\infty$ and $s_j = 1/k_j$ for $j = 0, \ldots, +\infty$. Since, for all s, $\operatorname{coker}(D, \mathbb{C}[[x]]_s/\mathbb{C}\{x\}) = 0$ the number of independent solutions of D in $\mathbb{C}[[x]]_s/\mathbb{C}\{x\}$ is equal to the index $\chi(D, \mathbb{C}[[x]]_s/\mathbb{C}\{x\})$ of D in $\mathbb{C}[[x]]_s/\mathbb{C}\{x\}$. It is then constant for $s_{j+1} < s \le s_j$, $j = 0, \ldots, r$. Hence, a series solution is at least s_1-Gevrey; if it is s-Gevrey with s satisfying $s_{j+1} < s \le s_j$, it is also s_j-Gevrey and, in particular, if it is s-Gevrey with $s < s_r$ it is convergent. □

Comments 4.2.8 (On the Examples of Section 1.1.2, p. 3)

▷ The Euler operator $\mathcal{E} = x^2 d/dx + 1$ and its homogeneous variant

$$\mathcal{E}_0 = x^3 \frac{\mathrm{d}^2}{\mathrm{d}x^2} + (x^2 + x)\frac{\mathrm{d}}{\mathrm{d}x} - 1$$

are irregular singular at 0. They have same indices and same irregularity as indicated on Fig. 4.3 below.

Moreover, the indices satisfy

$$\chi(\mathscr{E}, \mathbb{C}[[x]]_s) = \begin{cases} \chi(\mathscr{E}, \mathbb{C}[[x]]) & \text{for } s \geq 1 \\ \chi(\mathscr{E}, \mathbb{C}\{x\}) & \text{for } s < 1. \end{cases}$$

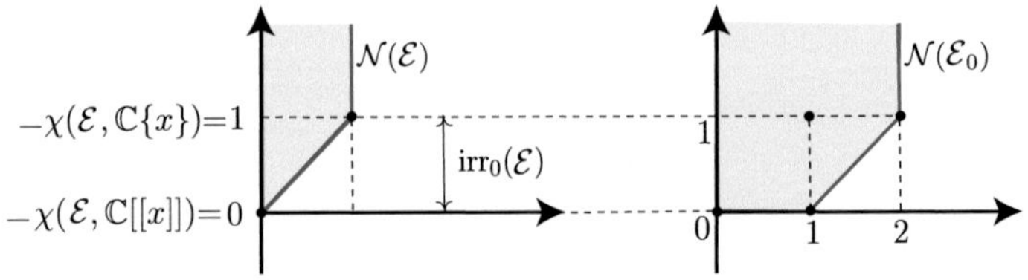

Fig. 4.3

The unique non-zero slope of the Newton polygons $\mathscr{N}(\mathscr{E})$ and $\mathscr{N}(\mathscr{E}_0)$ is equal to 1 since the Euler series is 1-Gevrey and s-Gevrey for no $s < 1$ (cf. Com. 1.2.3, p. 14).

▷ The exponential integral function satisfies $\mathscr{E}i(y) = 0$ where $\mathscr{E}i$ is the operator

$$\mathscr{E}i = x\frac{\mathrm{d}^2}{\mathrm{d}x^2} + (x+1)\frac{\mathrm{d}}{\mathrm{d}x}.$$

This operator is regular singular at 0 and irregular singular at infinity. One can check that the Newton polygon at 0 reduces to a horizontal slope. The picture below shows the Newton polygon $\mathscr{N}(\mathscr{E}i)$ at infinity. Recall that $\mathscr{N}(\mathscr{E}i)$ is the polygon symmetric, with respect horizontal axis, of the Newton polygon at 0 of the operator $\mathscr{E}i$ after the change of variable $z = 1/x$. Hence the change of signs in the indices. The series $\mathrm{Ei}(x)$ is 1-Gevrey.

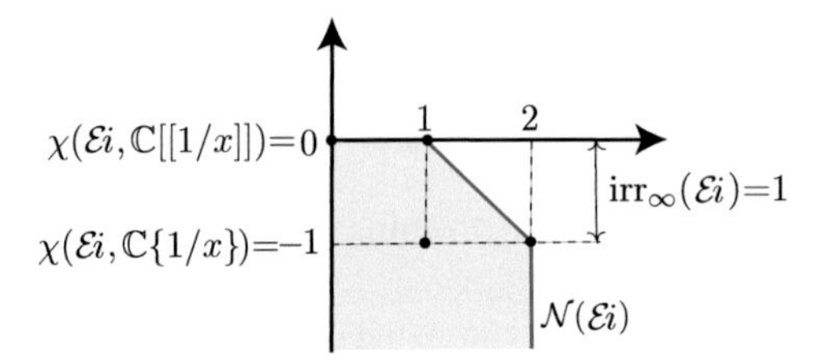

Fig. 4.4

▷ The hypergeometric operator

$$D_{3,1} = z\Big(z\frac{d}{dz}+4\Big) - z\frac{d}{dz}\Big(z\frac{d}{dz}+1\Big)\Big(z\frac{d}{dz}-1\Big)$$

is irregular singular at infinity. Its Newton polygon at infinity has a slope 0 and a slope $-\frac{1}{2}$. Its indices and its irregularity at infinity are indicated on figure below.

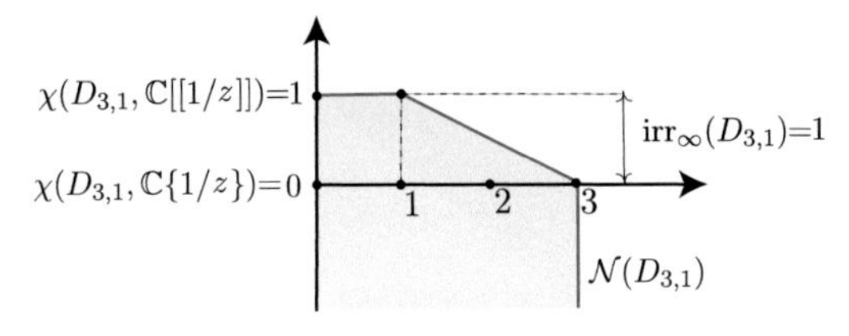

Fig. 4.5

The unique non-zero slope of the Newton polygon $\mathcal{N}(D_{3,1})$ is $-1/2$ at inifinity (hence, it is equal to $+1/2$ at 0 after the change of variable $x = 1/z$) and we saw that the hypergeometric series $\widetilde{g}(x)$ is 2-Gevrey.

4.3 Wild Analytic Continuation and Index Theorems

We sketch here another method to compute a larger variety of index theorems for D. For more details we refer to [LP97]. In that paper, indices are computed for D acting on a variety of spaces including the spaces considered before. The idea is to see each functional space as the 0-cohomology group of the sheaves $\mathscr{F}, \mathscr{F}^k$, and so on... on a convenient subset of the base space X, X^k and so on ... The admissible subsets U that are considered are finite unions of truncated narrow sectors (for a technical reason, sectors are assumed to be closed on their lower boundary) and, possibly, of a small disc centered at 0. The index $\chi(D, \mathbb{C}[[x]])$ of D acting on the space of formal series $\mathbb{C}[[x]]$ is assumed to be known, for instance, from a calculation as before (cor. 4.2.5 (i), p. 127). From the present viewpoint the situation is made more complicated by the fact that the base spaces are now varieties of real dimension 2. The cohomology groups $H^2(U;\mathscr{F}), H^2(U;\mathscr{F}^k)$ and so on... are non-zero groups but the cohomology groups H^i for $i \geq 1$ satisfy (cf. [LP97, Thms. 2.1 and 4.2]) the following sharp property.

Theorem 4.3.1 *For $i \geq 1$, the linear maps $D : H^i\big(U;\mathscr{F}\big) \longrightarrow H^i\big(U;\mathscr{F}\big)$ are isomorphisms for all admissible U.*
The same result is valid for $\mathscr{F}$ replaced by $\mathscr{F}^k, \mathscr{F}^{k_1,k_2}$, etc.

The technique is as follows. For small discs and narrow sectors, meaning small enough to contain no big point associated with D, the calculation is elementary and based on the isomorphism between the spaces of solutions for D and for a normal form D' of D over such domains. For a union of narrow sectors or of a small disc and narrow sectors the calculation follows from the use of Mayer-Vietoris sequences as follows (cf. [LP97, Lem. 3.5]).

Lemma 4.3.2 *Let $U = U_1 \cup U_2$ where U_1 and U_2 are either open or closed subsets of U and suppose U, U_1, U_2 and $U_1 \cap U_2$ are admissible subsets.*

If D has an index in $H^0(U_1;\mathscr{F})$, $H^0(U_2;\mathscr{F})$ and in $H^0(U_1 \cap U_2;\mathscr{F})$ then, it has an index in $H^0(U;\mathscr{F})$ given by

$$\begin{aligned}\chi\big(D, H^0(U;\mathscr{F})\big) = {} & \chi\big(D, H^0(U_1;\mathscr{F})\big) + \chi\big(D, H^0(U_2;\mathscr{F})\big) \\ & - \chi\big(D, H^0(U_1 \cap U_2;\mathscr{F})\big).\end{aligned}$$

The same result holds with $\mathscr{F}$ replaced by $\mathscr{F}^k, \mathscr{F}^{k_1,k_2}$ and so on ...

For a (non-exhaustive) list of indices which can be computed that way we refer to [LP97]. Let us just mention that the list includes indices of D acting on k-summable series over any k-wide arc I (cf. Def. 5.1.2, p. 135) or acting on multisummable series over any multi-arc $(I_1, I_2, \ldots, I_\nu)$ (cf. Sect. 7.7.1, p. 231).

These indices are formal meromorphic invariants of D as long as U does not contain a small disc about 0. Otherwise, their difference with $\chi(D, \mathbb{C}[[x]])$ are formal meromorphic invariants.

Chapter 5
Four Equivalent Approaches to k-Summability

Abstract This chapter deals with the simplest case of summability called k-summability ($k > 0$ either an integer or not) which applies to certain divergent series said *k-summable series* and it aims at being a detailed introduction to the subject. We present four approaches which show up to be equivalent characterizations of k-summable series. With each approach we discuss examples and we attach some applications fitting especially that point of view. The chapter includes sufficient conditions for a series solution of a differential equation to be k-summable. It also includes the proof of Nevanlinna's theorem at any level k.
We give extensive proofs for most of the results and we refer to the literature when the proofs are omitted. A good part of the chapter can be found in [Mal95] or, in [Bal94] and [Cos09] (mostly for the Borel-Laplace approach) and [LP97] (for an approach through wild analytic continuation). More references can be found in these papers and books.

Given a power series $\widetilde{f}(x)$ at 0 we know from the Borel-Ritt theorem that there are infinitely many functions asymptotic to $\widetilde{f}(x)$ on any given sector with an arbitrary opening. However, when $\widetilde{f}(x)$ satisfies an equation, these asymptotic functions do not satisfy the same equation in general. The main asymptotic expansion theorem fills in this gap on small enough sectors for series solutions of linear differential equations by asserting the existence of asymptotic solutions. However, the theorem does not guaranty uniqueness and consequently lets the situation under some indetermination.

The aim of a theory of summation on a given germ of sector (there might be some constraints on the size and the position of the sector) is to associate with any series an asymptotic function uniquely determined in a way as much natural as possible. What natural means depends on the category we want to consider. There is no known operator of summation applying to the algebra of all power series at one time and very little hope towards such a universal tool. For the theory to apply to series solutions of differential equations an eligible request is that the summation operator be a morphism of differential algebras from an algebra of power series (containing the series under consideration) into an algebra of asymptotic functions (containing the corresponding asymptotic solutions). Both algebras must be chosen carefully and correspondingly. To sum series solutions of a difference equation one should

M. Loday-Richaud, *Divergent Series, Summability and Resurgence II*, Lecture Notes in Mathematics, DOI 10.1007/978-3-319-29075-1_5

look for a summation operator being a difference morphism; to sum basic series, for a summation operator being a q-morphism and so on ... The simplest example is given by the usual summation of convergent power series from the algebra of convergent series into the algebra of germs of analytic functions at 0. Such a summation operator is indeed a morphism of differential algebras.

These questions were already widely considered by Euler. They have been developed at the end of the XIXth and the beginning of the XXth Century by mathematicians such as Borel, Birkhoff, Hardy and al. A cohomological point of view brought them an impulse in the late 1970's and 1980's mostly with the results of Y. Sibuya, B. Malgrange, J.-P. Ramis, J. Martinet, W. Balser and lately, C. Zhang for basic series, giving rise to the abstract notions of simple or multiple summability (called k-summability or multisummability). An extension of Borel's approach was almost simultaneously developed by J. Écalle, B. Braaksma, G. Immink, giving rise to the theory of resurgence and integral formulæ which apply to a variety of situations. The question of determining if any solution of a differential equation is k-summable for a convenient $k > 0$ is known as the Turrittin problem [Was76, p. 326]. In the 1980's, J.-P. Ramis and Y. Sibuya [RS89] (see also, [Lod90]) gave a negative answer by exhibiting an example that shows that series solutions of linear differential equations might be k-summable for no value of the parameter $k > 0$ (see below Exa. 7.1.1). They showed however that they are all, at worst, multisummable. The levels k_j entering the multisummability process are the levels of the equation (cf. Def. 3.3.4, p. 77). In the case of series solutions of linear difference equations J. Écalle noticed that some series are neither k-summable nor multisummable. He showed that one has to introduce a new concept named k^+-summability (cf. [Éca93, Im96]) for a simple level k as well as for multiple levels.

5.1 The First Approach: Ramis k-Summability

The problem is here addressed in the following form: to determine under which conditions the Taylor map

$$T_{s,I} : H^0(I; \mathscr{A}_s) \longrightarrow \mathbb{C}[[x]]_s$$

which, with a section of $\mathscr{A}_s$ over an arc I of S^1 (or of its universal cover $\check{S}^1 \simeq \mathbb{R}$), associates its s-Gevrey asymptotic expansion, could be inverted as a morphism of differential $\mathbb{C}$-algebras (recall $k = 1/s$). The answer is far from being straightforward and requires some restrictions both on I and $\mathbb{C}[[x]]_s$. This first approach is based on constraints for the asymptotic conditions themselves. It relies on the results of chapters 1 and 2.

Before to set up a definition let us observe the case of the Euler function.

Comment 5.1.1 (On the Euler Function (Exa. 1.1.4, p. 4))

Although the problem here addressed is independent of any equation, what can happen is well illustrated by the behavior of the solutions of the Euler equation.

We saw that $\mathrm{E}(x)$ is 1-Gevrey asymptotic at 0 to the Euler series $\widetilde{\mathrm{E}}(x)$ on any sector $\boldsymbol{\Delta}_I$ based on the arc $I =]-\frac{3\pi}{2}, \frac{3\pi}{2}[$ (cf. Coms. 1.2.9, p. 17). As before denote by $\mathrm{E}^-(x)$ and $\mathrm{E}^+(x)$ the two branches of $\mathrm{E}(x)$ on the half-plane $\boldsymbol{\Delta}_{-\pi} = \{x\,;\,\Re(x) < 0\}$; these branches are the respective analytic continuations of $\mathrm{E}_{-\pi+\varepsilon}(x)$ and $\mathrm{E}_{+\pi-\varepsilon}(x)$ as $\varepsilon > 0$ tends to 0. We saw that the functions $\mathrm{E}^-(x)$ and $\mathrm{E}^+(x)$ are distinct since otherwise $\widetilde{\mathrm{E}}(x)$ would admit an asymptotic expansion on a full neighborhood of 0 and then would be convergent. Recall the variation formula

$$\mathrm{E}^+(x) - \mathrm{E}^-(x) = 2\pi\mathrm{i}\exp(1/x)$$

The functions $\mathrm{E}^-(x)$ and $\mathrm{E}^+(x)$ are both 1-Gevrey asymptotic to $\widetilde{\mathrm{E}}(x)$ at 0 on the half-plane $\boldsymbol{\Delta}_{-\pi}$, and indeed, $\exp(1/x)$ is 1-Gevrey asymptotic to 0 on $\boldsymbol{\Delta}_{-\pi}$.

When the sector $\boldsymbol{\Delta}$ is narrow, that is, when $\boldsymbol{\Delta}$ is at most an open half-plane, then, $\mathrm{E}(x)$ provides always a 1-Gevrey asymptotic solution on $\boldsymbol{\Delta}$. However, when $\boldsymbol{\Delta} \subseteq \boldsymbol{\Delta}_{-\pi}$, the two solutions $\mathrm{E}^-(x)$ and $\mathrm{E}^+(x)$, and hence all the solutions, are 1-Gevrey asymptotic to $\widetilde{\mathrm{E}}(x)$. Existence is guaranteed, uniqueness fails.

When the sector $\boldsymbol{\Delta}$ is wide, that is, when $\boldsymbol{\Delta}$ contains a closed half-plane, then, either $\boldsymbol{\Delta}$ does not contain the closure $\overline{\boldsymbol{\Delta}}_{-\pi}$ of $\boldsymbol{\Delta}_{-\pi}$ and $\mathrm{E}(x)$ provides the unique 1-Gevrey asymptotic solution on $\boldsymbol{\Delta}$, or $\boldsymbol{\Delta}$ contains $\overline{\boldsymbol{\Delta}}_{-\pi}$ and there is no 1-Gevrey asymptotic solution on $\boldsymbol{\Delta}$. Uniqueness is guaranteed, existence can fail.

In conclusion, there is no good size for an open sector $\boldsymbol{\Delta}$ to guaranty both existence and uniqueness of s-Gevrey asymptotic solutions. We will see that this property remains valid for s-asymptotic functions, not necessarily solutions. Note also that the defect of uniqueness is an exponential function. More generally, flatness for solutions of linear differential equations is always related to exponential functions.

It is convenient to introduce the following definition.

Definition 5.1.2 (k-wide arc or sector)

▷ An arc I *(of* S^1 *or of its universal cover* $\check{S}^1$*) is said to be* k-wide *if it is bounded and either closed with opening* $|I| \geq \frac{\pi}{k}$ *or open with opening* $|I| > \frac{\pi}{k}$.

▷ A sector $\boldsymbol{\Delta}$ *is said to be* k-wide *if it is based on a k-wide arc.*

It follows from the Borel-Ritt theorem (Thm. 1.3.1 (ii), p. 22 and Cor. 1.3.4, p. 25) that the Taylor map

$$T_{s,I} : H^0(I;\mathscr{A}_s) \longrightarrow \mathbb{C}[[x]]_s$$

is surjective when the arc I is open with length $|I| \leq \pi/k$ and *a fortiori*, when I is closed with length $|I| < \pi/k$. Schematically, we can write

I open or closed but not k-wide $\Longrightarrow$ $T_{s,I}$ surjective

Consider now the injectivity of $T_{s,I}$. The example of the Euler function (cf. comment 5.1.1, p. 134) shows that the Taylor map $T_{s,I}$ may be not injective, at least, when I is small. For all I, the kernel of $T_{s,I}$ is the space $H^0(I;\mathscr{A}^{\leq -k})$.

Indeed, the left exactness of the functor $\Gamma(I;.) = H^0(I;.)$ applied to the short exact sequence

$$0 \to \mathscr{A}^{\leq -k} \longrightarrow \mathscr{A}_s \xrightarrow{T_s} \mathbb{C}[[x]]_s \to 0 \qquad (2.2)$$

implies the exactness of the sequence

$$0 \to H^0(I;\mathscr{A}^{\leq -k}) \longrightarrow H^0(I;\mathscr{A}_s) \xrightarrow{T_{s,I}} \mathbb{C}[[x]]_s.$$

A sufficient condition for $T_{s,I}$ to be injective is given by Watson's lemma[1].

Theorem 5.1.3 (Watson's Lemma)
Let Δ be an open sector with opening $|\Delta| = \pi/k$ and suppose that $f \in \mathscr{O}(\Delta)$ satisfies a global estimate of exponential type of order k on Δ, i.e., *there exist constants $C > 0$, $A > 0$ such that the following estimate holds for all $x \in \Delta$:*

$$|f(x)| \leq C \exp\left(-A/|x|^k\right)$$

Then, f is identically equal to 0 on Δ.

Roughly speaking, the lemma says:

"under a global estimate of exponential order k on Δ, the function f is too flat on a too wide sector to be possibly non 0".

For a proof, among the many possible references, quote [Mal95, p. 174, Lem. 1.2.3.3] or [Bal00, p. 75 Prop. 11].

In terms of sheaves Watson's lemma translates as follows.

Corollary 5.1.4 (Watson's Lemma) *The sections of $\mathscr{A}^{\leq -k}$ over any k-wide arc I are all trivial and consequently, the Taylor map $T_{s,I}$ is injective.*

Schematically (recall $s = 1/k$),

$$\boxed{k\text{-wide arc } I \implies H^0(I;\mathscr{A}^{\leq -k}) = 0 \implies T_{s,I} \text{ injective}}$$

Proof. It suffices to consider the case when I is compact. A section of $\mathscr{A}^{\leq -k}$ on I is represented by a finite and consistent collection of $f_j \in \overline{\mathscr{A}}^{\leq -k}(\Delta_j(R_j))$ where the sectors $\Delta_j(R_j)$ have radius R_j and cover the arc I. Let $R = \min_j(R_j)$. Then, the f_j's glue together into a function $f \in \overline{\mathscr{A}}^{\leq -k}(\Delta(R))$ where $\Delta(R)$ denotes the sector $\Delta(R) = \bigcup_j \Delta_j(R_j) \cap \{|x| < R\}$. Since $\Delta(R)$ contains I it is wider than π/k and Watson's lemma applies. □

[1] George Neville Watson (1886-1965) was an English mathematician. He is well known for his contribution to the theory of special functions, his lemma proved in 1918 and his Course of Modern Analysis written in collaboration with his advisor E. T. Whittaker, the so-called "Whittaker and Watson".

Comments 5.1.5 (Exponentials and Watson's Lemma)

▷ Choose $\boldsymbol{\Delta} = \{x\,;\, \pi/2 < \arg(x) < 3\pi/2\}$. Then, the exponential function $\exp(1/x)$ (which appears in the Euler example) belongs to $\overline{\mathscr{A}}^{\leq -1}(\boldsymbol{\Delta})$. The fact that this function be not zero is not conflicting with Watson's lemma. Indeed, denote $\theta = \arg(x)$; the best global estimate for $\exp(1/x)$ on $\boldsymbol{\Delta}$ is $\sup_{x\in\boldsymbol{\Delta}} |\exp(1/x)| = \sup_{\pi/2<\theta<3\pi/2} \exp(\cos\theta/|x|) = 1$ since $\cos\theta$ tends to 0 as θ tends to $\pm\pi/2$. Thus, $\exp(1/x)$ does not satisfy Watson's lemma hypothesis for $\boldsymbol{\Delta}$ and $k = 1$.

▷ On another hand, the exponential $\exp(1/x)$ satisfies Watson's estimate on any proper subsector of $\boldsymbol{\Delta}$. This shows that Watson's lemma is no more valid on a smaller sector; here, for $k = 1$ on a sector of opening less than π and for any $k > 0$, using an adequate exponential of order k, on a sector of opening less than π/k.

▷ *Euler series.* — We can now achieve our comment 5.1.1, p. 134 and show that when $\boldsymbol{\Delta}$ is a sector containing $\overline{\boldsymbol{\Delta}}_{-\pi}$ there exists no function (solution or not solution of the Euler equation) being 1-Gevrey asymptotic to the Euler series $\widetilde{\mathrm{E}}(x)$ on $\boldsymbol{\Delta}$.
Indeed, suppose $\boldsymbol{\Delta} =]\alpha,\beta[\times]0,R[$ with $\alpha < \pi/2 < 3\pi/2 < \beta$ and $f(x)$ be 1-Gevrey asymptotic to $\widetilde{\mathrm{E}}(x)$ on $\boldsymbol{\Delta}$. In restriction to $]\alpha, 3\pi/2[$, the function $f(x) - \mathrm{E}^+(x)$ is 1-Gevrey asymptotic to $\widetilde{\mathrm{E}}(x) - \widetilde{\mathrm{E}}(x) \equiv 0$; hence, it is 1-exponentially flat (cf. Prop. 1.2.17, p. 20) on a 1-wide sector and we can conclude by corollary 5.1.4, p. 136 of Watson's lemma that $f(x) = \mathrm{E}^+(x)$ on $]\alpha, 3\pi/2[$. Symmetrically, $f(x) = \mathrm{E}^-(x)$ on $]\pi/2, \beta[$. Hence, the contradiction since $\mathrm{E}^+ \neq \mathrm{E}^-$ on $]\pi/2, 3\pi/2[$.

The conditions on the arc I to insure either the injectivity or the surjectivity of the Taylor map $T_{s,I}$ are complementary and there is no intermediate condition insuring both injectivity and surjectivity. In such a situation a natural solution proposed by J.-P. Ramis in the early 80's to get both injectivity and surjectivity consisted in choosing for I a k-wide arc and restricting the space $\mathbb{C}[[x]]_s$ of s-Gevrey series into a smaller space.

Suppose we are given a power series $\widetilde{f}(x) = \sum_{n\geq 0} a_n x^n$ at 0.

Definition 5.1.6 (Ramis k-Summability)

▷ k-summability on a k-wide arc I (*recall* $s = 1/k$). — *The series* $\widetilde{f}(x)$ *is said to be* k-summable on I *if I is a k-wide arc and $\widetilde{f}$ belongs to the range of the Taylor map* $T_{s,I}$, i.e., *if there exists a section $f \in H^0(I;\mathscr{A}_s)$ which is s-Gevrey asymptotic to $\widetilde{f}$ on the large enough arc I.*

▷ k-summability in an $\arg(x) = \theta$ direction. — *The series $\widetilde{f}(x)$ is said to be k-*summable in the θ direction *if there exists a k-wide arc I bisected by θ on which $\widetilde{f}(x)$ is k-summable.*

▷ k-sum. — *The function f above, which is uniquely determined when it exists, is called* the k-sum of $\widetilde{f}(x)$ on I or in the θ direction *and we denote it by $f = \mathscr{S}_{k,I}(\widetilde{f})$ or by $f = \mathscr{S}_{k,\theta}(\widetilde{f})$.*

▷ k-summability. — *The series $\widetilde{f}(x)$ is said to be* k-summable *if it is k-summable in all directions but finitely many, called the* singular directions.

Notation 5.1.7 We denote by $\mathbb{C}\{x\}_{\{k,I\}}$ the set of all k-summable series on I and by $\mathbb{C}\{x\}_{\{k,\theta\}}$ the set of all k-summable series in the θ direction.

Notice that $\mathbb{C}\{x\}_{\{k,\theta\}} = \mathbb{C}\{x\}_{\{k,I\}}$ for I the closed arc bisected by θ with length π/k.

Remark 5.1.8 It follows from the definition that a series which is k-summable in all directions is necessarily convergent.

Comment 5.1.9 (On the Examples of Chapter 1)

From section 1.1.2, p. 3 we deduce:

▷ The Euler series $\widetilde{\mathrm{E}}(x)$ in example 1.1.4, p. 4 is 1-summable according to the definition above: precisely, it is 1-summable in all directions but the direction $\theta = \pi$.

▷ Since we have not yet proved that the hypergeometric series $\widetilde{g}(z)$ of example 1.1.6, p. 7 is a 2-Gevrey asymptotic expansion we cannot conclude about its hypothetical 1/2-summability.

▷ In example 1.1.7, p. 8, as for the Euler function, we can move the line of integration from $\mathbb{R}^+$ to the half-line d_θ with argument θ and get an estimate of the same type as estimate (1.10), p. 10, as long as $-\pi/2 < \theta < \pi/2$ (we leave that point as an exercise). This shows that the series $\widetilde{h}(z)$ is 1-summable in each θ direction satisfying $-\pi/2 < \theta < \pi/2$.

▷ Similarly, one can show that the series $\widehat{\ell}(z)$ of example 1.1.8, p. 10 is 1-summable in each θ direction satifying $-\pi/2 < \theta < \pi/2$.

The following proposition is straightforward.

Proposition 5.1.10 *With definitions as above, and especially $s = 1/k$, we can state:*

(i) *The sets $\mathbb{C}\{x\}_{\{k,I\}}$ of k-summable series on a k-wide arc I and $\mathbb{C}\{x\}_{\{k,\theta\}}$ of k-summable series in the θ direction are differential subalgebras of the Gevrey series space $\mathbb{C}[[x]]_s$;*

(ii) *For I a k-wide arc of S^1, the Taylor map*

$$\Gamma(I, \mathscr{A}_s) \xrightarrow{T_{s,I}} \mathbb{C}\{x\}_{\{k,I\}}$$

is an isomorphism of differential $\mathbb{C}$-algebras with inverse the summation map $\mathscr{S}_{k,I}$.

As in chapter 1 (cf. Prop. 1.2.13 and Cor. 1.2.14, p. 20) let us now observe the effect of a change of variable $x = t^r, r \in \mathbb{N}^*$. Let $I = (\alpha, \beta)$ be a k-wide arc. In accordance with the notation $\Delta^j_{/r}$ for sectors in section 1.2.2, p. 16, denote by $I^j_{/r}$ the arc

$$I^j_{/r} = \big((\alpha + 2j\pi)/r, (\beta + 2j\pi)/r\big)$$

so that when $\theta' = \arg(t)$ runs over $I_{/r} = I^0_{/r}$ then $\theta'' = \arg(\omega^\ell t)$ runs over $I^\ell_{/r}$ and $\theta = \arg(x = t^r)$ runs over I. Observe that $I^j_{/r}$ is kr-wide.

Proposition 5.1.11 (k-Summability in an Extension of the Variable)

The following two assertions are equivalent:

(i) *the series $\widetilde{f}(x)$ is k-summable on I with k-sum $f(x)$;*

(ii) *the series $\widetilde{g}(t) = \widetilde{f}(t^r)$ is kr-summable on $I_{/r}$ with kr-sum $g(t) = f(t^r)$.*

Proof. The equivalence is a direct consequence of definition 5.1.6, p. 137 of k-summability and of proposition 1.2.13, p. 19.

□

Given a series $\widetilde{g}(t)$ recall (cf. Sect. 1.2.2, p. 16) that r-rank reduction consists in replacing $\widetilde{g}(t)$ by the r series $\widetilde{g}_j(x), j=0,\ldots,r$ defined by

$$\widetilde{g}(t) = \sum_{j=0}^{r-1} t^j \widetilde{g}_j(t^r)$$

and that the series $\widetilde{g}_j(x)$ are given, for $j=0,\ldots,r-1$, by the relations

$$rt^j \widetilde{g}_j(t^r) = \sum_{\ell=0}^{r-1} \omega^{\ell(r-j)} \widetilde{g}(\omega^\ell t).$$

From corollary 1.2.14, p. 20 we can state:

Corollary 5.1.12 (k-Summability and Rank Reduction)
The following two properties are equivalent:
(i) *for $\ell=0,\ldots,r-1$ the series $\widetilde{g}(t)$ is k'-summable on $I^{\ell}_{/r}$ with k'-sum $g(t)$;*
(ii) *for $j=0,\ldots,r-1$ the r-rank reduced series $\widetilde{g}_j(x)$ is k'/r-summable on I with k'/r-sums $g_j(x)$ defined by the relation*

$$rx^{j/r} g_j(x) = \sum_{\ell=0}^{r-1} \omega^{\ell(r-j)} g(\omega^\ell x^{1/r}), \quad x^{1/r} \in I_{0/r}.$$

In particular, a series $\widetilde{g}(t)$ is k'-summable if and only if its associated r-rank reduced series are k'/r-summable.

With these results we may assume, without loss of generality, that k is small or large at convenience. In particular, we may assume that $k > 1/2$ so that closed arcs of length π/k are shorter than 2π and can be seen as arcs of S^1.

5.2 The Second Approach: Ramis-Sibuya k-Summability

Due to the quite simple integral formula defining the Euler function $f(x)$ we were able to prove, in accordance to Def. 5.1.6, p. 137, that the Euler series $\widetilde{\mathrm{E}}(x)$ is 1-summable in all directions but the direction $\theta = \pi$. However, to check s-asymptoticity on k-wide arcs is not an easy task in general (we refer for instance to our other examples in Sect. 1.1.2, p. 3) and it is worth to look for equivalent conditions in different form.

In this section, we discuss an alternate definition of k-summability, stated in the early 80' by J.-P. Ramis and Y. Sibuya, which is based on series seen as 0-cochains. In order to work on S^1 we assume that $k > 1/2$. This assumption does not affect the generality of the purpose as explained at the end of the previous section.

5.2.1 Definition

Let $\mathscr{I} = \{I_j\}_{j\in\mathbb{Z}/p\mathbb{Z}}$ be a "good" covering of S^1 (hence, without 3-by-3 intersections; cf. Def. 2.2.9, p. 53).
Its connected intersections 2-by-2 are the arcs $(\dot{I}_j = I_j \cap I_{j+1})$[2] and, given a sheaf $\mathscr{F}$ over S^1, a 1-cocycle of $\mathscr{I}$ with values in $\mathscr{F}$ is well defined by the data, for all $j \in \mathbb{Z}/p\mathbb{Z}$, of functions $\dot{\varphi}_j \in \mathscr{F}(\dot{I}_j)$.

Theorem 5.2.1 (Ramis-Sibuya Theorem)
*Suppose $\dot{\varphi} = (\dot{\varphi}_j)_{j\in\mathbb{Z}/p\mathbb{Z}}$ is a 1-cocycle of $\mathscr{I}$ with values in $\mathscr{A}^{\leq -k}$.
Then, there exist 0-cochains $(f_j \in \Gamma(I_j;\mathscr{A}))_{j\in\mathbb{Z}/p\mathbb{Z}}$ of $\mathscr{I}$ with coboundary $\dot{\varphi}$ and any such 0-cochain $(f_j)_{j\in\mathbb{Z}/p\mathbb{Z}}$ takes actually its values in $\mathscr{A}_s$,* i.e., $f_j \in \Gamma(I_j;\mathscr{A}_s)$ *for all j (recall that $s = 1/k$).*

The theorem says in particular that, under the condition that all the differences $-f_j + f_{j+1}$ are k-exponentially flat, all the f_j's are s-Gevrey asymptotic to a same s-Gevrey formal series $\widetilde{f}(x)$.

Proof. When $\dot{\varphi}$ is trivial ($\dot{\varphi}_j = 0$ for all j) then $\dot{\varphi}$ is the coboundary of any analytic function. Conversely, given any 0-cochain (f_j) which, by means of a refinement if necessary, we can assume to be a 0-cochain over a good covering the condition that its coboundary is trivial, i.e., $-f_j + f_{j+1} = 0$ for all j, implies that the functions f_j glue together into an analytic function f. The function f_j are, in particular, s-Gevrey asymptotic to f on I_j for any $s > 0$.

When $\dot{\varphi}$ is elementary (i.e., only one of its components is non zero; Def. 2.2.10, p. 54) a 0-cochain with values in $\mathscr{A}_s$ and coboundary $\dot{\varphi}$ is given by the Cauchy-Heine theorem 1.4.2 (ii), p. 27. The general case follows by additivity of cocycles. In all cases, there exists then a 0-cochain $(f_j)_{j\in\mathbb{Z}/p\mathbb{Z}}$ with values in $\mathscr{A}_s$ and coboundary $\dot{\varphi}$. Let $(g_j)_{j\in\mathbb{Z}/p\mathbb{Z}}$ be another 0-cochain of $\mathscr{I}$ with coboundary $\dot{\varphi}$. Then, the 0-cochain $(g_j - f_j)_{j\in\mathbb{Z}/p\mathbb{Z}}$ has a trivial coboundary and comes from an analytic function h: for all $j \in \mathbb{Z}/p\mathbb{Z}$, $g_j = f_j + h$ and then, like f_j, the function g_j belongs to $\mathscr{A}_s(I_j)$. □

The Ramis-Sibuya theorem admits the following corollary:

Corollary 5.2.2 *The natural injection $\mathscr{A}_s \hookrightarrow \mathscr{A}$ induces an isomorphism*

$$H^0(S^1;\mathscr{A}_s/\mathscr{A}^{\leq -k}) \xrightarrow{\ i\ } H^0(S^1;\mathscr{A}/\mathscr{A}^{\leq -k})$$

and, as a consequence (cf. *Cor. 2.1.30, p. 46*)*, the Taylor map induces an isomorphism*

$$H^0(S^1;\mathscr{A}/\mathscr{A}^{\leq -k}) \simeq \mathbb{C}[[x]]_s.$$

[2] There is an ambiguity with the notations when $p = 2$. In that case, the intersection $I_1 \cap I_2$ consists of two arcs which we denote indifferently by $\dot{I}_1$ and $\dot{I}_2$.

We can thus improve the characterization of s-Gevrey series given in section 2.2.3, p. 52 into a characterization free of Gevrey estimates (recall $s = 1/k$):

$$\left.\begin{array}{l} s\text{-Gevrey series} \\ \widetilde{f}(x) = \sum_{n\geq 0} a_n x^n \in \mathbb{C}[[x]]_s \end{array}\right\} \Longleftrightarrow \left\{\begin{array}{l} \text{(equivalence class of a)} \\ \text{0-cochain } (f_j) \text{ over } S^1 \\ \text{with values in } \mathscr{A} \\ \text{and coboundary } (f_j - f_\ell)_{j,\ell\in J} \\ \text{with values in } \mathscr{A}^{\leq -k} \end{array}\right.$$

This equivalence is a subsequent improvement with respect to the setting in section 2.2.3, p. 52 since to check that the 0-cochain is asymptotic in the sense of Poincaré is usually much simpler than to check its s-Gevrey asymptotics. While in section 2.2.3 it was sufficient to ask for the coboundary to be with values in $\mathscr{A}^{<0}$ it is now essential that the coboundary took its values in $\mathscr{A}^{\leq -k}$.

Definition 5.2.3 (k-Quasi-Sum)
Given $\widetilde{f}(x)$ an s-Gevrey series, the element $\varphi_0 \in H^0\big(S^1;\mathscr{A}/\mathscr{A}^{\leq -k}\big)$ associated with $\widetilde{f}(x)$ by the Taylor isomorphism of corollary 5.2.2, p. 140 is called the k-quasi-sum of $\widetilde{f}(x)$. *By extension, any 0-cochain (f_j) representing φ_0 is called* a k-quasi-sum of $\widetilde{f}(x)$.

With these results k-summability can be equivalently defined as follows.

Definition 5.2.4 (Ramis-Sibuya k-Summability) *An s-Gevrey series $\widetilde{f}(x)$ is said to be* k-summable with k-sum $f(x) \in H^0(I;\mathscr{A})$ on a k-wide arc I *if, in restriction to I, its k-quasi-sum φ_0 satisfies the condition*

$$\varphi_{0|_I}(x) = f(x) \mod \mathscr{A}^{\leq -k}.$$

Indeed, suppose φ_0 satisfies the condition above and let the 0-cochain (f_j) be a k-quasi-sum of $\widetilde{f}(x)$. We know by corollary 5.2.2, p. 140 of the Ramis-Sibuya theorem that all components f_j are s-Gevrey asymptotic to $\widetilde{f}(x)$. Hence, the same is true for $f(x)$ on the k-wide arc I and $f(x)$ fits definition 5.1.6, p. 137. Conversely, a k-sum $f(x)$ of $\widetilde{f}(x)$ in the sense of definition 5.1.6, can be completed into a k-quasi-sum (f_j) of $\widetilde{f}(x)$ using the Borel-Ritt theorem 1.3.1 (ii), p. 22 and proposition 1.2.17, p. 20. Thus, we can reformulated definition 5.2.4, p. 141 by saying:

The s-Gevrey series $\widetilde{f}(x)$ is k-summable with k-sum $f(x)$ on the k-wide arc I if there exists a k-quasi-sum of $\widetilde{f}(x)$ containing $f(x)$ as a component.

5.2.2 Applications to Differential Equations

As before we consider a linear differential operator with analytic coefficients at 0:

$$D = b_n(x)\frac{\mathrm{d}^n}{\mathrm{d}x^n} + b_{n-1}(x)\frac{\mathrm{d}^{n-1}}{\mathrm{d}x^{n-1}} + \cdots + b_0(x) \quad \text{with } b_n(x) \not\equiv 0.$$

▷ *The Maillet-Ramis theorem* (Thm. 4.2.7, p. 129) can be obtained as a consequence of the Ramis-Sibuya theorem as follows. Recall its statement:

Given $\widetilde{f}(x)$, solution of $Dy=0$, then, either $\widetilde{f}(x)$ is convergent or $\widetilde{f}(x)$ is s-Gevrey and $k=1/s$ is one of the levels $k_1<k_2<\cdots<k_r$ of $\widetilde{f}(x)$, i.e., *one of the non-zero slopes of the Newton polygon $\mathcal{N}_0(D)$ of D at* 0. *Other values of k are not optimal.*

Proof. Using the main asymptotic existence theorem (cor. 3.4.2 (i), p. 87) we can associate with $\widetilde{f}(x)$ a (non-unique) 0-cochain $\big(f_j(x)\big)_{j\in\mathbb{Z}/p\mathbb{Z}}$ consisting of asymptotic solutions of the equation $Dy=0$ over a good covering of S^1. The coboundary $\big(-f_j(x)+f_{j+1}(x)\big)$ consists of flat solutions; each such flat solution is equal to some linear combinations of all flat solutions of the equation. Now, solutions of the equation are flat if and only if the exponential factor they contain is flat. They are then flat of an order k which is one of the levels $k_1,k_2,\dots,k_r$. It follows that either the coboundary $\big(-f_j(x)+f_{j+1}(x)\big)$ is trivial (or cohomologous to trivial via flat functions) and the series $\widetilde{f}(x)$ is convergent or the coboundary takes its values in $\mathcal{A}^{\le -k_j}$ for a certain index j and the series $\widetilde{f}(x)$ is $1/k_j$-Gevrey according to the Ramis-Sibuya theorem (Thm. 5.2.1, p. 140) and proposition 1.2.10, p. 17.

If the series $\widetilde{f}(x)$ were $1/k$-Gevrey with $k_j<k<k_{j+1}$ then, by corollary 3.4.2 (ii), p. 87 of the main asymptotic existence theorem, the 0-cochain could be chosen with values in $\mathcal{A}_{1/k}$; by proposition 1.2.10, p. 17 its coboundary would be with values in $\mathcal{A}^{\le -k}$ and since it only comprises solutions it would necessarily take its values in $\mathcal{A}^{\le -k_{j+1}}$. If the series $\widetilde{f}(x)$ were $1/k$-Gevrey with $k_r<k$ then the coboundary would be trivial and $\widetilde{f}(x)$ convergent. Hence, the optimality of k is reached among the levels $k_1,k_2,\dots,k_r$. □

▷ *Summability properties*

Theorem 5.2.5 *Let a series $\widetilde{f}(x)$ be a solution of $Dy=0$ and suppose the equation has a unique level k associated with $\widetilde{f}(x)$* (cf. *Def. 3.3.4, p. 77*). *Then, $\widetilde{f}(x)$ is k-summable.*

Proof. We can assume that $k>1/2$. Prove that $\widetilde{f}(x)$ is k-summable in each θ direction which is anti-Stokes for no determining polynomial of the equation $Dy=0$.

Let $\bar{I}$ be the closed arc bisected by θ with length π/k. It suffices to prove that there exists a k-quasi-sum of $\widetilde{f}(x)$ with no discontinuity on $\bar{I}$. To this end, choose solutions $f_j(x)$ asymptotic to $\widetilde{f}(x)$ on open arcs $(I_j)_{j=1,\dots,m}$ that make a good covering of S^1. We can assume that the arcs $I_1,I_2,\dots,I_{m'}$ form a covering of $\bar{I}$, that the arc $I'=I_1\cup I_2\cup\cdots\cup I_{m'}$ contains no Stokes arc of $\widetilde{f}(x)$ and that the arcs $I_{m'+1},\dots,I_m$ have no intersection with $\bar{I}$. Applying lemma 3.4.3, p. 88, to I' we conclude like in the proof of corollary 3.4.2 (iii), p. 87, of the main asymptotic existence theorem that the natural map $H^0\big(I';\mathcal{V}\big)\longrightarrow H^0\big(I';\mathcal{V}/\mathcal{V}^{\le -k}\big)$ is surjective. Thus, by adding k-flat solutions, the 0-cochain $(f_j)_{j=1,\dots,m'}$ can be changed into a global solution $f(x)$ on I'. The 0-cochain $(f, f_{m'+1}, \dots, f_m)$ on $I', I_{m'+1}, \dots, I_m$ provides a k-sum of $\widetilde{f}(x)$ in the θ direction. □

Comments 5.2.6 (On Examples 1.1.4, p. 4, 1.1.5, p. 6 and 1.1.6, p. 7)

Theorem 5.2.5 applied to these examples yields the following results:

▷ The Newton polygon $\mathcal{N}(\mathcal{E})$ at the origin 0 of the Euler equation (1.1) (cf. Exa. 1.1.4, p. 4) and the Newton polygon at the origin 0 of the Euler equation $\mathcal{E}_0 y = 0$ in homogeneous form (cf. Exa. 2.1.27, p. 44) are drawn below.

The non-zero slopes reduce to a unique slope equal to 1. This implies that the exponentials in the formal solutions are all of degree 1. The fact that the horizontal length of the side of slope 1 is 1 means that there is only one such exponential (including multiplicity). The fact that $\mathcal{N}(\mathcal{E}_0)$ has one horizontal slope of length 1 means that there exists a one dimensional space of formal series solution of $\mathcal{E}_0 y = 0$ (possibly factored by a complex power of x; logarithms could also occur when the length is 2 and higher).

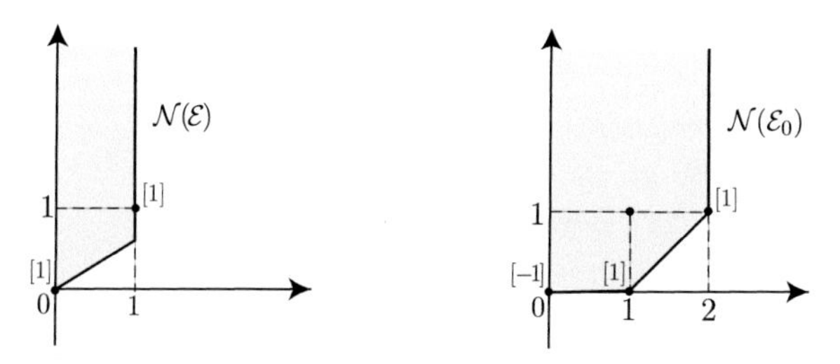

Fig. 5.1. Numbers inside brackets are the coefficients to take into account in the indicial or characteristic equations.

The Euler series $\widetilde{\mathrm{E}}(x)$ is the unique, up to multiplication by a constant, series solution of the Euler equation in homogeneous form (Exa. 2.1.27, p. 44). The exponent of the exponential is given by the characteristic equation associated with slope 1, i.e., the equation $r+1=0$ with solution $r_0 = -1$. Hence, the exponential $\mathrm{e}^{\int r_0/x^2} = \mathrm{e}^{1/x}$. The unique associated anti-Stokes direction is $\theta = \pi$.

Theorem 5.2.5 allows us to assert what we were already able to prove directly on this very simple example: the Euler series $\widetilde{\mathrm{E}}(x)$ is 1-summable in all directions but the direction $\theta = \pi$.

▷ The exponential integral $\mathrm{Ei}(z)$ has, at infinity, the same properties as the Euler function at 0 due to the formula $\mathrm{Ei}(z) = \mathrm{e}^{-z}\mathrm{E}(1/z)$.

▷ The Newton polygon at infinity of the generalized hypergeometric equation $D_{3,1}y = 0$ (Eq. (1.1.6), p. 7) drawn below has a horizontal slope of length 1 and a slope $1/2$ with horizontal length 2.

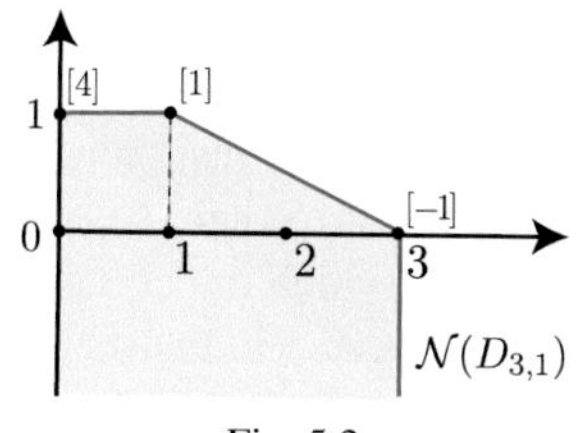

Fig. 5.2

It follows that the equation has a one dimensional space of formal series solutions (space generated by the "hypergeometric" series $\widetilde{g}(z)$); moreover, all exponentials are of degree 1/2. The characteristic equation associated with the slope 1/2 reads $r^2 - 1 = 0$ with solutions $r_{\pm 0} = \pm 1$ and the exponentials are $\exp(\int \pm z^{-1/2}) = \exp(\pm 2z^{1/2})$. The anti-Stokes directions are the directions $\theta = 2\pi \bmod 4\pi$. (We need to go to the Riemann surface of the logarithm since the slope is not an integer. After a ramification $z = t^2$ we could stay in the plane $\mathbb{C}$ of the variable t: the anti-Stokes directions would become $\theta = \pi \in S^1$.) The indicial equation associated with the horizontal slope reads $r + 4 = 0$ with solution $r_0 = -4$. Hence the factor $1/z^4$ in $\widetilde{g}(z)$.

Therefore, by theorem 5.2.5, the series $\widetilde{g}(z)$ is 1/2-summable with respect to the variable z with singular directions $\theta = 2\pi$ (mod. 4π), which we were not able to prove earlier.

Theorem 5.2.5 holds for systems in the following form.

Corollary 5.2.7 *Let $\mathrm{d}Y/\mathrm{d}x = B(x)Y$ be a differential system with a formal fundamental solution $\mathscr{Y}(x) = \widetilde{F}(x)x^L \mathrm{e}^{Q(1/x)}$ where $Q(1/x) = \bigoplus_{j=1}^{J} q_j(1/x)I_{n_j}$, the q_j's being distinct. Split the matrix $\widetilde{F}$ into column-blocks fitting the structure of Q:*

$$\widetilde{F}(x) = \left[\widetilde{F}_1(x)\ \widetilde{F}_2(x)\ \cdots\ \widetilde{F}_J(x)\right]$$

(for $j = 1, \dots, J$, the matrix $\widetilde{F}_j(x)$ has n_j columns). Fix j and suppose that the degrees of the polynomials $q_\ell - q_j$ for $\ell \neq j$ are all equal to k.
Then, the matrix $\widetilde{F}_j(x)$ (i.e., *any entry of $\widetilde{F}_j(x)$*) *is k-summable.*

Recall that the matrix $\widetilde{F}(x)$ satisfies the homological system (3.8), p. 76:

$$\frac{dF}{\mathrm{d}x} = B(x)F - F B_0(x)$$

which admits the polynomials $q_\ell - q_j$ for $j, \ell = 1, \dots, J$ as determining polynomials. Recall that $B_0(x)$ is the matrix of the normal form $\mathrm{d}Y/\mathrm{d}x = B_0(x)Y$ with the fundamental solution $\mathscr{Y}_0(x) = x^L \mathrm{e}^{Q(1/x)}$.

5.3 The Third Approach: Borel-Laplace Summation

Long before the other approaches, the method of summation by means of Borel and Laplace operators was introduced by Émile Borel in the "simplest" case of level one. Contrary to the previous approaches, instead of stating conditions of summmability, it provides integral formulæ for the sums. Note however that the integrand being not explicitly computable in general the sum is not really made explicit. This method serves as base for the theory of resurgence. A long introduction to Borel-Laplace summation in view to state resurgence is given in the first volume of this book [MS16, Chap 5]. There, D. Sauzin begins with the weaker concept of fine Borel-Laplace summation from which the classical Borel-Laplace summation, equivalent to 1-summability, is deduced. He insists on the effect of non linear operations which are useless in linear situations, for instance, to sum series that are solutions of linear differential equations.

The extension of the method to any level k is quite easy although more technical. In this section we define k-Borel-Laplace summability for any $k > 0$. Because of its importance we first sketch the case of level 1, deducing then the general case of level $k > 0$ by transmuting the formulæ with adequate ramifications of the variable x. We prove Nevanlinna's theorem and we use it to show the equivalence of k-Borel-Laplace summability with the notions of summability defined in the previous sections. Also, with Nevanlinna's theorem, the notion of k-fine Borel-Laplace summability (cf. Def. 5.3.11, p. 164) can be given an asymptotic definition, whereas in the first volume [MS16, Sect. 5.7] this notion is defined via Borel-Laplace integrals (and for $k = 1$). We prove Tauberian theorems to compare the processes of k-summability for various k; these theorems are strongly related to multisummability (cf. Prop. 7.2.4, p. 202). Finally we sketch a proof of the (summable)-resurgence of solutions of linear differential equations with the single level 1.

5.3.1 Definitions

Let us first recall the definition of Borel-Laplace summability in the classical case of level $k = 1$. As said before the definition at any level k will be deduced from the definition at level one by transmutation using an adequate ramification of the variable. One could aim at reducing the general case $k > 0$ to $k = 1$ by setting $x = t^{1/k}$ and taking t as new variable. This point of view is fully valuable and it is indeed the point of view chosen by B. Malgrange in [Mal95]. However, this method introduces non integer, and even non fractional, powers of x in general with connected problems. We prefer to continue with the initial variable x.

A definition of Borel-Laplace summability at infinity has been given in the first volume of this book [MS16, Sect. 5.9]. It can be translated at 0 as follows. For the convenience of the reader we begin with some recalls on Borel and Laplace operators.

Definition 5.3.1 (Classical Borel and Laplace Transforms)

(i) *Formal Borel and Laplace transforms.*

The (formal) Borel transform $\widehat{f} = \mathscr{B}(\widetilde{f})$ *of a series* $\widetilde{f}(x) = \sum_{n>0} a_n x^n$ *with valuation* 1 *is the series*

$$\boxed{\widehat{f}(\xi) = \sum_{n>0} a_n \frac{\xi^{n-1}}{\Gamma(n)}}$$

The (formal) Laplace transform $\widetilde{g} = \mathscr{L}(\widehat{g})$ *of a series* $\widehat{g}(\xi) = \sum_{n\geq 0} b_n \xi^n$ *is series*

$$\boxed{\widetilde{g}(x) = \sum_{n\geq 0} b_n \Gamma(1+n) x^{n+1}}$$

(ii) *Functional Borel and Laplace transforms in a given direction*

The Borel transform $\varphi_\theta = \mathscr{B}_\theta(f)$ *of a function* $f(x)$ in the θ direction *is defined, when the integral exists, by*

$$\boxed{\varphi_\theta(\xi) = \frac{1}{2\pi \mathrm{i}} \int_{\gamma_\theta} f(x) e^{\xi/x} \frac{dx}{x^2}}$$

with γ_θ *a Hankel-type contour in the* θ *direction at the origin* (cf. *Fig. 5.3, p. 146*).

The Laplace transform $f_\theta = \mathscr{L}_\theta(\varphi)$ *of a function* $\varphi(\xi)$ in the θ direction *is defined, when the integral exists, by*

$$\boxed{f_\theta(\varphi)(x) = \int_0^{e^{i\theta}\infty} \varphi(\xi)\,\mathrm{e}^{-\xi/x}\,\mathrm{d}\xi}$$

Formal Borel and formal Laplace operators on one hand, Borel and Laplace operators in a given θ direction on the other hand are inverse from each other.

The "Hankel" contour γ_θ is oriented positively. Let us observe that we need a contour that ends at 0 since the function is studied near the origin; if we worked at infinity we would take a usual Hankel-type contour at infinity, image of γ_θ by the change of variable $x = 1/z$. Notice that the tangents at 0 belong to the half-plane directed by $\theta + \pi$ so that $\mathrm{e}^{\xi/x}$ is flat on γ_θ at 0 when $\arg(x)$ is close to θ.

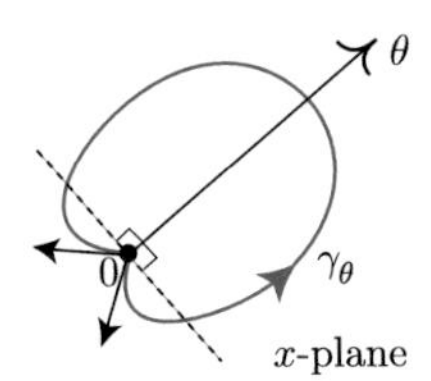

Fig. 5.3. The contour γ_θ

The above definitions are compatible with each other. Consider the monomial x^n. As an analytic function it follows from Hankel's formula for the Gamma function that its Borel transform is given by $\mathscr{B}_\theta(x^n) = \xi^{n-1}/\Gamma(n)$ in any θ direction. As a formal monomial (a formal series!) its formal Borel transform is "also" given by $\mathscr{B}(x^n) = \xi^{n-1}/\Gamma(n)$. The only difference is that the former one should be viewed as an analytic function when the latter one should be viewed as a series. The same result occurs for Laplace operators and it extends to convergent series. Indeed, let $\widetilde{f}(x) = \sum_{n>0} a_n x^n$ be a convergent series; $\widetilde{f}(x)$ is then the Taylor series of its sum $f(x)$. The Borel transform of $f(x)$ in the θ direction reads $\mathscr{B}_\theta(f)(\xi) = \sum a_n \mathscr{B}_\theta(x^n) = \sum a_n \xi^{n-1}/\Gamma(n)$ (one can commute $\sum$ and $\int$). It is then the sum of the formal Borel series of $\widetilde{f}(x)$ and it does not depend on the direction.

Observe that the formal Borel transform applies to series without constant term. With the constant 1 it would be natural to associate the Dirac distribution δ at 0. This is necessary in certain situations, for instance, when one needs to work with convolution algebras (δ is then a neutral element). For our purpose, this is unnecessary and we assume that our series have no constant term.

Remarks 5.3.2 (about Notation)

1. Following the use in resurgence theory, we typically denote by

▷ $\widetilde{f}(x)$, with a tilde, a formal series in the Laplace plane (plane of the initial variable x).

▷ $\widehat{f}(\xi)$, with a hat, its (formal) Borel transform in the Borel plane (with variable ξ). This is coherent with the use of denoting with a hat a Fourier transform since a Borel transform is of Fourier type.

▷ When the series $\widehat{f}(\xi)$ is convergent we denote by $\varphi(\xi)$ its sum; if it is summable in the θ direction we denote its sum in the θ direction by $\varphi_\theta(\xi)$ or even simply by $\varphi(\xi)$ if the direction can be understood without ambiguity.

▷ As much as possible we use latin letters such as $x, f, \dots$ in the Laplace plane and the corresponding greek letters $\xi, \varphi, \dots$ in the Borel plane.

2. When working at the origin, contrary to what happens at infinity, all directional operators (Borel, Laplace introduced above, k-Borel, k-Laplace and accelerators below) to be used in a given summation process follow a same θ direction. Once the direction is clearly made explicit, in view of simplicity, there is no harm to skip the index θ and to denote the operators in the θ direction by $\mathscr{B}$ and $\mathscr{L}$ instead of $\mathscr{B}_\theta$ and $\mathscr{L}_\theta$.

3. Sometimes, one uses different notation for the formal Borel operator, say $\widehat{\mathscr{B}}$ or $\widetilde{\mathscr{B}}$, and for the functional Borel operator $\mathscr{B}_\theta$ or $\mathscr{B}$ if one skips the direction θ when clearly identified. However, from the remark above about monomials and convergent series there is no necessity for introducing two different notations, at least when functions and series are clearly identified by the context. Note however that we do not denote the same way a convergent series (seen as a formal object) and its sum (seen as an analytic function).

Here are some of the basic actions of the Borel transform either formal or not.

$$\begin{aligned}
\frac{1}{x} f(x) &\xrightarrow{\mathscr{B}} \frac{\mathrm{d}}{\mathrm{d}\xi}\varphi(\xi) \quad \text{(assume } f(x)/x \text{ has no constant term),}\\
x^2 \frac{\mathrm{d}}{\mathrm{d}x} f(x) &\xrightarrow{\mathscr{B}} \xi\varphi(\xi),\\
f(x)g(x) &\xrightarrow{\mathscr{B}} \varphi * \psi(\xi) = \int_0^\xi \varphi(\xi-\eta)\psi(\eta)\,\mathrm{d}\eta.
\end{aligned}$$

The formulæ are valid both for series $f(x)$ and $g(x)$ ($\varphi(\xi)$ and $\psi(\xi)$ are then their formal Borel transform) or for functions ($\varphi(\xi)$ and $\psi(\xi)$ are then their functional

Borel transform in some direction). In the third formula, when $\varphi(\xi)$ and $\psi(\xi)$ are formal series the integral is taken term by term.
Observe that an ordinary product is changed into a convolution product so that, in the Borel plane, the multiplication law is convolution. The list can be read from the right to the left as well in order to get the corresponding actions of the Laplace transform when it exists.

The Borel-Laplace summabillity is defined as follows.

Definition 5.3.3 (Borel-Laplace Summability)

A series $\widetilde{f}(x) = \sum_{n>0} a_n x^n$ *is said to be* Borel-Laplace summable in the θ_0 direction *if the following two conditions are satisfied:*

(i) *The series* $\widetilde{f}(x)$ *is* 1*-Gevrey,* i.e., *its Borel transform* $\widehat{f}(\xi)$ *converges* $\big($*recall that* $\widehat{f}(\xi) = \sum_{n>0} a_n \xi^{n-1}/\Gamma(n)\big)$.

(ii) *The sum* $\varphi(\xi)$ *of the Borel series* $\widehat{f}(\xi)$ *can be analytically continued to an infinite open sector* $\sigma = \sigma_{\theta_1,\theta_2}(\infty)$, *where* $\theta_1 < \theta_0 < \theta_2$, *with exponential growth of order 1 at infinity. Denoting again by* φ *this analytic continuation and taking a narrower sector* σ *if necessary we can assume that there exist constants* $A, K > 0$ *such that*

$$\big|\varphi(\xi)\big| \leq K \exp\big(A|\xi|\big) \quad \textit{on } \sigma.$$

The Borel-Laplace summability is an open condition with respect to the direction. Indeed, according to the previous definition, if the series $\widetilde{f}(x)$ is Borel-Laplace summable in the θ_0 direction then it is Borel-Laplace summable in all directions θ satisfying $\theta_1 < \theta < \theta_2$ and one can consider the Laplace integrals

$$f_\theta(x) = \int_0^{e^{i\theta}\infty} \varphi(\xi)\, e^{-\xi/x}\, d\xi \quad \text{for all } \theta \in]\theta_1, \theta_2[.$$

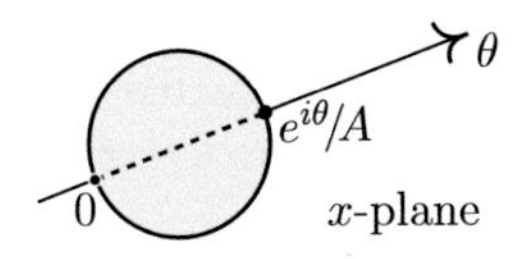

Fig. 5.4. Borel disc $\mathscr{D}_\theta(A)$

From the general properties of Laplace integrals all functions $f_\theta(x)$ are analytic on an open disc $\mathscr{D}_\theta(A)$ with diameter the segment $[0, e^{i\theta}/A]$. Such a disc is called *a Borel disc* in the θ direction (cf. Fig. 5.4).

One proves (Prop. 5.3.7, p. 152) that the functions $f_\theta(x)$ for $\theta_1 < \theta < \theta_2$ glue together into an analytic function $f(x)$ defined on the domain $\bigcup_\theta \mathscr{D}_\theta(A)$. The domains of definition of $\varphi(\xi)$ called *champaign cork* and of $f(x)$ called *Borel croissant* are sketched in Fig. 5.5.

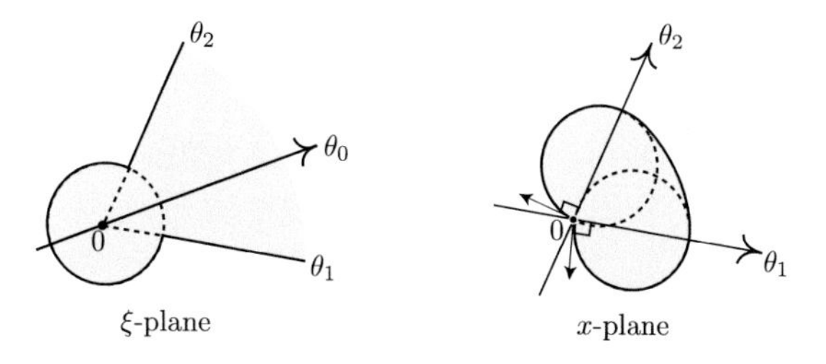

Fig. 5.5. Champaign cork in the ξ-plane and Borel croissant in the x-plane

We can now set up the following definition of the Borel-Laplace sum of $\widehat{f}(x)$ in the θ_0 direction.

Definition 5.3.4 (Borel-Laplace Sum)
Under the conditions of Def. 5.3.3, p. 148 the Borel-Laplace sum of $\widetilde{f}(x)$ in the θ_0 direction *is the function $f(x)$ obtained by gluing together the functions $f_\theta(x)$ for all θ in $]\theta_1, \theta_2[$ or, at least, all θ in a sub-arc of directions neighboring θ_0.*

Observe that, for any small $\varepsilon > 0$, there exists a sector $\Delta = \Delta_{\theta_1-\pi/2+\varepsilon,\theta_2+\pi/2-\varepsilon}$ which is based on the arc $I_{\theta_1,\theta_2}(\varepsilon) =]\theta_1 - \pi/2+\varepsilon, \theta_2+\pi/2-\varepsilon[$ and included in the domain $\bigcup_\theta \mathscr{D}_\theta(A)$; for ε small enough the arc $I_{\theta_1,\theta_2}(\varepsilon)$ contains the closed arc $[\theta_0 - \pi/2, \theta_0+\pi/2]$ with length π bisected by θ_0. A condition usually asked for a function to be a sum of a series is that the function be linked to the series by an asymptotic condition. Although we did not say anything of the kind yet this is indeed the case here: the function $f(x)$ is 1-Gevrey asymptotic to $\widetilde{f}(x)$ on Δ for all ε, a result which we will prove in the general case of level $k > 0$.

As already mentioned in Rem. 5.3.2, p. 147, to obtain a Borel-Laplace sum in the θ_0 direction at the origin one has to consider Laplace integrals in direction θ_0 and neighboring ones. Things go differently at infinity since a Borel transform changes the orientation and so, to get Borel-Laplace sums in direction θ_0 one has to take Laplace integrals in direction $-\theta_0$ and neighboring ones.

Despite the fact that this definition comes with explicit integral formulæ for the sum of the series $\widetilde{f}(x)$ this sum remains somehow indeterminate since an explicit calculation of $\varphi(\xi)$ in terms of classical functions is, in general, out of reach and even impossible.

Definitions 5.3.1, p. 145, 5.3.3, p. 148 and 5.3.4, p. 149 can be extended to any level $k > 0$ as follows. The classical Borel and Laplace operators correspond to $k = 1$. Denote temporarily by $\mathscr{B}_1$ or $\mathscr{B}_{1,\theta}$ the classical Borel operators defined as above and generally, by $\mathscr{B}_k$ or $\mathscr{B}_{k,\theta}$ the k-Borel operators. Denote by $\mathscr{L}_{1,\theta}$ the Laplace operator and generally, by $\mathscr{L}_{k,\theta}$ the k-Laplace operator in direction θ. The operators of level k are transmuted from those of level one by means of ramifications according to the following schemes. Let the ramification ρ_k be defined by $\rho_k(x) = t^{1/k}$ and its inverse $\rho_{1/k}$ by $\rho_{1/k}(\tau) = \xi^k$.

The k-Borel operators in the θ direction are defined by the following commutative diagram:

$$\begin{array}{ccc} f(x) & \xrightarrow{\mathscr{B}_{k,\theta}} & \mathscr{B}_{k,\theta}(f)(\xi) = \psi(\xi^k) \\ \rho_k \big\downarrow & & \big\uparrow \rho_{1/k} \\ f(t^{1/k}) & \xrightarrow{\mathscr{B}_{1,k\theta}} & \psi(\tau) \end{array}$$

where

$$\psi(\tau) = \frac{1}{2\pi\mathrm{i}} \int_{\gamma_{k\theta}} f(t^{1/k}) e^{\tau/t} dt/t^2 = \frac{1}{2\pi\mathrm{i}} \int_{\gamma'_\theta} f(x) e^{\tau/x^k} k\, dx/x^{k+1},$$

the path γ'_θ being deduced from the "Hankel" contour $\gamma_{k\theta}$ by the ramification $t = x^k$ as shown on Fig. 5.6.

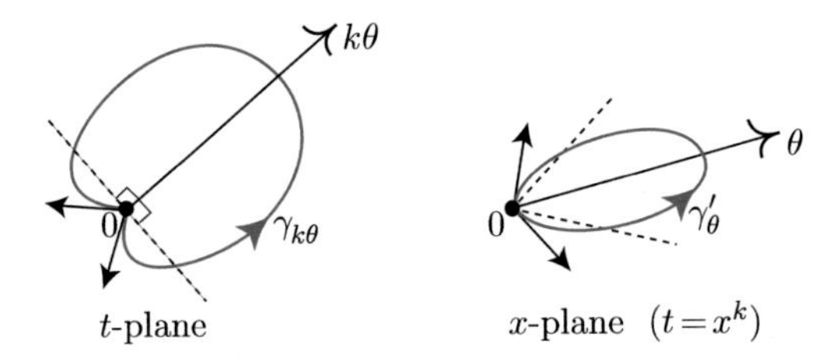

Fig. 5.6. The "Hankel" contour $\gamma_{k\theta}$ compared to γ'_θ when $k > 1$

The formal k-Borel operators are obtained by applying these formulæ to the monomials $f(x) = x^n$. One obtains $\mathscr{B}_k(x^n) = \xi^{n-k}/\Gamma(n/k)$. In accordance to the fact that the usual 1-Borel transform applies to series with valuation $k_0 \geq 1$, the k-Borel transform applies to series with valuation $k_0 \geq k$.

The k-Laplace operators are defined by the following commutative diagram:

$$\begin{array}{ccc} \varphi(\xi) & \xrightarrow{\mathscr{L}_{k,\theta}} & \mathscr{L}_{k,\theta}(\varphi)(x) = g(x^k) \\ \rho_k \big\downarrow & & \big\uparrow \rho_{1/k} \\ \varphi(\tau^{1/k}) & \xrightarrow{\mathscr{L}_{1,k\theta}} & g(t). \end{array}$$

where

$$g(t) = \int_0^{\mathrm{e}^{ik\theta}\infty} \varphi(\tau^{1/k})\, \mathrm{e}^{-\tau/t}\, \mathrm{d}\tau = \int_{\xi=0}^{\mathrm{e}^{i\theta}\infty} \varphi(\xi)\, \mathrm{e}^{-\xi^k/t}\, \mathrm{d}(\xi^k).$$

We can then state the following definitions generalizing for any $k > 0$ the classical definitions of Borel and Laplace transforms stated above with $k = 1$.

Definition 5.3.5 (k-Borel and k-Laplace Transforms)

(i) *Formal k-Borel and k-Laplace transforms.*

The (formal) k-Borel transform $\widehat{f} = \mathscr{B}_k(\widetilde{f})$ *of a series* $\widetilde{f}(x) = \sum_{n \geq k_0} a_n x^n$ *with valuation* $k_0 \geq k$ *is the series*

$$\boxed{\widehat{f}(\xi) = \sum_{n \geq k_0} a_n \frac{\xi^{n-k}}{\Gamma(n/k)}}$$

The (formal) k-Laplace transform $\widetilde{g} = \mathscr{L}_k(\widehat{g})$ *of a series* $\widehat{g}(\xi) = \sum_{n \geq 0} b_n \xi^n$ *is the series*

$$\boxed{\widetilde{g}(x) = \sum_{n \geq 0} b_n \Gamma(1 + n/k) x^{n+k}}$$

(ii) *Functional k-Borel and k-Laplace transforms in a given direction.*

The k-Borel transform *of a function* $f(x)$ in the θ direction *is defined, when the integral exists, by*

$$\boxed{\mathscr{B}_{k,\theta}(f)(\xi) = \frac{1}{2\pi \mathrm{i}} \int_{\gamma_\theta} f(x) e^{\xi^k/x^k} \frac{k dx}{x^{k+1}}}$$

with γ_θ *a Hankel-type contour like drawn in Fig. 5.6, p. 150.*

The k-Laplace transform *of a function* $\varphi(\xi)$ in the θ direction *is defined, when the integral exists, by*

$$\boxed{\mathscr{L}_{k,\theta}(\varphi)(x) = \int_{\xi=0}^{\mathrm{e}^{i\theta}\infty} \varphi(\xi) \mathrm{e}^{-\xi^k/x^k} \mathrm{d}(\xi^k)}$$

Like in the classical case when $k = 1$, the formal k-Borel and formal k-Laplace operators on one hand, the k-Borel and k-Laplace operators in a given θ direction on the other hand are inverse from each other.

These operators are also coherent: $\mathscr{B}_k(x^n) = \xi^{n-k}/\Gamma(n/k)$ as a formal series and $\mathscr{B}_{k,\theta}(x^n) = \xi^{n-k}/\Gamma(n/k)$ in any θ direction as a function. And again, when the θ direction is clearly identified, we allow us to skip the index θ (cf. Rem. 5.3.2, p. 147).

Notice that the formal k-Borel transform applies to series with valuation $\geq k$ like in the classical case with $k = 1$.

Definition 5.3.3, p. 148, can be generalized as follows.

Definition 5.3.6 (k-Borel-Laplace Summability)

Let k_0 be a positive integer satisfying $k_0 \geq k > 0$.

A series $\widetilde{f}(x) = \sum_{n>k_0} a_n x^n$ is said to be k-Borel-Laplace summable in the θ_0 direction *if the following two conditions are satisfied:*

(i) *The series $\widetilde{f}(x)$ is a Gevrey series of order k, that is, its k-Borel transform $\widehat{f}(\xi) = \sum_{n>k_0} a_n \xi^{n-k}/\Gamma(n/k)$ converges.*

(ii) *The sum $\varphi(\xi)$ of the Borel series $\widehat{f}(\xi)$ can be analytically continued to an infinite open sector $\sigma = \sigma_{\theta_1,\theta_2}(\infty)$, where $\theta_1 < \theta_0 < \theta_2$, with exponential growth of order k at infinity. Denoting again by φ this analytic continuation and taking a narrower sector σ if necessary we can assume that there exist constants $A, K > 0$ such that*

$$\left|\varphi(\xi)\right| \leq K \exp\left(A|\xi|^k\right) \quad \textit{on } \sigma. \tag{5.1}$$

Under the action of the ramification $\rho_{1/k}$ the Borel disc $\mathscr{D}_\theta(A) = \{x\,;\,\Re(\mathrm{e}^{ik\theta}/t) > A\}$ in the $k\theta$ direction is changed into the *Fatou petal* $\mathscr{P}_\theta(A) = \{x\,;\,\Re(\mathrm{e}^{ik\theta}/x^k) > A\}$ in the θ direction as shown in Fig. 5.7.

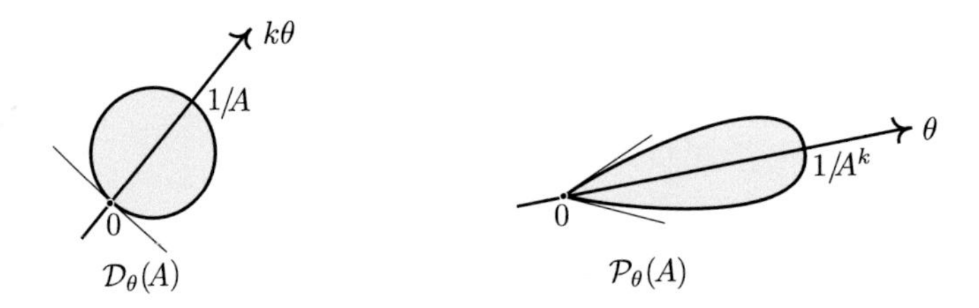

Fig. 5.7. Borel disc vs Fatou petal

Proposition 5.3.7 *With notation as in Def. 5.3.6, p. 152, set*

$$f_\theta(x) = \int_{\xi=0}^{\mathrm{e}^{i\theta}\infty} \varphi(\xi)\,\mathrm{e}^{-\xi^k/x^k}\,\mathrm{d}(\xi^k) \quad \textit{for all } \theta \textit{ satisfying } \theta_1 < \theta < \theta_2.$$

The functions f_θ glue together into a function $f(x)$ defined and analytic on $\bigcup_\theta \mathscr{P}_\theta(A)$.

Proof. Setting $\zeta = \xi^k$ and $t = x^k$ the integral reduces to an usual Laplace transform in the $k\theta$ direction. It is then sufficient to prove the result for $k = 1$. Hence, we assume that $k = 1$.

For any $\theta \in]\theta_1, \theta_2[$ the function $f_\theta(x)$ is analytic on $\mathscr{D}_\theta(A)$. Indeed, the integrand is analytic. For x in the open set $\mathscr{D}_\theta(A)$ there exists $B > A$ such that x belongs to $\mathscr{D}_\theta(B)$ too. Then, on $\mathscr{D}_\theta(B)$, φ satisfies $|\varphi(\xi)\mathrm{e}^{-\xi/x}| \leq K\exp\left(-(B-A)|\xi|\right)$ and $\left|\frac{\mathrm{d}}{\mathrm{d}x}\left(\varphi(\xi)\mathrm{e}^{-\xi/x}\right)\right| \leq \frac{K}{|x^2|}|\xi|\exp\left(-(B-A)|\xi|\right)$. We can thus apply Lebesgue's dominated convergence theory to conclude on a neighborhood of x.

Prove now that the f_θ's glue together into an analytic function on $\bigcup_\theta \mathscr{D}_\theta(A)$.

Consider two directions $\theta' < \theta''$ in $]\theta_1, \theta_2[$ such that, say, $\theta'' - \theta' \leq \pi/4 < \pi/2$ and apply Cauchy's theorem to $\varphi(\xi)\mathrm{e}^{-\xi/x}$ along the boundary γ_R, oriented positively, of a sector of radius R limited by the lines $\theta = \theta'$ and $\theta = \theta''$. Denote by C_R the arc of circle from $R\mathrm{e}^{i\theta'}$ to $R\mathrm{e}^{i\theta''}$. Then,

$$\int_0^{R\mathrm{e}^{i\theta'}} + \int_{C_R} - \int_0^{R\mathrm{e}^{i\theta''}} \varphi(\xi)\mathrm{e}^{-\xi/x}\mathrm{d}\xi = 0$$

and

$$f_{\theta'}(x) - f_{\theta''}(x) = \lim_{R\to+\infty} \int_{C_R} \varphi(\xi)\mathrm{e}^-\xi/x\mathrm{d}\xi.$$

We must prove that this limit is 0, for x in some open subset of $\mathscr{D}_{\theta'}(A) \cap \mathscr{D}_{\theta''}(A)$. Denoting $x = |x|\,\mathrm{e}^{i\omega}$ we can write

$$\begin{aligned}\Big|\int_{C_R} \varphi(\xi)\mathrm{e}^{-\xi/x}\mathrm{d}\xi\Big| &\leq \int_{\theta'}^{\theta''} K\exp\big(-(\Re(\mathrm{e}^{i\theta}/x) - A)|\xi|\big)\,|\xi\,\mathrm{e}^{i\theta}|\,\mathrm{d}\theta \\ &= KR\int_{\theta'}^{\theta''} \exp\big(-(\cos(\theta-\omega)/|x| - A)R\big)\,\mathrm{d}\theta.\end{aligned}$$

Now choose $x \in \mathscr{D}_{\theta'}(A) \cap \mathscr{D}_{\theta''}(A)$ such that $\theta' < \omega < \theta''$. Such x form an open set $\mathscr{D}_{\theta',\theta''}$ and it suffices to prove that $f_{\theta'}$ and $f_{\theta''}$ coincide on $\mathscr{D}_{\theta',\theta''}$. From the inequality $|\theta - \omega| < \pi/4$ for all θ from θ' to θ'' we deduce the estimate

$$\Big|\int_{C_R} \varphi(\xi)\mathrm{e}^{-\xi/x}\mathrm{d}\xi\Big| \leq K(\theta'' - \theta')R\exp\Big(-\big(\tfrac{1}{|x|\sqrt{2}} - A\big)R\Big)$$

and the integral tends to 0 as R tends to infinity as soon as $|x| < 1/(A\sqrt{2})$. It results

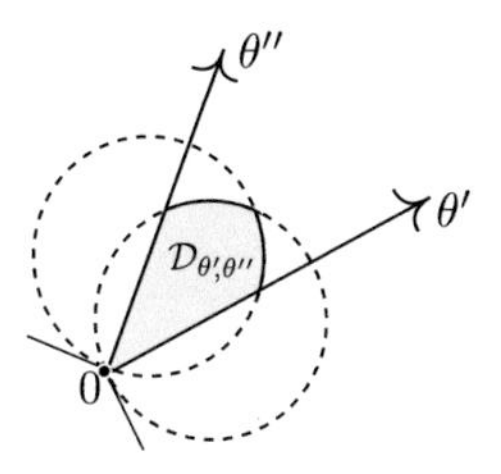

Fig. 5.8. Domain $\mathscr{D}_{\theta',\theta''}$ with $\theta'' - \theta' < \pi/4$

that the Laplace integrals $f_{\theta'}(x)$ in the θ' direction and $f_{\theta''}(x)$ in the θ'' direction coincide on the (non empty) open set $\mathscr{D}_{\theta',\theta''} \cap \{|x| < 1/(A\sqrt{2})\}$ and they are, then, analytic continuations of each other. $\square$

Definition 5.3.8 (k-Borel-Laplace Sum) *Under the conditions of Def. 5.3.6, p. 152,* the k-Borel-Laplace sum of $\widetilde{f}(x)$ in the θ_0 direction *is the function $f(x)$ obtained by gluing together the functions $f_\theta(x)$ defined in Prop. 5.3.7, p. 152 for all θ in a subarc of directions neighboring θ_0.*

Notice that the domain of definition of $f(x)$ contains an open sector with vertex 0, with opening larger than π/k and bisected by θ_0.

As expected, the sum $f(x)$ of the series $\widetilde{f}(x)$ satisfies k-Gevrey asymptotic conditions to $\widetilde{f}(x)$ in the θ_0 direction. This property is proved in the next two sections as well as the connected question of the equivalence between k-summability and k-Borel-Laplace summability (cf. Prop. 5.3.12, p. 165).

5.3.2 Nevanlinna's Theorem and Summability

We begin with the proof of Nevanlinna's theorem which we use to prove the equivalence of k-summability and k-Borel-Laplace summability.

Assume we are given a direction θ_0 issuing from 0 which, by means of a rotation, we assume to be $\theta_0 = 0$ and let us first describe the curves and domains we will be concerned with. We consider two copies of $\mathbb{C}$, one which we call the *Laplace plane* with coordinate x and the other one, called *Borel plane*, with coordinate ξ.

We fix $k > 0$ and we introduce two new copies of $\mathbb{C}$ with coordinates $Z = 1/x^k$ and $\zeta = \xi^k$ respectively.

▷ In the x-plane we consider for any $\ell > 0$, the domain (*Borel disc* when $k = 1$, *Fatou petal* when $k \neq 1$; cf. Fig. 5.9) defined by

$$\boldsymbol{\Delta}_\ell = \left\{x \in \mathbb{C}\,;\, \Re\left(\frac{1}{x^k}\right) > \ell^k \text{ and } \left|\arg(x)\right| < \frac{\pi}{2k}\right\}.$$

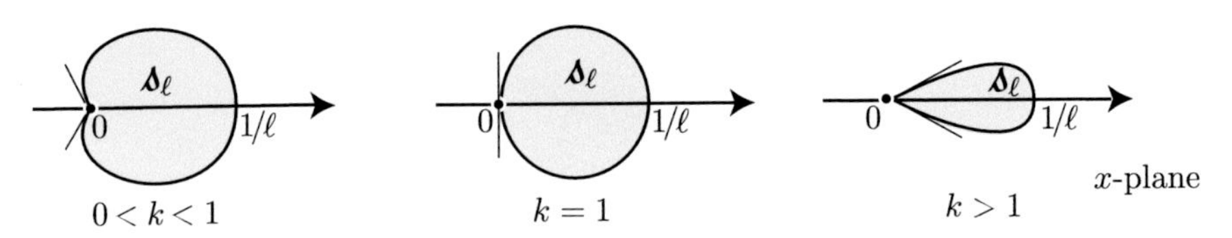

Fig. 5.9. Fatou petals and Borel disc

Note that when $k < 1/2$ the domain $\boldsymbol{\Delta}_\ell$ has an angle larger than 2π at 0. It lives then on the Riemann surface of logarithm with similar shape. If wished it is still possible however to assume $k \geq 1/2$ by means of a change of variable $x = t^p$ with an integer p such that $kp \geq 1/2$.

▷ In the Z-plane, we consider the image $\mathcal{S}_\ell$ of Δ_ℓ by the map $Z = 1/x^k$. This is the half-plane $\{Z\,;\Re(Z) > \ell^k\}$ (cf. Fig. 5.10, p. 155).

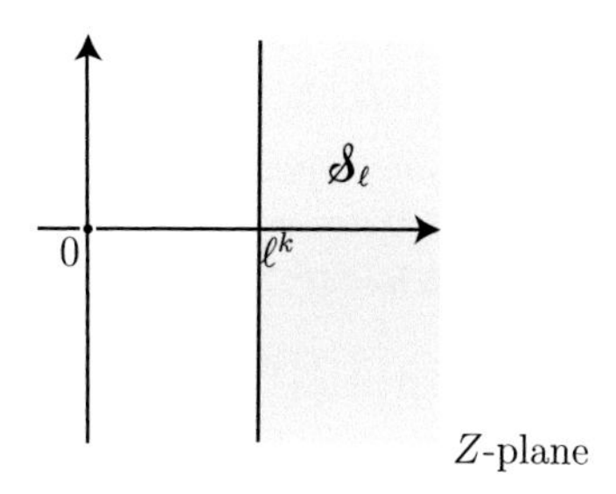

Fig. 5.10. Half-plane $\mathcal{S}_\ell$

▷ In the ζ-plane, for $B > 0$, we consider the domain $\Sigma_B = D(0,B^k) \cup \Sigma'_B$ union of the open disc $D(0,B^k)$ with center 0 and radius B^k and of the set Σ'_B of points in $\mathbb{C}$ at a distance less than B^k of the line $[B^k, +\infty[$ (cf. Fig. 5.11, p. 155).

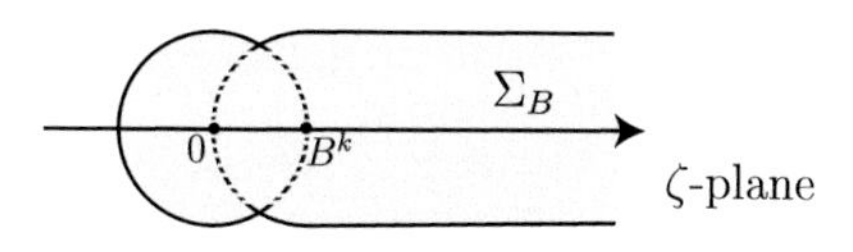

Fig. 5.11. Domain Σ_B

▷ In the ξ-plane, we consider for $B > 0$, the domain $\sigma_B = D(0,B) \cup \sigma'_B$ union of the disc $D(0,B)$ with center 0 and radius B and of the image σ'_B of Σ'_B by the map $\xi = \zeta^{1/k}$ for the choice of the principal determination of the kth-root (cf. Fig. 5.12, p. 155).

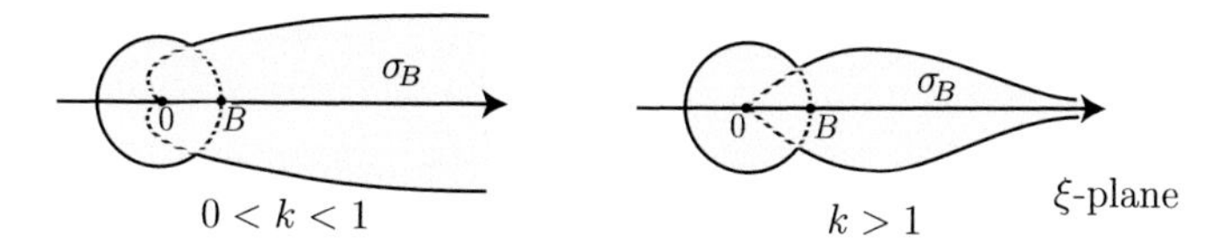

Fig. 5.12. Domain σ_B

All these domains depend on k. If necessary, we make k explicit by denoting $\Delta_0(k), \Delta_\ell(k)$, etc.

Theorem 5.3.9 (Nevanlinna's Theorem)

Let $k>0$. Suppose the series $\widetilde{f}(x)=\sum_{n\geq k_0} a_n x^n$ has valuation $k_0\geq k$ and denote by $\widehat{f}(\xi)=\sum_{n\geq k_0}\frac{a_n}{\Gamma(n/k)}\xi^{n-k}$ its associated k-Borel series.

Then, the following two assumptions are equivalent:

(i) *The k-Borel series $\widehat{f}(\xi)$ is convergent and its sum $\varphi(\xi)$ can be analytically continued to a domain σ_B with global exponential growth of order k at infinity, that is, there exist constants $A,K>0$ such that*

$$\big|\varphi(\xi)\big|\leq K\exp\big(A|\xi|^k\big)\quad \textit{for all}\ \ \xi\in\sigma_B. \tag{5.2}$$

(ii) *There exist $\ell_0>0$ and $f(x)\in\mathscr{A}(\boldsymbol{\Delta}_{\ell_0})$ asymptotic to $\widetilde{f}(x)$ with global k-Gevrey estimates on $\boldsymbol{\Delta}_{\ell_0}=\boldsymbol{\Delta}_{\ell_0}(k)$, that is, there exist constants $C,D>0$ such that for all $x\in\boldsymbol{\Delta}_{\ell_0}$ and $N\in\mathbb{N}^*$*

$$\Big|f(x)-\sum_{n=k_0}^{N-1}a_nx^n\Big|\leq C\Big(\frac{N}{k}\Big)^{N/k}\mathrm{e}^{-N/k}\frac{|x|^N}{D^N}. \tag{5.3}$$

Moreover, the functions $f(x)$ and $\varphi(\xi)$ are k-Laplace and k-Borel transforms of each other (cf. *Def. 5.3.5, p. 151): given $\ell>\ell_0$ they satisfy*

$$\begin{aligned} f(x) &= \textstyle\int_0^{+\infty}\varphi(\xi)\,\mathrm{e}^{-\xi^k/x^k}\,\mathrm{d}(\xi^k)\quad \textit{for all}\ \ x\in\boldsymbol{\Delta}_\ell\\ \varphi(\xi) &= \tfrac{1}{2\pi\mathrm{i}}\textstyle\int_{\Re(1/u^k)=\ell^k}f(u)\,\mathrm{e}^{\xi^k/u^k}\,\mathrm{d}(1/u^k)\quad \textit{for all}\ \ \xi>0.\end{aligned}$$

Observe that condition (ii) is not a mere asymptotic condition on the sector $\boldsymbol{\Delta}_{\ell_0}$ with opening π/k but a uniform one.

Proof. ▷ (i) *implies* (ii) is proved for level $k=1$, for instance, in [Mal95, p. 182] or in volume I of this book [MS16, Thm. 5.20]. In volume III it is referred to as "the easy part of Nevanlinna's theorem" ([Dela16, Prop. 3.13]. The proof below is valid for any level $k>0$.

Assume that $\varphi(\xi)$ satisfies (i) and set $f(x)=\int_0^{+\infty}\varphi(\xi)\mathrm{e}^{-\xi^k/x^k}\,\mathrm{d}(\xi^k)$. We have to prove that $f(x)$ is analytic and satisfies condition (5.3) on some Fatou petal $\boldsymbol{\Delta}_{\ell_0}$.

From condition (5.2), p. 156, we know that $f(x)$ is defined and analytic for any x satisfying $\Re(1/x^k)>A$; hence, adding the condition $|\arg(x)|<\pi/(2k)$, on any Fatou petal $\boldsymbol{\Delta}_\ell$ with $\ell^k>A$. Choose $\ell_0^k>A$. Set $\zeta=\xi^k$ and denote by $\zeta^{1/k}$ the principal k^{th} root of ζ. The Laplace integral reads $f(x)=\int_0^{+\infty}\varphi(\zeta^{1/k})\mathrm{e}^{-\zeta/x^k}\,\mathrm{d}\zeta$ and from Cauchy's theorem we can integrate as well on a path γ_b homotopic to $[0,+\infty[$ in Σ_B and defined as follows: the path γ_b follows a straight line from 0 to b^k and continues along a horizontal line from b^k to $+\infty$. Choose b with argument, say, $\beta=\pi/(4k)$ and $|b|>0$ so small that the whole path γ_b is included in Σ_B.

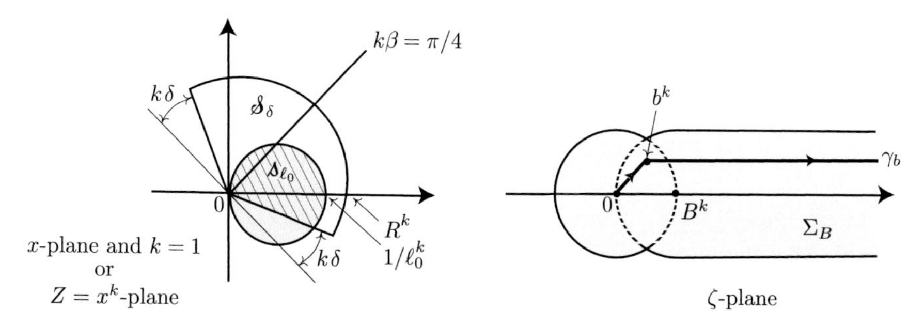

Fig. 5.13. The path γ_b and the domain $\Delta_{\ell_0} \cap \mathcal{S}_\delta$ (hachured)

Decompose $f(x)$ into $f(x) = f^b(x) + g^b(x)$ where

$$f^b(x) = \int_{\zeta=0}^{b^k} \varphi(\zeta^{1/k})\,\mathrm{e}^{-\zeta/x^k}\,\mathrm{d}\zeta \quad \text{and} \quad g^b(x) = \int_{\zeta=b^k}^{+\infty} \varphi(\zeta^{1/k})\,\mathrm{e}^{-\zeta/x^k}\,\mathrm{d}\zeta.$$

Now, choose a sector $\mathcal{S}_\delta$ in the Laplace plane (variable x) bisected by β, with radius $R > 1/\ell_0$ and opening $\pi/k - 2\delta$. Suppose $0 < \delta < \pi/(4k)$ so that $\mathcal{S}_\delta$ contains the upper part $\Delta_{\ell_0} \cap \{x\,;\,\Im(x) \geq 0\}$ of Δ_{ℓ_0}.

From lemma 1.3.2, p. 24, we know that, then, $f^b(x)$ satisfies a global s-Gevrey asymptotic condition like (5.3), p. 156, on $\mathcal{S}_\delta$: there exist constants $C', D' > 0$ such that, for all $N \in \mathbb{N}$ and $x \in \mathcal{S}_\delta$,

$$\Big| f^b(x) - \sum_{n=k_0}^{N-1} a_n x^n \Big| \leq C' \Big(\frac{N}{k}\Big)^{N/k} \mathrm{e}^{-N/k} \frac{|x|^N}{D'^N}.$$

Prove that $g^b(x)$ is k-exponentially flat on $\mathcal{S}_\delta \cap \Delta_{\ell_0}$.

Indeed, parameterizing the second part of γ_b by $\zeta = b^k + \tau$, we can write

$$g^b(x) = \mathrm{e}^{-b^k/x^k} \int_0^{+\infty} \varphi\big((b^k + \tau)^{1/k}\big)\,\mathrm{e}^{-\tau/x^k}\,\mathrm{d}\tau.$$

Hence, using assumption (5.2), p. 156, we obtain

$$|g^b(x)| \leq K\mathrm{e}^{A|b^k|}\,\mathrm{e}^{-\Re(b^k/x^k)} \int_0^{+\infty} \mathrm{e}^{-(\Re(1/x^k)-A)\tau}\,\mathrm{d}\tau.$$

However, if x belongs to $\mathcal{S}_\delta$ then $|\arg(b^k/x^k)| < \pi/2 - k\delta$, which implies that there exists a constant $c > 0$ such that $-\Re(b^k/x^k) \leq -c/|x^k|$. If x belongs to Δ_{ℓ_0} then the integral satisfies $|\int_0^{+\infty} \mathrm{e}^{-(\Re(1/x^k)-A)\tau}\,\mathrm{d}\tau| \leq \int_0^{+\infty} \mathrm{e}^{-(\ell_0^k - A)\tau}\,\mathrm{d}\tau = 1/(\ell_0^k - A)$. Thus, on $\mathcal{S}_\delta \cap \Delta_{\ell_0}$, the function $g^b(x)$ satisfies a global condition of exponential decay of order k: there exist constants $c, h > 0$ such that $|g^b(x)| \leq h\mathrm{e}^{-c/|x|^k}$. This implies, like in proposition 1.2.17, p. 20, that $g^b(x)$ is asymptotic to 0 and satisfies a global

s-Gevrey asymptotic condition on $\mathcal{S}_\delta \cap \Delta_{\ell_0}$: there exist constants $C'', D'' > 0$ such that, for all $N \in \mathbb{N}$ and $x \in \mathcal{S}_\delta \cap \Delta_{\ell_0}$,

$$|g^b(x)| \le C'' \Big(\frac{N}{k}\Big)^{N/k} \mathrm{e}^{-N/k} \frac{|x|^N}{D''^N}.$$

In conclusion, taking $C = \max(C', C'')$ and $D = \min(D', D'')$, we have proved that $f(x)$ satisfies condition (5.3), p. 156, on $\mathcal{S}_\delta \cap \Delta_{\ell_0}$.

Symmetrically, choosing $\overline{b}$ instead of b and the path $\gamma_{\overline{b}}$ symmetric from γ_b with respect to the real axis, we prove that $f(x)$ satisfies condition (5.3), p. 156, on $\overline{\mathcal{S}}_\delta \cap \Delta_{\ell_0}$ where $\overline{\mathcal{S}}_\delta$ stands for the symmetric sector of $\mathcal{S}_\delta$ with respect to the real axis. However, Δ_{ℓ_0} is included in $\mathcal{S}_\delta \cup \overline{\mathcal{S}}_\delta$. Hence, the conclusion on all of Δ_{ℓ_0}. This achieves the proof of that part.

▷ (ii) *implies* (i). This part was first proved by F. Nevanlinna[3] [Nev19, pp. 44–45]. Since any monomial x^n satisfies the theorem the series $\widetilde{f}(x) = \sum_{n \ge k_0} a_n x^n$ can be supposed to have valuation $k_0 > k$ (we exclude valuation equal to k when k is an integer).

From condition (5.3), p. 156, the series $\widehat{f}(\xi)$ converges with radius at least D (see Prop. 1.2.10, p. 17). Hence, its sum $\varphi(\xi)$ defines an analytic function on the Riemann surface of logarithm when $0 < |\xi| < D$.

Again set $Z = 1/x^k$ and $\zeta = \xi^k$ and denote by $Z^{1/k}$ and $\zeta^{1/k}$ the principal k^{th}-roots of Z and ζ. The function $F(Z) = f(1/Z^{1/k})$ is analytic on the half-plane $\Pi_{\ell_0^k} = \{\Re(Z) > \ell_0^k\}$.

Choose $\ell > \ell_0$ and set

$$\phi(\zeta) = \frac{1}{2\pi \mathrm{i}} \int_{\ell^k + i\mathbb{R}} F(U)\, \mathrm{e}^{\zeta U}\, \mathrm{d}U \quad \text{for all } \zeta > 0.$$

This formula makes sense. Indeed, in the new variables, condition 5.3, p. 156, becomes: for all $N \ge k_0$ and all $Z \in \Pi_{\ell_0^k}$, the function $F(Z)$ satisfies

$$\begin{aligned} |R_N(Z)| \equiv \Big| F(Z) - \textstyle\sum_{n=k_0}^{N-1} \frac{a_n}{Z^{n/k}} \Big| &\le C \Big(\tfrac{N}{k}\Big)^{N/k} \mathrm{e}^{-N/k} \frac{1}{(D|Z|^{1/k})^N} \\ &\le C' \Gamma(N/k) \frac{1}{(D'|Z|^{1/k})^N} \quad \text{(using Stirling formula)} \end{aligned} \tag{5.4}$$

for any $D' < D$ jointly with a convenient $C' > 0$. In particular, there exist constants $M_0, M_1 > 0$ such that

$$|F(Z)| \le \frac{M_1}{|Z|^{k_0/k}} \le M_0 \quad \text{for all } Z \in \Pi_{\ell_0^k} \tag{5.5}$$

[3] Frithiof Nevanlinna (1894-1977) was a Finn mathematician and Rolf's elder brother.

and since $k_0/k > 1$ this implies that $F(Z)$ belongs to $L^1(\ell^k + i\mathbb{R})$. Hence, its Fourier integral $\phi(\zeta)$ exists and is continuous with respect to $\zeta \in \mathbb{R}$.

We have to prove that:

1. The function $F(Z)$ can be written in the form

$$F(Z) = \int_0^{+\infty} \phi(\zeta) e^{-Z\zeta} \, \mathrm{d}\zeta.$$

2. The Borel series $\widehat{f}(\xi)$ converges to $\phi(\xi^k)$ for $0 < \xi < D$; hence the analytic continuation of $\phi(\xi^k)$ by $\varphi(\xi)$ to the disc $|\xi| < D$ (0 might be a branch point for the series $\phi(\zeta)$ itself).

3. The function $\phi(\zeta)$ can be analytically continued to Σ'_D.

4. The function $\phi(\zeta)$ has exponential growth of order 1 at infinity on Σ'_D.

Prove now the four steps.

1. Given $Z \in \Pi_{\ell^k}$ we enclose it in a domain Ω limited by the vertical line at ℓ^k and an arc of a circle centered at 0 with radius R as drawn in Fig. 5.14, p. 159.

By Cauchy's integral formula we can write

$$F(Z) = \frac{1}{2\pi \mathrm{i}} \int_{\partial\Omega} F(U) \frac{\mathrm{d}U}{Z-U},$$

the boundary $\partial\Omega$ of Ω being oriented clockwise.

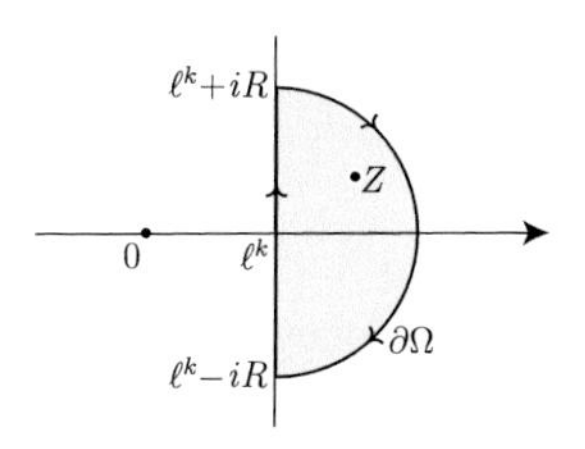

Fig. 5.14

From $|F(Z)| \leq M_1/|Z|^{k_0/k}$ (Estimate (5.5), p. 158) we deduce that the integral along the half-circle tends to zero as R tends to infinity. Hence, $F(Z) = \frac{1}{2\pi \mathrm{i}} \int_{\ell^k + i\mathbb{R}} F(U) \frac{\mathrm{d}U}{Z-U}$. Write $\frac{1}{Z-U} = \int_0^{+\infty} \mathrm{e}^{(U-Z)\zeta} \, \mathrm{d}\zeta$ so that

$$F(Z) = \frac{1}{2\pi \mathrm{i}} \int_{\ell^k + i\mathbb{R}} F(U) \int_0^{+\infty} \mathrm{e}^{(U-Z)\zeta} \, \mathrm{d}\zeta dU.$$

Fubini's theorem can be applied to the iterated integral since, using again estimate (5.5), we obtain $\left| F(U) \mathrm{e}^{-(Z-U)\zeta} \right| \leq \frac{M_1}{|U|^{k_0/k}} \mathrm{e}^{-(\Re(Z) - \ell^k)\zeta}$ with $\Re(Z) - \ell^k > 0$.

Hence, the followed formulae

$$F(Z)=\int_0^{+\infty}\phi(\zeta)\,\mathrm{e}^{-Z\zeta}\,\mathrm{d}\zeta \ \text{ with } \ \phi(\zeta)=\frac{1}{2\pi\mathrm{i}}\int_{\ell^k+i\mathbb{R}}F(U)\,\mathrm{e}^{U\zeta}\,\mathrm{d}U.$$

Moreover, $\phi(\zeta)$ is independent of $\ell>\ell_0$ (apply Cauchy's theorem to $F(U)\,\mathrm{e}^{U\zeta}$ along a rectangle with vertical sides at $\Re(Z)=\ell^k$ and $\Re(Z)=\ell'^k, \ell'^k\neq\ell^k$, and let the horizontal sides go to infinity).

2. From $F(U)=\sum_{n=k_0}^{N-1}a_n/U^{n/k}+R_N(U)$ and $\phi(\zeta)=\frac{1}{2\pi\mathrm{i}}\int_{\ell^k+i\mathbb{R}}F(U)\,\mathrm{e}^{U\zeta}\,\mathrm{d}U$ for all $\zeta>0$ we can write

$$\phi(\zeta)=\sum_{n=k_0}^{N-1}\frac{a_n}{\Gamma(n/k)}\,\zeta^{n/k}+\phi_N(\zeta) \text{ with } \ \phi_N(\zeta)=\frac{1}{2\pi\mathrm{i}}\int_{\ell^k+i\mathbb{R}}R_N(U)\,\mathrm{e}^{U\zeta}\,\mathrm{d}U.$$

By condition (5.3), p. 156, we have

$$|\phi_N(\zeta)|\leq\frac{C}{2\pi}\Big(\frac{N}{k}\Big)^{N/k}\frac{\mathrm{e}^{-N/k}}{D^N}\,\mathrm{e}^{\ell^k\zeta}\Big|\int_{\ell^k+i\mathbb{R}}\frac{\mathrm{d}U}{|U|^{N/k}}\Big|$$

while

$$\ell^{N-k}\Big|\int_{\ell^k+\mathbb{R}}\frac{\mathrm{d}U}{|U|^{N/k}}\Big|=\int_{\mathbb{R}}\frac{\mathrm{d}\tau}{\sqrt{1+\tau^2}^{N/k}}\leq\int_{\mathbb{R}}\frac{\mathrm{d}\tau}{\sqrt{1+\tau^2}^{k_0/k}}<+\infty.$$

Hence, there exists a constant $C_0>0$ such that

$$|\phi_N(\zeta)|\leq C_0\Big(\frac{N}{k}\Big)^{N/k}\mathrm{e}^{-N/k+\ell^k\zeta}\,\frac{1}{(\ell D)^N}.$$

Take $0<\zeta<D^k$ and consider the right-hand side as a function of ℓ.
The function $y(\ell)=\frac{\mathrm{e}^{\ell^k\zeta}}{\ell^N}$ for $\ell>0$ reaches its minimal value at $\ell_1(N)=\big(\frac{N}{k}\big)^{1/k}\frac{1}{\zeta^{1/k}}$ and $y(\ell_1(N))=\big(\frac{k}{N}\big)^{N/k}\mathrm{e}^{N/k}\zeta^{N/k}$. Choose $n_0=n_0(\zeta)$ so large that

$$\ell_1=\Big(\frac{n_0}{k}\Big)^{1/k}\frac{1}{\zeta^{1/k}}>\ell_0$$

For $N\geq n_0(\zeta)$ one has also $\ell_1(N)>\ell_0$ and since $\phi_N(\zeta)$ does not depend on $\ell>\ell_0$, we can take $\ell=\ell_1(N)$. Then, $\phi_N(\zeta)$ satisfies $|\phi_N(\zeta)|\leq C_0\frac{\zeta^{N/k}}{D^N}$ and tends to 0 as N tends to infinity.
Hence, $\phi(\xi^k)=\sum_{n\geq k_0}\frac{a_n}{\Gamma(n/k)}\xi^n=\varphi(\xi)$ for $0<\xi<D$ which proves that $\widehat{f}(\xi)$ is the Taylor series of $\phi(\xi^k)$ at 0 and that $\phi(\xi^k)$ can be analytically continued by $\varphi(\xi)$ to the disc $\{\xi\,;\,|\xi|<D\}$.

From now on, we denote $\varphi(\xi)=\phi(\xi^k)$.

3. Given $\zeta_0 \geq D^k$, prove that the Taylor series of $\phi(\zeta)$ at ζ_0 converges with radius D^k and converges to $\phi(\zeta)$ for ζ real.

Prove first that $\phi(\zeta)$ is infinitely derivable for $\zeta > 0$. Given $N \in \mathbb{N}^*$, let $m \in \mathbb{N}$ satisfy $k(N+1) < m \leq k(N+1)+1$. Write

$$\phi(\zeta) = \sum_{n=k_0}^{m} \frac{a_n}{\Gamma(n/k)} \zeta^{n/k} + \frac{1}{2\pi \mathrm{i}} \int_{\ell^k + i\mathbb{R}} R_{m+1}(U)\, \mathrm{e}^{U\zeta}\, \mathrm{d}U$$

and look at the νth derivative of the integrand for $1 \leq \nu \leq N$. From estimate (5.4), p. 158, we can write

$$\left| R_{m+1}(U) U^{\nu} \mathrm{e}^{U\zeta} \right| \leq C' \Gamma\big((m+1)/k\big) \frac{\mathrm{e}^{\ell^k \zeta}}{D'^{m+1}} |U|^{\nu-(m+1)/k} \leq \frac{C''}{|U|^{1+1/k}} \tag{5.6}$$

where the constant C'' is independent of ζ so long as ζ stays bounded. By Lebesgue's theorem we can then conclude that $\phi(\zeta)$ can be derivated N times under the sign of integration for any $\zeta > 0$.

To estimate the Nth derivative when $\zeta \geq \zeta_0 \geq D^k$ we write

$$\frac{\partial^N}{\partial \zeta^N} \phi(\zeta) = J_N + I_N$$

where

$$\begin{aligned} J_N &= \sum_{n=k_0}^{m} \frac{a_n}{\Gamma(n/k)} \frac{n}{k} \Big(\frac{n}{k} - 1\Big) \cdots \Big(\frac{n}{k} - N + 1\Big) \zeta^{n/k-N}, \\ I_N &= \frac{1}{2\pi \mathrm{i}} \int_{\ell^k + i\mathbb{R}} R_{m+1}(U) U^N \mathrm{e}^{U\zeta}\, \mathrm{d}U. \end{aligned}$$

From the Gevrey condition (5.3), p. 156 or (5.4), p. 158 (cf. Prop. 1.2.10, p. 17) and the fact that $|Z|^{1/k} \geq 1/\gamma$ it follows that

$$\frac{|a_n|}{\Gamma(n/k)} \leq \frac{C'}{D'^n} \Big(\frac{\gamma}{D'} \frac{\Gamma((n+1)/k)}{\Gamma(n/k)} + 1 \Big)$$

for all $n \in \mathbb{N}^*$, and since $\frac{\Gamma((n+1)/k)}{\Gamma(n/k)}$ behaves like $\left(\frac{n}{k}\right)^{1/k}$ as n tends to infinity we can conclude that, for all $B = D - \varepsilon < D'$ there exists a constant $C'' > 0$ such that

$$\frac{|a_n|}{\Gamma(n/k)} \leq \frac{C''}{B^n} \quad \text{for all } n \in \mathbb{N}^*.$$

This implies that $|J_N| \leq C''\frac{m}{k}\Big(\frac{m}{k}-1\Big)\cdots\Big(\frac{m}{k}-N+1\Big)\zeta^{-N}\sum_{n=k_0}^{m}\Big(\frac{\zeta^{1/k}}{B}\Big)^n$

and, since $\sum_{n=k_0}^{m}(\frac{\zeta^{1/k}}{B})^n \leq m\frac{\zeta^{m/k}}{B^m} \leq k(N+1+1/k)\frac{\zeta^{1+1/k}}{B^{Nk+k+1}}\zeta^N$ if $B^k \leq \zeta$, that

$$|J_N| \leq \frac{kC''}{B^{k+1}}\Gamma\Big(N+3+\frac{1}{k}\Big)\frac{\zeta^{1+1/k}}{B^{kN}} \quad \text{when } B^k \leq \zeta.$$

From estimates (5.6), p. 161 and (5.4), p. 158, we obtain

$$\begin{aligned}
|I_N| &\leq \frac{1}{2\pi}\Big|\int_{\ell^k+i\mathbb{R}}|R_{m+1}(U)|\,|U|^N \mathrm{e}^{\ell^k\zeta}\,\mathrm{d}U\Big| \\
&\leq \frac{1}{2\pi}C'\Gamma\big((m+1)/k\big)\frac{\mathrm{e}^{\ell^k\zeta}}{D'^{\,m+1}}\max\Big(\frac{1}{\ell},\frac{1}{\ell^2}\Big)\int_{\mathbb{R}}\frac{dT}{\sqrt{1+T^2}^{\,1+1/k}} \\
&\leq C'''\Gamma(N+1+2/k)\frac{\mathrm{e}^{\ell^k\zeta}}{D'^{kN}} \quad \text{for a convenient } C'''>0 \\
&\leq C'''\Gamma(N+1+2/k)\frac{\mathrm{e}^{\ell^k\zeta}}{B^{kN}}.
\end{aligned}$$

Recall that $k(N+1)<m\leq k(N+1)+1$ so that $1<m-kN+1-k\leq 2$. Hence, the term $\max\big(\frac{1}{\ell},\frac{1}{\ell^2}\big)$ and the power of the integrand.

Adding these two estimates we see that, for all positive $B<D$, there exists a constant $\alpha_B>0$ such that

$$\Big|\frac{\partial^N\phi}{\partial\zeta^N}(\zeta)\Big| \leq \alpha_B\Gamma(N+3+2/k)\frac{\mathrm{e}^{\ell^k\zeta}}{B^{kN}} \quad \text{for all } \zeta\geq B^k. \tag{5.7}$$

Hence, the Taylor series of $\phi(\zeta)$ at ζ_0,

$$\sum_{N\geq 0}\frac{1}{\Gamma(1+N)}\frac{\partial^N\phi}{\partial\zeta^N}(\zeta_0)(\zeta-\zeta_0)^N,$$

converges for $|\zeta-\zeta_0|<B^k$. Making B tend to D, we can conclude that it converges for $|\zeta-\zeta_0|<D^k$ and consequently, the Taylor series of $\phi(\zeta)$ at ζ_0 has a radius of convergence at least equal to D^k.

To prove that this Taylor series converges to $\phi(\zeta)$ we write the Taylor-Lagrange formulas

$$\begin{aligned}
\phi(\zeta) &= \sum_{p=0}^{n-1}\frac{1}{\Gamma(1+p)}\frac{\partial^p\phi}{\partial\zeta^p}(\zeta_0)(\zeta-\zeta_0)^p+\psi_n(\zeta), \\
\psi_n(\zeta) &= \frac{1}{\Gamma(1+n)}\frac{\partial^n\phi}{\partial\zeta^n}\big(\zeta_0+\theta(\zeta-\zeta_0)\big)(\zeta-\zeta_0)^n, \quad 0<\theta<1.
\end{aligned}$$

For $\zeta_0 \geq D^k$ and $\zeta \geq B^k$, then $\zeta_0 + \theta(\zeta - \zeta_0) > B^k$ and we can apply estimate (5.7), p. 162, to $\frac{\partial^n \phi}{\partial \zeta^n}(\zeta_0 + \theta(\zeta - \zeta_0))$ so that

$$\left|\psi_n(\zeta)\right| \leq \alpha_B \frac{\Gamma(n+3+2/k)}{\Gamma(1+n)} \mathrm{e}^{\ell^k \max(\zeta,\zeta_0)} \frac{|\zeta - \zeta_0|^n}{B^{kn}}$$

and $\psi_n(\zeta)$ tends to 0 as n tends to infinity as soon as $\max(B^k, \zeta_0 - B^k) < \zeta < \zeta_0 + B^k$. Therefore, the sum of the Taylor series of ϕ at any $\zeta_0 \geq B^k$ coincides with $\phi(\zeta)$ on the interval $\max(B^k, \zeta_0 - B^k) < \zeta < \zeta_0 + B^k$. This proves that $\phi(\zeta)$ admits an analytic continuation to $\Sigma'_{B\varepsilon} \cap \{\Re(\zeta) > B^k\}$. Since the two intervals $]0, D^k[$ and $]\max(B^k, \zeta_0 - B^k), \zeta_0 + B^k[$ for $\zeta_0 = D^k$, for instance, overlap this analytic continuation fit the analytic continuation by $\varphi(\zeta^{1/k})$ on $D(0, D^k) \cap \Sigma'_B \cap \{\Re(\zeta) > B^k\}$.

Letting now B tend to D allows us to extend the analytic continuation of $\phi(\zeta)$ up to Σ'_D. Hence, the analytic continuation of $\varphi(\xi)$ to the full domain σ_D.

4. Suppose $0 < B < D$ be given. Since $\varphi(\xi)$ is analytic in the disc $\mathscr{D}(0, D)$ it is bounded on the smaller disc $\mathscr{D}(0, B)$. It results that $\phi(\zeta)$ is bounded in $\mathscr{D}(0, B^k)$ and it suffices to prove the exponential estimate in the discs $\mathscr{D}(\zeta_0, B^k)$ for $\zeta_0 \geq D^k$.

The analytic continuation of ϕ to the disc $\mathscr{D}(\zeta_0, D^k)$ is given by the Taylor series

$$\phi(\zeta) = \sum_{n \geq 0} \frac{1}{\Gamma(1+n)} \frac{\partial^n \phi}{\partial \zeta^n}(\zeta_0)(\zeta - \zeta_0)^n.$$

Apply estimate (5.7), p. 162 to $\frac{\partial^n \phi}{\partial \zeta^n}(\zeta_0)$ with $B' > B$.

It follows that, on the disc $\mathscr{D}(\zeta_0, B^k)$, the function ϕ satisfies

$$\begin{aligned}
|\phi(\zeta)| &\leq \alpha_{B'} \sum_{n \geq 0} \frac{\Gamma(n+3+2/k)}{\Gamma(1+n)} \frac{|\zeta - \zeta_0|^n}{B'^{kn}} \mathrm{e}^{\ell^k \zeta_0} \\
&\leq \alpha_{B'} \sum_{n \geq 0} \frac{\Gamma(n+3+2/k)}{\Gamma(1+n)} \left(\frac{B^k}{B'^k}\right)^n \mathrm{e}^{\ell^k B^k} \mathrm{e}^{\ell^k \Re \zeta} < +\infty
\end{aligned}$$

(write $\zeta_0 = (\zeta_0 - \Re(\zeta)) + \Re(\zeta)$ and $|\zeta_0 - \Re(\zeta)| < B^k$). The estimate being valid for all $\ell > \ell_0$ we can conclude that there exist constants $K > 0$, $A > \ell_0^k$ such that

$$|\phi(\zeta)| \leq K \mathrm{e}^{A|\zeta|} \quad \text{on } \Sigma'_B.$$

Hence the result which achieves the proof of the second part of the theorem and the rest of the statement.

□

Remarks 5.3.10

▷ The proof of "(ii) implies (i)" shows that one can choose any $B < D$ and that $\varphi(\xi)$ is actually analytic on σ_D. Thus, condition (ii) implies that $\varphi(\xi)$ has exponential growth of order k on σ_D (meaning that there exist exponential estimates in restriction to any proper subdomain σ_B of σ_D). On the contrary starting from (i) with a constant B there is no similar control on D.

▷ Condition (ii) of Nevanlinna's theorem implies that $f(x)$ is a section of the sheaf $\mathscr{A}_s$ of s-Gevrey asymptotic functions: $f \in H^0(I\,;\mathscr{A}_s)$ where I is the open arc $I =]-\pi/(2k), +\pi/(2k)[$ centered at $\theta_0 = 0$ with length π/k (recall $s = 1/k$). Indeed, in any θ direction belonging to I, there exists in Δ_{ℓ_0} an open sector containing the direction θ on which f is s-Gevrey asymptotic to $\widetilde{f}$. Conversely, the existence of local s-Gevrey asymptotic conditions in any direction $\theta \in I$ does not imply a global estimate on Δ_{ℓ_0} since the constants might be unbounded when the direction θ approaches the boundaries of I. In other words, condition (ii) of Nevanlinna's theorem is stronger than the mere s-Gevrey asymptoticity of $f(x)$ on I. Also, one should notice that f does not belong to the presheaf $\overline{\mathscr{A}}_s(I)$ since there exists, in general, no sector $\mathscr{S}_{I,R}$ based on I with radius R contained in Δ_{ℓ_0}, that is, no sector on which f is even defined and analytic (cf. Exa. 2.1.25, p. 41).

▷ The statement of Nevanlinna's theorem asserts in particular that, given a series $\widetilde{f}(x)$, if a function $f(x)$ exists which satisfies estimate (5.3), p. 156, on a Fatou domain $\Delta_{\ell_0}(k)$ then it is unique. This corroborates the fact that estimate (5.3) is stronger than s-Gevrey asymptoticity on I which is insufficient to guaranty uniqueness (cf. Com. 5.1.1 on Euler example, p. 134). Moreover, for $f(x)$ to exist, it suffices that the Borel transform of $\widetilde{f}(x)$ could be analytically continued to a tubular neighborhood of $\mathbb{R}^+$ (in the $\theta_0 = 0$ direction) with exponential growth of order k.

These remarks justify the following definition.

Definition 5.3.11 (k-fine summability) *A series $\widetilde{f}(x)$ is said to be* k-fine summable *with sum $f(x)$ in the θ_0 direction if one of the equivalent two conditions is satisfied.*

(a) *There exists a Fatou petal $\Delta_{\ell_0}(k)$ bisected by θ_0 with opening π/k on which $f(x)$ satisfies, for all N and some constants $C, D > 0$, a global estimate*

$$\Big|f(x) - \sum_{n=k_0}^{N-1} a_n x^n\Big| \le C\Big(\frac{N}{k}\Big)^{N/k} \mathrm{e}^{-N/k} \frac{|x|^N}{D^N}$$

(b) *The Borel transform of $\widetilde{f}(x)$ converges and its sum $\varphi(\xi)$ can be analytically continued to a tubular neighborhood σ_{B,θ_0} of the line d_{θ_0} from 0 to infinity in the θ_0 direction with global exponential growth of order k: there exist constants $A, K > 0$ such that*

$$|\varphi(\xi)| \le K \exp\big(A|\xi|^k\big) \quad \textit{for all} \ \ \xi \in \sigma_{B,\theta_0}$$

The sum $f(x)$ of $\widetilde{f}(x)$ is then the Laplace transform of φ in the θ_0 direction.

The case $k = 1$ was introduced in volume I [MS16, Sect. 5.7] with definition (b).

Clearly, k-summability in the θ_0 direction implies k-fine summability in the same θ_0 direction with "same" sum (same but the defintion domain which is larger for the k-sum). The converse is false: uniform asymptoticity on an open sector with opening π/k does not imply asymptoticity on a sector with wider opening.

Proposition 5.3.12 *k-Borel-Laplace summability in a given θ_0 direction is equivalent to k-summability in the same θ_0 direction.*

Proof. By means of a rotation we assume again that $\theta_0 = 0$.

▷ *k-Borel-Laplace summability implies k-summability.*

Let $\widetilde{f}(x) = \sum_{n\geq k_0} a_n x^n$ with $k_0 \geq k$ be a k-Borel-Laplace summable series in the θ_0 direction. This means that its k-Borel transform $\widehat{f}(\xi) = \sum_{n\geq k_0} \frac{a_n}{\Gamma(n/k)} \xi^{n-k}$ converges with some radius $R > 0$; its sum $\varphi(\xi)$ can be analytically continued to an unlimited sector $\sigma(I)$ based on an open arc, say $I =]-\alpha, +\alpha[$, containing the $\theta_0 = 0$ direction where $\varphi(\xi)$ satisfies the following inequality for some positive constants A and K:

$$\left|\varphi(\xi)\right| \leq K \exp\left(A|\xi^k|\right).$$

Denote by $\mathscr{D} = D(0,R) \cup \sigma(I)$ the "champaign cork" union of the disc of convergence of $\widehat{f}(\xi)$ and the sector $\sigma(I)$ and choose two directions β and $-\beta$ satisfying the conditions $\beta < \pi/(2k)$ and $-\alpha < -\beta < 0 < \beta < \alpha$.

Consider first the β direction. There exists a tubular neighborhood σ_B of the line d_β from 0 to infinity in the β direction which is contained in $\mathscr{D}$ and where condition (i) of Nevanlinna's theorem holds. Nevanlinna's theorem implies then that the k-Laplace integral $f_\beta(x) = \mathscr{L}_{k,\beta}(\varphi)(x)$ of $\varphi(\xi)$ in the β direction is k-Gevrey asymptotic to $\widetilde{f}(x)$ on a Fatou petal $\Delta_\ell(k,\beta)$ bisected by β with opening π/k.
By symmetry the same result holds for the k-Laplace integral $f_{-\beta}(x) = \mathscr{L}_{k,-\beta}(\varphi)(x)$ of $\varphi(\xi)$ in the $-\beta$ direction on a Fatou petal $\Delta_\ell(k,-\beta)$ in the $-\beta$ direction with opening π/k.

From the choice of β, the intersection $\Delta_\ell(k,\beta) \cap \Delta_\ell(k,-\beta)$ of these two Fatou petals is non empty and, from Prop. 5.3.7, p. 152, the functions $f_\beta(x)$ and $f_{-\beta}(x)$ glue together into an analytic function $f(x)$ satisfying a k-Gevrey asymptotic estimate on the union $\Delta_\ell(k,\beta) \cup \Delta_\ell(k,-\beta)$. This union contains a sector bisected by $\theta_0 = 0$ with opening larger than π/k. This shows that $f(x)$ is the k-sum of $\widetilde{f}(x)$ in the θ_0 direction (cf. Def. 5.1.6, p. 137). Hence $\widetilde{f}(x)$ is k-summable in the θ_0 direction.

▷ *k-summability implies k-Borel-Laplace summability.*

Suppose we are given a k-summable series $\widetilde{f}(x) = \sum_{n\geq k_0} a_n x^n$ with $k_0 \geq k$ in the $\theta_0 = 0$ direction. This means that there exist an analytic function $f(x)$ and constants $C, D > 0$ such that

$$\left|f(x) - \sum_{n=k_0}^{N-1} a_n x^n\right| \leq C\Gamma(N/k)\, D^N |x|^N$$

on a sector $\mathcal{s} = \{x\,;\,|\arg(x)| < \pi/(2k) + \delta$ and $|x| < r\}$ bisected by $\theta_0 = 0$ with opening larger than π/k. In any θ direction satisfying $-\delta \le \theta \le \delta$ one can draw a Fatou petal $\mathcal{s}_\ell(k,\theta)$ bisected by θ, with opening π/k and contained in $\mathcal{s}$. The function $f(x)$ satisfies Nevanlinna's condition (ii) on each of these Fatou petals with same constants C and D. Nevanlinna's theorem implies then that the Borel series $\widehat{f}(\xi)$ of $\widetilde{f}(x)$ is convergent; its sum $\varphi(\xi)$ can be analytically continued to any tubular neighborhood of the lines d_θ with thickness D and moreover, it satisfies on each of these neighborhoods an exponential estimate of order k with same constants. The union of these tubular neighborhoods jointly with the disc of convergence of $\widetilde{f}(\xi)$ contains a champaign cork neighborhood of $\mathbb{R}^+$ (in the $\theta_0 = 0$ direction) on which $\varphi(\xi)$ has exponential growth of order k. This shows that the series $\widetilde{f}(x)$ is k-Borel-Laplace summable in the θ_0 direction and achieves the proof of the proposition. □

Comments 5.3.13 (On Examples 1.1.4. p. 4, 1.1.7, p. 8, and 1.1.8, p. 10)

▷ The Borel transform $\varphi(\xi) = 1/(\xi+1)$ of the Euler series $\widetilde{\mathrm{E}}(x)$ (example 1.1.4) has exponential growth of order one (and even less) in all directions. However, the sum of the Borel series cannot be continued up to infinity in the $\theta = \pi$ direction because of the pole $\xi = -1$ of φ and indeed, we saw that the Euler series $\widetilde{\mathrm{E}}(x)$ is 1-summable in all directions but the $\theta = \pi$ direction.

▷ The Borel series $\widehat{h}(\zeta)$ of example 1.1.7 has as sum the function $\varphi(\zeta) = 1/(\mathrm{e}^{-\zeta} - 2)$ which has exponential growth of order 1 in all directions. However, the function $\varphi(\zeta)$ has a line of poles $\zeta_n = -\ln(2) + 2ni\pi$, $n \in \mathbb{Z}$. Hence, we can now conclude that the series $h(z)$ is 1-summable in all directions but $\theta = \arg(-\ln(2) + 2ni\pi)$ for all $n \in \mathbb{Z}$ and their closure $\theta = \pm\pi/2$. In particular, it is not 1-summable in the sense of definition 5.1.6 (point 4), p. 137, which requires 1-summability in all directions but finitely many. This shows that solutions of difference equations, even when they are mild, can be not summable.

▷ The Borel series $\widehat{\ell}(\zeta)$ of example 1.1.8 has as sum the function $\varphi(\zeta) = \mathrm{e}^{-\zeta + \mathrm{e}^{-\zeta} - 1}$ which is an entire function with exponential growth of order 1 in all directions $\Re(\zeta) \ge 0$ and exponential growth of no order in the directions belonging to $\Re(\zeta) < 0$. Hence, the series $\widetilde{h}(z)$ is 1-summable in all directions in $\Re(z) > 0$ and not 1-summable in the other directions.

5.3.3 Tauberian Theorems

The Tauberian theorems we have in mind wish at comparing various k-sums of a given series in a given direction when several ones exist (cf. [Mal95, Théorème 2.4.2.2] and [Bal94, Thms. 2.1 and 2.2]). We begin with the following result.

Theorem 5.3.14 (Pre-Tauberian Theorem) *Given* $0 < k_1 < k_2$ *define* κ_1 *by*

$$1/\kappa_1 = 1/k_1 - 1/k_2.$$

Suppose we are given two closed arcs I_1 *and* $\widehat{I}_1$ *with same middle point* θ_0 *and respective length* $|I_1| = \pi/k_1$ *and* $|\widehat{I}_1| = \pi/\kappa_1$. *Given a formal power series* $\widetilde{f}(x) \in \mathbb{C}[[x]]$ *denote by* $\widehat{g}(\xi) = \mathscr{B}_{k_2}(\widetilde{f})(\xi)$ *its* k_2*-Borel transform* (cf. *Def. 5.3.1, p. 145*).

The following two assertions are equivalent.

(i) *The series* $\widetilde{f}(x)$ *is* k_1*-summable on* I_1 *with* k_1*-sum* $f(x)$*;*

(ii) *The series* $\widehat{g}(\xi)$ *is* κ_1*-summable on* $\widehat{I_1}$ *and its* κ_1*-sum* $g(\xi)$ *can be analytically continued to an unlimited open sector* $\widehat{\sigma}$*, containing* $\widehat{I_1}\times]0,+\infty[$*, with exponential growth of order* k_2 *at infinity.*

Moreover, $\mathscr{B}_{k_2}(f)(\xi) = g(\xi)$ *and* $\mathscr{L}_{k_2}(g)(x) = f(x)$ *in the* θ_0 *direction and the neighboring directions.*

Proof. The theorem being true (and empty) for monomials we can assume that the series $\widetilde{f}(x) = \sum_{n\geq k_0} a_n x^n$ has valuation $k_0 > k_2$.

▷ *Prove that* (i) *implies* (ii). We proceed as in the proof of Nevanlinna's theorem. By assumption, the series $\widetilde{f}(x)$ has a k_1-sum $f(x)$ on the closed arc I_1, hence, on a larger open arc. Thus, there exists a closed arc I_1' containing I_1 in its interior, there exist $r_0 > 0$ and constants $A, C > 0$ such that the estimate

$$\Big| f(x) - \sum_{n=k_0}^{N-1} a_n x^n \Big| \leq C N^{N/k_1} A^N |x|^N \tag{5.8}$$

holds for all $N \in \mathbb{N}^*$ and all x in the sector $\mathscr{S}_1' = I_1' \times]0, r_0]$.

For convenience, we normalize the Borel transform $\mathscr{B}_{k_2}$ into the classical Borel transform $\mathscr{B}_1$ of level 1. To this end, set $Z = 1/x^{k_2}$, $\zeta = \xi^{k_2}$ and $R_0 = 1/r_0^{k_2}$. In the coordinate Z, the sector $\mathscr{S}_1' = I_1' \times]0, r_0]$ is changed into a sector $\mathscr{S}_1 = J_1 \times [R_0, +\infty[$ with opening larger than π (indeed, $k_2 > k_1$). The series $\widetilde{f}(x)$ becomes the series $\widetilde{F}(Z) = \widetilde{f}(1/Z^{1/k_2})$ and the function $f(x)$ the function $F(Z) = f(1/Z^{1/k_2})$. In the coordinate ζ, the function $g(\xi)$ becomes $G(\zeta) = g(\zeta^{1/k_2})$.

Suppose first that $1/k_1 < 2/k_2$ (hence, $k_2\pi/(2k_1) < \pi$). Recall that, by assumption, we have $1/k_2 < 1/k_1$ (hence, $k_2\pi/(2k_1) > \pi/2$). After performing a rotation to normalize the direction θ_0 to 0 we get the picture in Fig. 5.15 where $J_1 = [-\omega_1, \omega_1]$ with $\omega_1 = k_2\big(\pi/(2k_1) + \varepsilon\big)$ (suppose ε chosen so small that $\omega_1 < \pi$).

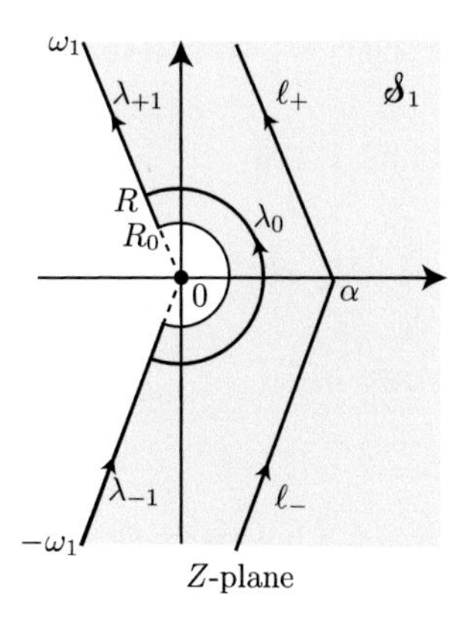

ω_1 ω Σ $-\omega$ $-\omega_1$

ζ-plane

Fig. 5.15

Denote by $g(\xi)=\mathscr{B}_{k_2,\theta_0}(f)(\xi)$ the k_2-Borel transform of $f(x)$ in direction $\theta_0=0$. The Borel path to define $G(\zeta)=g(\zeta^{1/k_2})$ can be chosen as the boundary

$$\partial\mathscr{S}_1=\partial_{-1}\cup\partial_0\cup\partial_{+1}$$

of $\mathscr{S}_1$ where ∂_0 denotes the part of the boundary which is a circular arc of radius R_0 and $\partial_{\pm 1}$ the two straight lines of $\partial\mathscr{S}_1$. Assume that $\partial\mathscr{S}_1$ is oriented as in Fig. 5.15. From Cauchy's theorem the path $\partial\mathscr{S}_1$ can equivalently be deformed into its homothetic $\lambda=\lambda_{-1}\cup\lambda_0\cup\lambda_{+1}$ where λ_0 has radius $R>R_0$ or into a broken line $\ell=\ell_-\cup\ell_+$ passing through a large enough $\alpha>0$ as shown in Fig. 5.15. Thus, γ being any of the Borel paths above, $G(\zeta)$ reads as

$$G(\zeta)=\frac{1}{2\pi\mathrm{i}}\int_\gamma F(U)\,\mathrm{e}^{\zeta U}\,dU. \tag{5.9}$$

It suffices to prove that

1. the function $G(\zeta)$ is defined and holomorphic on the unlimited open sector $\Sigma=J\times]0,+\infty[$ where $J=]-\omega,+\omega[$ with $\omega=k_2(\pi/(2\kappa_1)+\varepsilon)$;

2. the function $G(\zeta)$ has exponential growth of order one at infinity on Σ and has $F(Z)$ as Laplace transform;

3. there exist constants $C',A'>0$ such that the following estimate holds for all N and all $\zeta\in\Sigma'=J'\times]0,1/R_0[$ where $J'=[-\omega',+\omega']$, $\omega'=\omega-k_2\varepsilon/2$:

$$\Big|G(\zeta)-\sum_{n=k_0}^{N-1}\frac{a_n}{\Gamma(n/k_2)}\zeta^{n/k_2-1}\Big|\le C'N^{N/\kappa_1}A'^N|\zeta|^{N/k_2-1}. \tag{5.10}$$

Notice that $\Sigma'\Subset\Sigma$ (Def. 1.0.2, p. 2) and $\Sigma'\supsetneq[-k_2\pi/(2\kappa_1),+k_2\pi/(2\kappa_1)]\times]0,1/R_0[$.

1. In the variable Z, estimate (5.8), p. 167, reads

$$\Big|F(Z)-\sum_{n=k_0}^{N-1}\frac{a_n}{Z^{n/k_2}}\Big|\le CN^{N/k_1}\frac{A^N}{|Z|^{N/k_2}}\quad\text{for all } Z\in\mathscr{S}_1 \tag{5.11}$$

which, taking $N=k_0$, implies that there exists constants $M_0,M_1>0$ such that

$$|F(Z)|\le\frac{M_1}{|Z|^{k_0/k_2}}\le M_0,\quad\text{for all } Z\in\mathscr{S}_1. \tag{5.12}$$

In the integral of formula (5.9), p. 168, choose $\gamma=\partial\mathscr{S}_1$ and, for $j=\pm1,0$, denote

$$G_j(\zeta)=1/(2\pi\mathrm{i})\int_{\partial_j}F(U)\,\mathrm{e}^{\zeta U}\,dU$$

so that $G(\zeta)=G_{-1}(\zeta)+G_0(\zeta)+G_{+1}(\zeta)$.

The term $G_0(\zeta)$ is a Riemann integral and determines a holomorphic function for all ζ. From estimate (5.12), p. 168, and the fact that $k_0/k_2>1$ the function $F(U)$ is Lebesgue integrable on $\partial_{-1}\cup\partial_{+1}$ and, consequently, the functions $G_{\pm1}(\zeta)$ are defined and holomorphic on the half-planes $\Re(\zeta\, \mathrm{e}^{\pm i\omega_1}) < 0$ respectively. Thus, the function $G(\zeta)$ is defined and holomorphic on the sector Σ, intersection of these two half-planes.

2. Denote $G_\pm(\zeta) = 1/(2\pi \mathrm{i}) \int_{\ell_\pm} F(U) \mathrm{e}^{\zeta U} dU$ so that $G(\zeta) = G_-(\zeta) + G_+(\zeta)$ for all $\zeta \in \Sigma$. Parameterizing the paths $\ell_\pm$ by $U = \alpha + u\mathrm{e}^{\pm i\omega_1}$ we deduce from estimate (5.12), p. 168, that $G_\pm(\zeta)$ satisfies

$$|G_\pm(\zeta)| \leq M_1 \frac{\mathrm{e}^{\alpha\Re(\zeta)}}{2\pi} \int_0^{+\infty} \frac{1}{|\alpha + u\mathrm{e}^{\pm i\omega_1}|^{k_0/k_2}} du \quad \text{for all } \zeta \in \Sigma.$$

Thus, there exists a constant $c > 0$ such that $|G(\zeta)| \leq c\, \mathrm{e}^{\alpha|\zeta|}$ for all $\zeta \in \Sigma$, and this proves the exponential growth of order 1 of $G(\zeta)$ at infinity on Σ.

Prove that the Laplace transform $\mathscr{L}(G)(Z)$ in the $\theta_0 = 0$ direction is equal to $F(Z)$ on the half-plane $\{Z\,;\Re(Z) > R_0\}$. By definition, $\mathscr{L}(G)(Z)$ reads

$$\mathscr{L}(G)(Z) = \frac{1}{2\pi \mathrm{i}} \int_0^{+\infty} \Big(\int_{\partial\mathscr{S}_1} F(U)\mathrm{e}^{\zeta U} dU \Big) \mathrm{e}^{-\zeta Z} d\zeta$$

and the function $F(U)\mathrm{e}^{\zeta(U-Z)}$ is in $L^1(\partial\mathscr{S}_1 \times \mathbb{R}^+)$ when $\Re(Z) > R_0$. Indeed, parameterizing $\partial_{\pm 1}$ by $U = (R_0 + V)\mathrm{e}^{\pm i\omega_1}$ and ∂_0 by $U = R_0 \mathrm{e}^{i\theta}$ provides the estimates

$$\left| F(U)\mathrm{e}^{\zeta(U-Z)} \right| \leq \begin{cases} M_1\, \mathrm{e}^{\zeta(R_0 \cos\omega_1 - \Re(Z))} / |R_0 + V|^{k_0/k_2} & \text{for } (U,\zeta) \in \partial_\pm \times R^+ \\ M_0\, \mathrm{e}^{\zeta(R_0 - \Re(Z))} & \text{for } (U,\zeta) \in \partial_0 \times R^+. \end{cases}$$

By Fubini's theorem we can then write

$$\begin{aligned} \mathscr{L}(G)(Z) &= \frac{1}{2\pi \mathrm{i}} \int_{\partial\mathscr{S}_1} \Big(F(U) \int_0^{+\infty} \mathrm{e}^{\zeta(U-Z)} d\zeta \Big) dU \\ &= \frac{1}{2\pi \mathrm{i}} \int_{\partial\mathscr{S}_1} -\frac{F(U)}{U-Z} dU \quad (\partial\mathscr{S}_1 \text{ turns negatively around } Z) \\ &= F(Z) \quad \text{(by Cauchy's formula).} \end{aligned}$$

3. Use now the path λ with unknown radius R to be made explicit later. We want to estimate the quantity

$$\left| G(\zeta) - \sum_{n=k_0}^{N-1} \frac{a_n}{\Gamma(n/k_2)} \zeta^{n/k_2 - 1} \right| \leq Q_{-1} + Q_0 + Q_{+1}$$

where $Q_j = 1/(2\pi) \left| \int_{\lambda_j} \big(F(U) - \Sigma_{k_0}^{N-1} a_n/U^{n/k_2}\big) \mathrm{e}^{\zeta U} dU \right|$, $j = \pm 1, 0$ for all $\zeta \in \Sigma'$.

To estimate $Q_{\pm 1}$ use inequality (5.11), p. 168 to write

$$Q_{\pm 1} \leq \frac{C}{2\pi} N^{N/k_1} A^N \int_R^{+\infty} \frac{\mathrm{e}^{\Re(\zeta U)}}{|U|^{N/k_2}} d|U|$$

where $\Re(\zeta U) = |\zeta U|\cos(\theta' \pm \omega_1)$ and $|\theta' = \arg(\zeta)| \leq \omega'$. For $\zeta = |\zeta|\,\mathrm{e}^{i\theta'}$ in Σ then $\theta' + \omega_1$ satisfy $\pi/2 < |\theta' \pm \omega_1| < 3\pi/2$. Since Σ' is a proper subsector of Σ there exists $c' > 0$ such that $\cos(\theta' + \omega_1) \leq -c'$ for all $\zeta \in \Sigma'$ and therefore, $\Re(\zeta U) \leq -c'|\zeta U|$ for all $\zeta \in \Sigma'$ and $U \in \lambda_{\pm 1}$. Using this estimate and the change of variable $V = c'|\zeta|\,|U|$ in the latter integral we obtain

$$\begin{aligned} Q_{\pm 1} &\leq \frac{C}{2\pi} N^{N/k_1} A^N (c'|\zeta|)^{N/k_2 - 1} \int_{c'|\zeta|R}^{+\infty} \frac{\mathrm{e}^{-V}}{V^{N/k_2}} dV \\ &\leq \frac{C}{2\pi} N^{N/k_1} A^N (c'|\zeta|)^{N/k_2 - 1} \frac{1}{(c'|\zeta|R)^{N/k_2}} \int_0^{+\infty} \mathrm{e}^{-V} dV \\ &= \frac{C}{2\pi} N^{N/k_1} A^N (c'|\zeta|)^{-1} \frac{1}{R^{N/k_2}}. \end{aligned}$$

For each $\zeta \in \Sigma'$ choose $R = N/|\zeta|$ (then, $R > R_0$) and denote by C_1' the constant $C_1' = C/(2\pi c')$. It follows that Σ', $Q_{\pm 1}$ satisfies on Σ' the estimate

$$Q_{\pm 1} \leq C_1' N^{N/\kappa_1} A^N |\zeta|^{N/k_2 - 1} \quad \text{(recall } 1/\kappa_1 = 1/k_1 - 1/k_2\text{)}.$$

To estimate Q_0 parameterize λ_0 by $U = R\mathrm{e}^{i\theta}$ with $R = N/|\zeta|$ to obtain

$$Q_0 \leq \frac{C}{2\pi} N^{N/k_1} A^N \frac{1}{R^{N/k_2 - 1}} \int_{-\omega_1}^{\omega_1} \mathrm{e}^{\Re(N\mathrm{e}^{i\theta})} d\theta \leq \frac{C}{2\pi} N^{N/\kappa_1} N A^N |\zeta|^{N/k_2 - 1} \mathrm{e}^N 2\omega_1.$$

Choosing $A_2' > A\mathrm{e}$ (so that $NA^N\mathrm{e}^N < \mathrm{Cst}.A_2'^N$) and $C_2' = \mathrm{Cst}.C\omega_1/\pi$ we obtain

$$Q_0 \leq C_2' N^{N/\kappa_1} A_2'^N |\zeta|^{N/k_2 - 1} \quad \text{on } \Sigma'.$$

By adding these estimates and choosing $A' = A_2'$ and $C' = 2C_1' + C_2'$ it follows that estimate (5.10), p. 168 is satisfied for all $N \geq 1$ and all $\zeta \in \Sigma'$.

Suppose now that $1/k_1 \geq 2/k_2$ (hence, $k_2\pi/(2k_1) \geq \pi$).

We observe that when ω_1 passes the value π the expression of ω is changed from $\omega = -\pi/2 + \omega_1 = k_2\big(\pi/(2\kappa_1) + \varepsilon\big)$ to $\omega = 3\pi/2 - \omega_1 = \pi - k_2\big(\pi/(2\kappa_1) + \varepsilon\big)$. Hence, as ω_1 increases through the value π (also k_2/k_1 and k_2/κ_1 increase) the value of ω first increases up to $\pi/2$ (when $\omega_1 = \pi$) and then decreases. The sector Σ is no more large enough to prove the κ_1-summability of $g(\xi)$. We can pass through that difficulty by breaking $\mathcal{S}_1$ into finitely many subsectors of opening less than 2π. To this end, choose some θ_j directions and closed arcs $J_{1,j}$ of length less than 2π whose interiors make a covering of J_1. From Cauchy's theorem the Borel transforms of $F(Z)$ at the various θ_j directions are analytic continuations of each others and we can apply the previous proof to each arc $J_{1,j}$ taking now the Borel transform in the θ_j direction. This ends the proof of that part.

▷ *Prove that* (ii) *implies* (i). Again we can restrict the study to the case when $1/k_1 < 2/k_2$ (hence, $k_2\pi/(2k_1) < \pi$).

We use the same notations as before.

Assume conditions 1, 2 and 3 are satisfied. Denote by a the type of exponential growth of $G(\zeta)$ on Σ' and, up to increasing the value of R_0 (hence, up to shrinking the interval $]0,1/R_0[$), suppose $R_0 > a$.

Choose a θ_1 direction in Σ' (i.e., $|\theta_1| \leq \omega'$) and set

$$F_{\theta_1}(Z) = \int_0^{+\infty e^{i\theta_1}} G(\zeta)\,\mathrm{e}^{-\zeta Z} d\zeta.$$

Condition 2 says that $G(\zeta)$ has exponential growth, say, of type a and it follows that the above definition of $F_{\theta_1}(Z)$ defines a holomorphic function $F_{\theta_1}(Z)$ on the half-plane $\Re(\mathrm{e}^{i\theta_1}Z) > a$ bisected by $-\theta_1$ at the distance a of 0. By Cauchy's theorem the functions $F_{\theta_1}(Z)$ for the various values of $\theta_1 \in \Sigma'$ are analytic continuations from each other and we denote by $F(Z)$ the function they define on the open sector $\mathscr{S}_1$ union of the half-planes associated with all $\theta_1 \in \Sigma'$. Observe that, since the opening of Σ' is larger than $k_2\pi/\kappa_1$, the opening of $\mathscr{S}_1$ is larger than $k_2\pi/k_1$.

By means of a rotation we can assume that $\theta_1 = 0$ and use estimate (5.10), p. 168, for $\zeta > 0$. Given $0 < \beta < \varepsilon/2$, denote by Π_β the sector

$$\Pi_\beta = \{Z\,;\, |\arg Z| \leq \pi/2 - \beta \;\text{ and }\; \Re(Z) \geq R_0 > a\}.$$

Notice that the condition on β implies that the sector $\mathscr{S}'_1$ union of the Π_β's associated with the various directions $\theta_1 \in \Sigma'$ has opening more than $k_2\pi/k_1$ (and this is also the case for $\mathscr{S}'_1 \Subset \mathscr{S}_1$).

Prove that estimate (5.11), p. 168:

$$\left| F(Z) - \sum_{n=k_0}^{N-1} \frac{a_n}{Z^{n/k_2}} \right| \leq C N^{N/k_1} \frac{A^N}{|Z|^{N/k_2}}$$

(there exist $A, C > 0$) holds for all N and all Z in Π_β, for, the constants involved are valid for any choice of θ_1 in Σ'.

Since $\int_0^{+\infty} \zeta^{n/k_2-1} e^{-\zeta Z} d\zeta = \Gamma(n/k_2)/Z^{n/k_2}$ we can write

$$F(Z) - \sum_{n=k_0}^{N-1} \frac{a_n}{Z^{n/k_2}} = \int_0^{+\infty} \Big(G(\zeta) - \sum_{n=k_0}^{N-1} \frac{a_n}{\Gamma(n/k_2)} \zeta^{n/k_2-1} \Big) e^{-\zeta Z} d\zeta$$

and then,

$$\left| F(Z) - \sum_{n=k_0}^{N-1} \frac{a_n}{Z^{n/k_2}} \right| \leq P_1 + P_2 + P_3$$

where

$$\begin{cases} P_1 = \int_0^{1/R_0} \left| G(\zeta) - \sum_{n=k_0}^{N-1} \frac{a_n}{\Gamma(n/k_2)} \zeta^{n/k_2-1} \right| |e^{-\zeta Z}| d\zeta, \\ P_2 = \int_{1/R_0}^{+\infty} |G(\zeta) e^{-\zeta Z}| d\zeta, \\ P_3 = \sum_{n=k_0}^{N-1} \frac{|a_n|}{\Gamma(n/k_2)} \int_{1/R_0}^{+\infty} |\zeta^{n/k_2-1} e^{-\zeta Z}| d\zeta. \end{cases}$$

From estimate (5.10), p. 168, we obtain, on Π_β,

$$\begin{aligned} P_1 &\leq C' A'^N N^{N/\kappa_1} \int_0^{+\infty} \zeta^{N/k_2-1} e^{-\zeta \Re(Z)} d\zeta \\ &= C' \frac{A'^N}{\Re(Z)^{N/k_2}} N^{N/\kappa_1} \Gamma(N/k_2) \\ &\leq C' \frac{A'^N}{(|Z| \cos(\beta))^{N/k_2}} N^{N/\kappa_1} \Gamma(N/k_2) \quad \text{since } \Re(Z) \geq |Z| \cos\beta \text{ on } \Pi_\beta \\ &\leq C_1 \frac{A_1^N}{|Z|^{N/k_2}} N^{N/\kappa_1} N^{N/k_2} = C_1 \frac{A_1^N}{|Z|^{N/k_2}} N^{N/k_1} \quad \text{for larger constants } A_1, C_1 > 0. \end{aligned}$$

From condition 2 we obtain, on Π_β,

$$\begin{aligned} P_2 &\leq c \int_{1/R_0}^{+\infty} e^{(a-\Re(Z))\zeta} d\zeta \\ &\leq \frac{c e^{-(\Re(Z)-a)/R_0}}{(\Re(Z)-a)} \\ &\leq \frac{c e^{a/R_0}}{R_1 - a} \cdot \frac{n^n e^{-n} R_0^n}{(|Z| \cos\alpha)^n} \quad \text{for all } n > 0 \text{ and using } \Re(Z) \geq |Z| \cos\beta \\ &\leq C_2 \frac{A_2^N}{|Z|^{N/k_2}} N^{N/k_1} \end{aligned}$$

by taking $n = N/k_2$, $A_2 = \big(R_0/(ek_2 \cos\beta)\big)^{1/k_2}$ and $C_2 = ce^{a/R_0}/(R_1 - a)$ and using the inequality $N^{1/k_2} < N^{1/k_1}$.

From estimate (5.10), p. 168, we deduce (see Prop. 1.2.10, p. 17) that there exist constants A'' and $C'' > 0$ such that, for all n,

$$\left| \frac{a_n}{\Gamma(n/k_2)} \right| \leq C'' n^{n/\kappa_1} A''^n. \tag{5.13}$$

It follows that P_3 satisfies

$$\begin{aligned} P_3 &\leq \sum_{n=k_0}^{N-1} C'' n^{n/\kappa_1} A''^n \int_{1/R_0}^{+\infty} \frac{1}{R_0^{n/k_2-1}} (R_0\zeta)^{N/k_2-1} e^{-\zeta\Re(Z)} d\zeta \quad \text{since } R_0\zeta \geq 1 \\ &\leq C'' N \max\left(A''^N, A''^{k_0}\right) R_0^{(N-n)/k_2} N^{N/\kappa_1} \frac{\Gamma(N/k_2)}{(|Z|\cos\beta)^{N/k_2}} \\ &\leq C_3 \frac{A_3^N}{|Z|^{N/k_2}} N^{N/k_1} \quad \text{as before with large enough constants } A_3 \text{ and } C_3. \end{aligned}$$

Adding these three estimates we obtain

$$\left| F(Z) - \sum_{n=k_0}^{N-1} \frac{a_n}{Z^{n/k_2}} \right| \leq C \frac{A^N}{|Z|^{N/k_2}} N^{N/k_1} \quad \text{on } \Pi_\alpha \tag{5.14}$$

by setting $A = \max(A_1, A_2, A_3)$ and $C = C_1 + C_2 + C_3$. The constants A and C are independent of $\theta_1 \in \Sigma'$. Henceforth, estimate (5.14), p. 173 is valid for all $Z \in \mathcal{S}_1'$ and this proves the k_1-summability of $f(x)$ in the θ_0 direction since the opening of $\mathcal{S}_1'$ is larger than $k_2\pi/k_1$. This achieves the proof of the theorem. □

The Tauberian theorems of J. Martinet and J.-P. Ramis [MarR89, Prop. 4.3][4] are now easy corollaries of this theorem.

Corollary 5.3.15 (Martinet-Ramis Tauberian Theorem 1)

Let $0 < k_1 < k_2$ and let $I_1 \supseteq I_2$ be, respectively, a k_1-wide and a k_2-wide arc of S^1. Set $s_2 = 1/k_2$.

If a series $\widetilde{f}(x)$ is both s_2-Gevrey and k_1-summable on I_1 then it is k_2-summable on I_2 and the two sums agree on I_2..

Observe that the assertion is not trivial since, according to definition 5.1.6, p. 137, being k_2-summable on I_2, compared to being k_1-summable on I_1, is a stronger condition to be satisfied on the smaller arc I_2.

Proof. It is sufficient to prove the theorem when I_1 and I_2 are closed of length π/k_1 and π/k_2 respectively. Let θ_1 and θ_2 be the bisecting directions of I_1 and I_2; they satisfy $|\theta_1 - \theta_2| \leq \pi/k_1 - \pi/k_2 = \pi/k$. A k_1-sum $f_1(x)$ of $\widetilde{f}(x)$ exists on a larger open arc $I_{1,\varepsilon}$ containing I_1 and lives in a sector $\Delta_{1,\varepsilon}$ based on $I_{1,\varepsilon}$. By theorem 5.3.14, p. 166, the k_2-Borel transform $g(\xi)$ of $f_1(x)$ in the θ_1 direction lives on an unbounded sector σ of opening $\pi/k + \varepsilon$ bisected by θ_1 and has there exponential growth of order k_2. Moreover, $g(\xi)$ is the unique function s-Gevrey asymptotic to $\widehat{g}(\xi)$ on σ since the opening of σ is larger than $s\pi = \pi/k$. On another hand, since $\widetilde{f}(x)$ is s_2-Gevrey, the formal Borel transform $\widehat{g}(\xi)$ of $\widetilde{f}(x)$ is convergent. Its sum in the usual sense and the unique s-Gevrey asymptotic function $g(\xi)$ must necessarily agree. Denote by σ_c the union of the sector σ with the disc of convergence of $\widehat{g}(\xi)$ and keep the notation $g(\xi)$ for the function $g(\xi)$ continued to σ_c. The domain σ_c

[4] Caution: the notation s in that article corresponds to our κ_1.

contains the θ_2 direction. Using definition 5.3.8, p. 153, we can then conclude that $\widetilde{f}(x)$ is k_2-summable in the θ_2 direction.

In addition, from theorem 5.3.14, p. 166, we know that $f_1(x)$ is the analytic continuation of the k_2-Laplace transform of $g(\xi)$ in the θ_1 direction. On another hand, it follows from Cauchy's theorem that the Laplace transforms of $g(\xi)$ in the θ_1 and θ_2 directions are analytic continuations from each other. Hence, the k_2-sum $f_2(x)$ of $\widetilde{f}(x)$ coincides with $f_1(x)$ on I_2 and the proof is achieved. □

Corollary 5.3.16 (Martinet-Ramis Tauberian Theorem 2)

Let $0 < k_1 < k_2$ *and* $s_2 = 1/k_2$.

If a series $\widetilde{f}(x)$ *is both* k_1*-summable and* s_2*-Gevrey then it is convergent. A fortiori, if* $\widetilde{f}(x)$ *is* k_1*- and* k_2*-summbable for* $k_1 \neq k_2$ *then,* $\widetilde{f}(x)$ *is convergent.*

Proof. Any closed arc of length π/k_2 can be included in an arc of length π/k_1 on which $\widetilde{f}(x)$ is k_1-summable. Henceforth, by the previous corollary, $\widetilde{f}(x)$ is k_2-summable in all directions and it follows that it is convergent (cf. Rem. 5.1.8, p. 138). □

Example 5.3.17 (Leroy Series)

The Leroy series

$$\widetilde{\mathrm{L}}(x) = \sum_{n\geq 0}(-1)^n n!\, x^{2n+2}$$

is the series deduced from the Euler series $\widetilde{\mathrm{E}}(x) = \sum_{n\geq 0}(-1)^n n!\, x^{n+1}$ (cf. Exa. 1.1.4, p. 4) by substituting x^2 for x. The Leroy series is thus divergent and $1/2$-Gevrey ($k_2 = 2, s_2 = 1/2$). It satisfies the Leroy equation

$$x^3 y' + 2y = 2x^2.$$

Show that $\widetilde{\mathrm{L}}(x)$ is both 1- and 2-summable in each θ direction satisfying $|\theta| < \pi/4 \bmod \pi$ with "same" sums (meaning the same on the intersection of their domains of definition). We choose the directions θ satisfying $|\theta| < \pi/4$, the case when $|\theta - \pi| < \pi/4$ being similar. From the 1-summability of the Euler series $\widetilde{\mathrm{E}}(x)$ in any θ direction satisfying $|\theta| < \pi$ we deduce that $\widetilde{\mathrm{L}}(x)$ is 2-summable in any θ direction satisfying $|\theta| < \pi/2$ with a 2-sum $\mathrm{E}(x^2)$ defined on $|\arg(x)| < 3\pi/4$. In particular, $\mathrm{E}(x^2)$ is $1/2$- and then also 1-Gevrey asymptotic to $\widetilde{\mathrm{L}}(x)$ on $|(\arg(x)| < 3\pi/4$. Since the sector $|\arg(x)| < 3\pi/4$ is wider than π this shows that $\mathrm{E}(x^2)$ is also the (unique) 1-sum of $\widetilde{\mathrm{L}}(x)$ on $|(\arg(x)| < 3\pi/4$ (cf. Def. 5.1.6, p. 137).

We have thus showed that the Leroy series satisfies the Tauberian theorem 5.3.15, p. 173, taking $k_1 = 1, k_2 = 2, I_1 = I_2 = \{\theta\,;\,|\theta| < 3\pi/4\}$ and that, moreover, the 1-sum and the 2-sum when they both exist, agree.

Show that $\widetilde{\mathrm{L}}(x)$ is not 1-summable in any θ direction satisfying $\pi/4 \leq \theta \leq 3\pi/4 \bmod \pi$. Indeed, suppose $\pi/4 < \theta < 3\pi/4 \bmod \pi$. The 1-Borel transform $\mathscr{B}_1\widetilde{\mathrm{L}}(\xi)$ of $\widetilde{\mathrm{L}}(x)$ satisfies an equation obtained from the Leroy equation by substituting ξ (multiplication by ξ) for $x^2 d/dx$ and $d/d\xi$ (derivation w.r.t ξ) for $1/x$ and so, it satisfies the equation

$$\xi Y + 2\frac{dY}{d\xi} = 2.$$

After noticing that $\mathscr{B}_1\widetilde{\mathrm{L}}(0) = 0$ we observe that $\mathscr{B}_1\widetilde{\mathrm{L}}(\xi)$ is the Taylor series of the entire function $\Phi(\xi) = \exp(-\xi^2/4)\int_0^\xi \mathrm{e}^{t^2/4}dt$ solution of the same equation. Because of the condition $\pi/4 < \theta < 3\pi/4$ the function $\Phi(\xi)$ has exponential growth of order exactly 2 in the θ direction.

Hence, it cannot be applied a Laplace transform and the series $\widetilde{\mathrm{L}}(x)$ is not 1-summable in the θ direction, and therefore it is not 1-summable in all θ directions belonging to $\pi/4 \le \theta \le 3\pi/4 \bmod \pi$. This property is compatible with the Tauberian theorem 5.3.16, p. 174, since, as the series $\widetilde{\mathrm{L}}(x)$ is divergent, it cannot be both 1- and 2- summable in almost all directions.

5.3.4 The Borel-Laplace Summability and the Summable-Resurgence

We saw in theorem 5.2.5, p. 142 that solutions of linear differential equations with a unique level k are k-summable in any non anti-Stokes direction. In this section, we investigate more properties of such solutions called resurgence and summable-resugence in the case when $k = 1$. These notions of resurgence and summable-resurgence are precisely defined and developed in volume I, [MS16] and in [Sau05] in the case of a discrete and a one-dimensional lattice of singular points for their Borel transform, respectively; see also [CNP93]. They were introduced by J. Écalle [Éca81, Éca85] in a very general setting where they apply to a wide class of series, among which solutions of non linear ordinary differential equations or of difference equations. The case of series that are solutions of linear differential equations presents some specificity. These induce only a finite number of singular points in the Borel plane and, because of the linearity, it is unnecessary to complete this finite set of singular points into a lattice. Moreover, the theory can be given a direct approach in the Laplace plane (that is, the plane of the initial variable x) which is made explicit in [LR11, Sect. 4.4].

The aim of this section is to show that series that are solutions of a linear differential equation with the unique level one are summable-resurgent, and in particular, resurgent. Another proof, based on majorant series, is given in [LR11] which allows a better description of the singularities in the Borel plane. It is however more technical.

Be aware of the fact that being summable-resurgent, the definition of which we recall below, is a *stronger* requirement than both being summable and resurgent. Indeed, whereas summability requires exponential estimates on sectors in the Borel plane at infinity, summable-resurgence requires similar estimates on any sector of some universal covering of the Borel plane deprived of certain points.

Let D be a linear differential operator with meromorphic (convergent) coefficients and let us expand it with respect to the derivation $\delta = x^2\frac{\mathrm{d}}{\mathrm{d}x}$:

$$D = b_n(x)\delta^n + b_{n-1}(x)\delta^{n-1} + \cdots + b_0(x).$$

Without loss of generality we may assume that the coefficients $b_j(x)$ are all analytic in x with minimal valuation equal to 0 (otherwise replace D by $x^{n_0}D$ for a convenient value of $n_0 \in \mathbb{Z}$). This implies that its Newton polygon $\mathscr{N}_0(D)$ at 0 is bounded from below by the first diagonal of the axis with a non empty intersection with this diagonal.

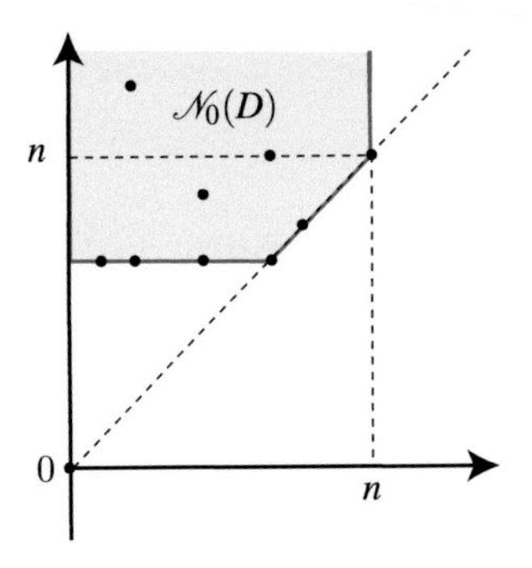

Fig. 5.16

Suppose moreover that the Newton polygon $\mathcal{N}_0(D)$ has two slopes, equal to 0 and 1, and that $\widetilde{f}(x)$ is a series solution of the equation $Dy = 0$. The other formal solutions are linear combinations of log-series $\widetilde{f}_j(x)x^{\lambda_j}$ including power series, or of log-exp-series $\widetilde{f}_j(x)x^{\lambda_j}\mathrm{e}^{q_j(1/x)}$ where $\widetilde{f}_j(x) \in \mathbb{C}[[x]][\ln x]$, $\lambda_j \in \mathbb{C}$ and where $q_j(1/x) = -a_j/x$ with $a_j \neq 0$ is a non-zero monomial of degree 1 in $1/x$.

Under these conditions the coefficients of D must satisfy $b_0(0) = 0$ and $b_n(0) \neq 0$ (cf. proof of Prop. 3.3.18, p. 84). Recall that the Stokes values of $\widetilde{f}(x)$ are, up to sign, the leading coefficients a_j of the determining polynomials $q_j(1/x)$ (cf. Def. 3.3.5, p. 77). They indicate the anti-Stokes directions for $\widetilde{f}(x)$ as well as for any series $\widetilde{f}_j(x)$ appearing in a log-series solution; we remind however that log-exp-series are associated with other Stokes values.

From theorem 5.2.5, p. 142, we know that $\widetilde{f}(x)$ is 1-summable in all non anti-Stokes directions. From the Borel-Laplace point of view this means that, given any line d_θ from 0 to infinity in a θ direction which is not anti-Stokes, its associated Borel series $\widehat{f}(\xi)$ is convergent and can be continued to a sector neighboring d_θ with exponential growth of order 1.

We prove below that such properties can be extended to a much larger domain.

▷ Suppose first that the coefficients $b_j(x)$ of D are algebraic. A priori, this means that they are polynomials in x and $1/x$. Actually, because of the above assumptions, they are polynomials in x alone. Let $\nu \geq 0$ be their maximal degree and consider the operator $D' = x^{-\nu}D$.

The operator D' reads

$$D' = B_n(1/x)\delta^n + B_{n-1}(1/x)\delta^{n-1} + \cdots + B_0(1/x)$$

with coefficients $B_j(1/x) = x^{-\nu}b_j(x)$ that are polynomials in $1/x$ of maximal degree ν. Denote $B_j(1/x) = (1/x^\nu)(\gamma_j + O(x))$.

By means of a Borel transform, D' is changed into the linear differential operator (cf. Sect. 3.3.3.2, p. 83)

$$\Delta' = B_n\Big(\frac{\mathrm{d}}{\mathrm{d}\xi}\Big)\xi^n + B_{n-1}\Big(\frac{\mathrm{d}}{\mathrm{d}\xi}\Big)\xi^{n-1} + \cdots + B_0\Big(\frac{\mathrm{d}}{\mathrm{d}\xi}\Big)$$

and the Borel transform $\widehat{f}(\xi)$ of $\widetilde{f}(x)$ satisfies the equation $\Delta'\widehat{y} = 0$. Observe that the derivatives in the operator Δ' are to the left whereas the powers $\xi^n, \xi^{n-1}, \dots$ of the variable ξ are positioned to the right.

Lemma 5.3.18 *The set $\mathscr{S}$ of the singular points of the equation $\Delta'\widehat{y} = 0$ is the set of the Stokes values of $\widetilde{f}(x)$ in the equation $Dy = 0$.*

Proof. By construction, the operator Δ' has order ν. Taking into account the relations $b_0(0) = 0$ and, $\frac{\mathrm{d}^k}{\mathrm{d}\xi^k}\xi^\ell = \xi^\ell\frac{\mathrm{d}^k}{\mathrm{d}\xi^k} +$ "lower order terms", it is of the form

$$\Delta' = \Big(\sum_{j=1}^{n} \gamma_j \xi^j\Big)\frac{\mathrm{d}^\nu}{\mathrm{d}\xi^\nu} + \text{"lower order terms"}.$$

The condition $b_n(0) \neq 0$ means $\gamma_n \neq 0$, hence the polynomial $\xi \mapsto \sum_{j=1}^n \gamma_j \xi^j$ has degree n. Therefore, the singular points of the equation $\Delta'\widehat{y} = 0$ are the zeroes of the polynomial $\sum_{j=1}^n \gamma_j \xi^j = 0$ which obviously vanishes for $\xi = 0$. However, this polynomial is also, up to a power of ξ, the 1-characteristic polynomial of D and we saw (cf. Sect. 3.3.3.3, p. 84) that the non-zero Stokes values associated with $\widetilde{f}(x)$ are given by the roots of the various characteristic polynomials. In our case, since the Newton polygon of D has no other slopes than 0 and 1, there is no other characteristic polynomial than the 1-characteristic polynomial and we can conclude that the singular points of the equation $\Delta'\widehat{y} = 0$ are all the Stokes values of the equation $Dy = 0$ including 0. □

It follows from the Cauchy-Lipschitz theorem and Grönwall's lemma that $\widehat{f}(\xi)$ can be analytically continued along any path drawn in $\mathbb{C}$ which avoids the (finite) set $\mathscr{S}$ of the singular points of the equation $\Delta'\widehat{y} = 0$. The domain to which extend the sum of the convergent series $\widehat{f}(\xi)$ is then the Riemann surface, named $\mathscr{R}_{\mathscr{S}}$, which consists of (the terminal end of) all homotopy classes in $\mathbb{C} \setminus \mathscr{S}$ of paths issuing from 0 and bypassing all points of $\mathscr{S}$. Only homotopically trivial paths are allowed to turn back to 0. The surface $\mathscr{R}_{\mathscr{S}}$ looks very much like the universal covering of $\mathbb{C} \setminus \mathscr{S}$ but the fact that 0 is not a branch point in the first sheet (we always start with a convergent power series $\widehat{f}(\xi)$). We denote by $\varphi(\xi)$ the sum of $\widehat{f}(\xi)$ as well as its analytic continuation to $\mathscr{R}_{\mathscr{S}}$.

The above property is named *resurgence* and we can state:

Lemma 5.3.19 *The series $\widetilde{f}(x)$ is* resurgent with singular support the set $\mathscr{S}$ *of the Stokes values associated with $\widetilde{f}(x)$ in the equation $Dy = 0$. Said otherwise, the Borel transform $\widehat{f}(\xi)$ of $\widetilde{f}(x)$ is convergent and can be analytically continued to the surface $\mathscr{R}_{\mathscr{S}}$.*

Let $\Delta_{\mathscr{R}_{\mathscr{S}}}$ denote a sector of $\mathscr{R}_{\mathscr{S}}$, pull back of a sector $\sigma = I \times]R, +\infty[$ of $\mathbb{C}$ (polar coordinates) where $I = \{\theta_1 < \theta < \theta_2\,; |\theta_2 - \theta_1| < 2\pi\}$ is a bounded arc of directions. Suppose moreover that σ contains no singular point of the equation (that is, $(I \times]R, +\infty[) \cap \mathscr{S} = \emptyset$). One can reach $\Delta_{\mathscr{R}_{\mathscr{S}}}$ from 0 following a $\mathscr{C}^1$-path γ of finite length in $\mathbb{C} \setminus \mathscr{S}$.

Lemma 5.3.20 *The series $\widetilde{f}(x)$ is* summable-resurgent *on $\mathscr{R}_{\mathscr{S}}$, that is, its Borel transform $\varphi(\xi)$ has exponential growth of order 1 at infinity on any sector $\Delta_{\mathscr{R}_{\mathscr{S}}}$ with bounded opening on $\mathscr{R}_{\mathscr{S}}$.*

Proof. This is a direct consequence of proposition 3.3.18, p. 84. If the opening of $\Delta_{\mathscr{R}_{\mathscr{S}}}$ were not bounded we could turn infinitely many times around ∞ adding exponential terms at each turn. Hence, the necessity to bound the opening of $\Delta_{\mathscr{R}_{\mathscr{S}}}$. □

▷ Now, consider the general case when D has meromorphic, not necessarily algebraic, coefficients.

Theorem 5.3.21 *Let D be a linear differential operator with meromorphic (convergent) coefficients at 0 and suppose D has the unique level $k = 1$. Then, any series $\widetilde{f}(x)$ solution of $Dy = 0$ is summable-resurgent with singular support the Stokes values of $\widetilde{f}(x)$ in the equation $Dy = 0$* (cf. *Def. 3.3.5, p. 77*).

Proof. The algebraisation theorem of Birkhoff (see [Bir09] or [Sib90, thm. 3.3.1]) says that to any linear differential operator D with meromorphic coefficients at 0 there exists a meromorphic transformation which changes D into an operator D' with polynomial coefficients. The two equations $Dy = 0$ and $D'y = 0$ have the same determining polynomials, hence the same set $\mathscr{S}$ of Stokes values associated with a series $\widetilde{f}(x)$ in $Dy = 0$ or associated with its image after meromorphic transformation in $D'y = 0$. The operator D' is relevant of lemma 5.3.20, p. 178. The Borel transform of a convergent series is an entire function with exponential growth of order 1 at infinity.

We thus have to prove that, if a function $\varphi(\xi)$ is defined on all of $\mathscr{R}_{\mathscr{S}}$ with exponential growth of order 1 at infinity and $g(\xi)$ is an entire function with exponential growth of order 1 at infinity then $g * \varphi$ is well defined on all of $\mathscr{R}_{\mathscr{S}}$ and has exponential growth of order 1 at infinity.

Since g and φ are both analytic near the origin 0 their convolution product is well defined near 0 by the integral $g * \varphi(\xi) = \int_0^\xi g(\xi - t)\varphi(t)\,\mathrm{d}y$. Given any path γ from 0 to ξ in $\mathbb{C} \setminus \mathscr{S}$ (but its starting point 0), the integral $\int_\gamma g(\xi - t)\varphi(t)\,\mathrm{d}y$ is well defined and determines the analytic continuation of $g * \varphi$ along γ. Changing the path γ into a homotopic one does not affect the result according to Cauchy's theorem. Hence, $g * \varphi$ is well defined on all of $\mathscr{R}_{\mathscr{S}}$. The fact that this function has exponential growth of order 1 at infinity follows from the fact that the convolution of exponentials $\mathrm{e}^{A\xi}$ and $\mathrm{e}^{B\xi}$ of order 1 is itself a combination of exponentials of order 1.

□

Remarks 5.3.22

▷ The previous theorem is valid for *all series* appearing in a formal fundamental solution of a linear differential equation with meromorphic coefficients and unique level $k = 1$ even those that are factored by a complex power of x, an integer power of $\log(x)$ or an exponential. This follows from the fact that these series are themselves solution of a (another) linear differential equation of the same type. This is easily seen on systems: in a formal fundamental solution $\widetilde{F}(x)x^L e^Q(1/x)$ the factor $\widetilde{F}(x)$ is solution of the homological system, itself with meromorphic coefficients and unique level 1. In each case, the set $\mathscr{S}$ to be considered is the set of the Stokes values associated with the selected series (cf. Def. 3.3.5 (vi)).

▷ The definition of resurgent or summable-resurgent series given above fit the definitions given in volume I [MS16, Part II] but the fact that we consider here a finite set $\mathscr{S}$ of singular points in the Borel plane instead of a lattice. This is specific to the linearity. Indeed, applying a Borel transform to a linear differential equation results in a convolution equation which contains only convolution products of resurgent functions (the unknown functions) by entire functions (the Borel transforms of the convergent coefficients in the initial equation). There is no convolution powers or products of (non entire) resurgent functions. Recall that the convolution product of two functions with a singular point at a and b respectively generates a singular point at $a+b$; hence, the necessity of considering monoïds or lattices of singular points in general. In the linear case, convolution products with entire functions generate no new singular point and all the information is concentrated at the initial singular points in $\mathscr{S}$.

▷ Resurgent series associated with a given lattice of singular points Ω (singular for their Borel transform) form an algebra, called *resurgence algebra*, in which one can perform *alien calculus* by means of new derivatives, called *alien derivatives*, related to each singular point in Ω. Resurgence algebras and alien calculus are the subject of volume I [MS16, Part II]. There, the definition of the algebra is given via the convolution product in the Borel plane, that is, after applying a Borel transform, and it requires to continue convolution products all over the resurgence surface $\mathscr{R}_\Omega$. The lattice Ω is assumed to be discrete in $\mathbb{C}$. Alien derivatives are defined via an average of some analytic continuations along various paths.

▷ In the linear case, that is, when one wants to treat only series satisfying linear differential equations, one can give definitions of an adequate resurgent algebra and alien derivatives directly in the Laplace plane as shown in [LR11, Sect. 4.4.1], escaping so to the analytical continuation of convolution products. Observe that, most often, the lattice generated in the Borel plane by the set $\mathscr{S}$ of the singular points related to a given linear differential equation is dense in $\mathbb{C}$; hence, it does not fit the hypothesis of being a discrete lattice as assumed above. Alien derivatives can be seen as the components of the logarithm of the Stokes automorphisms graded by the adjoint action of the exponential torus [MS16, Sect. 2.2.3.2]. Examples of such calculations are found in volume I [MS16, Part I, Exa. 2.43 to 2.46] and in [LR11,

Exa. 4.5 and 4.6]. They can be performed without the assumption of a unique level equal to 1.

5.4 The Fourth Approach: Wild Analytic Continuation

The fourth definition of k-summability deals with wild analytic continuation, that is, continuation of the series in the infinitesimal neighborhood of 0 (cf. Sect. 3.6, p. 113).

5.4.1 k-Wild-Summability

A Gevrey series is a germ at 0 of the sheaf $\mathscr{F}$. We call *wild analytic continuation* any of its continuations as sections of $\mathscr{F}$. A series $\widetilde{f}(x)$ which is k-summable on a k-wide arc I can be wild analytically continued to a domain containing the disc $D(0,k)$ and the sector $\{(\theta,k')\,;\,\theta\in I \text{ and } 0<k'\leq+\infty\}$. These conditions are not quite sufficient to characterize k-summable series on I since the set of global sections of $\mathscr{F}$ over the open disc $D(0,k)$ is isomorphic to

$$\mathbb{C}[[x]]_{s+} := \varprojlim_{\varepsilon\to 0+} \mathbb{C}[[x]]_{s+\varepsilon} = \bigcap_{\varepsilon>0} \mathbb{C}[[x]]_{s+\varepsilon} \supset \mathbb{C}[[x]]_s$$

and is thus, bigger than $\mathbb{C}[[x]]_s$. As for the set of global sections of $\mathscr{F}$ over the closed disc $\overline{D}(0,k)$, it is isomorphic to

$$\mathbb{C}[[x]]_{s-} := \varinjlim_{\varepsilon\to 0+} \mathbb{C}[[x]]_{s-\varepsilon} = \bigcup_{\varepsilon>0} \mathbb{C}[[x]]_{s-\varepsilon} \subset \mathbb{C}[[x]]_s$$

and smaller than $\mathbb{C}[[x]]_s$ (cf. Prop. 3.6.3, p. 116).

The right domain lies between $D(0,k)$ and $\overline{D}(0,k)$. It can be made explicit in the sheaf space $(X^k,\mathscr{F}^k)$ since, indeed, the set of global sections of $\mathscr{F}^k$ over the closure $\overline{D}(0,\{k,0\})$ in X^k of the open disc $D(0,\{k,0\})$ is isomorphic to $\mathbb{C}[[x]]_s$ (cf. Prop. 3.6.5, p. 119).

We can then state the following new definition of k-summability:

Definition 5.4.1 (k-Wild-Summability)

Let I be a k-wide arc of S^1 (cf. *Def. 5.1.2, p. 135*).

A series $\widetilde{f}(x)=\sum_{n\geq 0}a_nx^n$ is k-wild-summable *on I if it can be wild analytically continued to a domain containing the closed disc $\overline{D}(0,\{k,0\})$ and the sector $I\times\,]0,+\infty]$.*

We call such a domain a k-sector in X^k.

The definition above, being the exact translation of Ramis-Sibuya definition of k-summability, we can state:

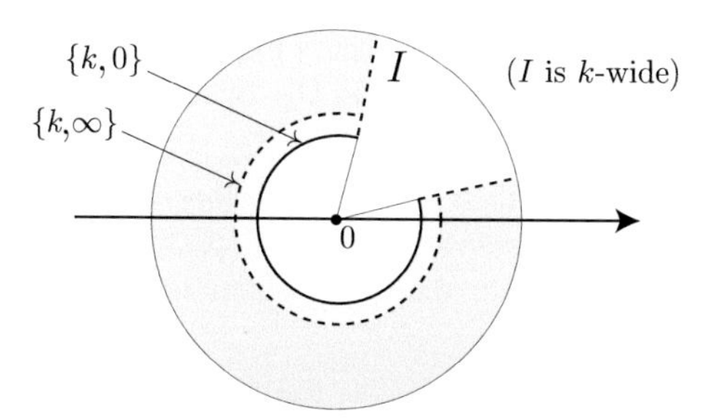

Fig. 5.17. Domain for a k-sum in X^k (in white)

Proposition 5.4.2 *k-summability on I and k-wild-summability on I are equivalent.*

5.4.2 Applications

1. *Tauberian Theorems* — Let us revisit the Martinet-Ramis tauberian theorem 1 (Cor. 5.3.15, p. 173) from the point of view of wild analytic continuation.

Consider a k_1-summable series $\widetilde{f}(x)$ on I_1. In the viewpoint of wild analytic continuation this property translates in the space $(X^{k_1}, \mathscr{F}^{k_1})$ (cf. Sect. 3.6.2 p. 116) as the condition that the series $\widetilde{f}(x)$ admits a continuation as a section of the sheaf $\mathscr{F}^{k_1}$ to the k_1-sector $\boldsymbol{\Delta}_{k_1,I_1} = \overline{D}(0, \{k_1, 0\}) \cup \big(I_1 \times]0, +\infty]\big)$ (see Fig. 5.17). The fact that it is k_2-summable on I_2 has a similar interpretation in the space $(X^{k_2}, \mathscr{F}^{k_2})$. To interpret both we need to work in the space $(X^{k_1,k_2}, \mathscr{F}^{k_1,k_2})$.

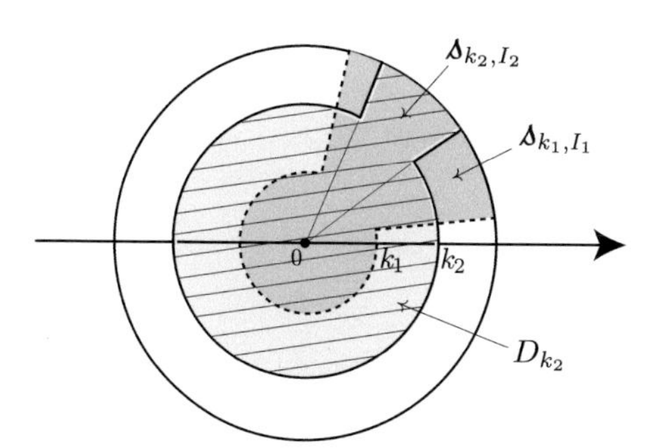

Fig. 5.18. $\boldsymbol{\Delta}_{k_1,I_1}$ (in green) and $\boldsymbol{\Delta}_{k_2,I_2}$ (hachured) in X^{k_1,k_2}

The Tauberian theorem says that, given $k_1 < k_2$ and $I_2 \subset I_1$, the fact that $\widetilde{f}(x)$ be k_1-summable on I_1 (i.e., that it can be continued to the k_1-sector $\boldsymbol{\Delta}_{k_1,I_1}$) and that it be also s_2-Gevrey (i.e., that it can be continued to the disc $D_{k_2} = \overline{D}(0, \{k_2, 0\})$) implies that it can be continued to the k_2-sector $\boldsymbol{\Delta}_{k_2,I_2}$. Clearly, the sector $\boldsymbol{\Delta}_{k_2,I_2}$ is included in $\Delta_{k_1,I_1} \cup D_{k_2}$. The theorem asserts that, on the intersection $\Delta_{k_1,I_1} \cap D_{k_2}$, the two continuations agree. This is true on D_{k_1} since there is a unique continuation of $\widetilde{f}(x)$ to D_{k_1} (cf. Cor. 5.2.2 p. 140). On $\left(D_{k_2} \setminus D_{k_1}\right) \cap \Delta_{k_1,I_1}$ the compatibility of the two continuations means that their difference belongs to $H^0\left(I_1, \mathscr{A}^{\leq -k_1}/\mathscr{A}^{\leq -k_2}\right)$. The relative Watson's lemma below (Thm. 7.2.1 p. 200) asserts that such a space reduces to the null section. Hence, the two continuations agree and define a k_2-sum of $\widetilde{f}(x)$ on I_2.

2. *Functions of k-summable series* — Let us prove that analytic functions of k-summable series are k-summable series.

Proposition 5.4.3 *Let be given a k-wide arc I and r series $\widetilde{f}_1(x), \dots, \widetilde{f}_r(x)$ which are k-summable on I with k-sums $f_1(x), \dots, f_r(x)$ respectively.*
Assume that $\widetilde{f}_1(0) = \cdots = \widetilde{f}_r(0) = 0$.
If $g(x, y_1, \dots, y_r)$ is an analytic function on a neighborhood of 0 in $\mathbb{C}^{r+1}$ then, the series $g(x, \widetilde{f}_1(x), \dots, \widetilde{f}_r(x))$ is k-summable on I with k-sum $g(x, f_1(x), \dots, f_r(x))$.

Proof. According to Prop. 1.2.6, p. 16, the expression $g(x, \widetilde{f}_1(x), \dots, \widetilde{f}_r(x))$ determines a well-defined s-Gevrey series, hence, a germ at 0 of the sheaf $\mathscr{F}^k$ which can be continued to the closed disc $\overline{D}(0, \{k, 0\})$.

The series $\widetilde{f}_1(x), \dots, \widetilde{f}_r(x)$ being k-summable on I and vanishing at $x = 0$ can be continued to the sector $I \times]0, +\infty]$ with values in an arbitrary small neighborhood of 0. The function g being holomorphic on a neighborhood of 0 the series $g\left(x, \widetilde{f}_1(x), \dots, \widetilde{f}_r(x)\right)$ can also be continued to the sector $I \times]0, +\infty]$ with analytic continuation $g\left(x, f_1(x), \dots, f_r(x)\right)$. □

3. *Summability of solutions of differential equations* — Let $\widetilde{f}(x)$ be a series satisfying a linear differential equation (or system).

The fences to the wild analytic continuation of $\widetilde{f}(x)$ in X as well as in any space $X^k, X^{k_1,k_2}, \dots$ are the big points of the exponentials $\exp(q_j(1/x))$ for all $j \in J$ appearing in a formal fundamental solution. Indeed, when a direction passes a big point it exits the definition domain of the associated exponential and flat terms become undefined.

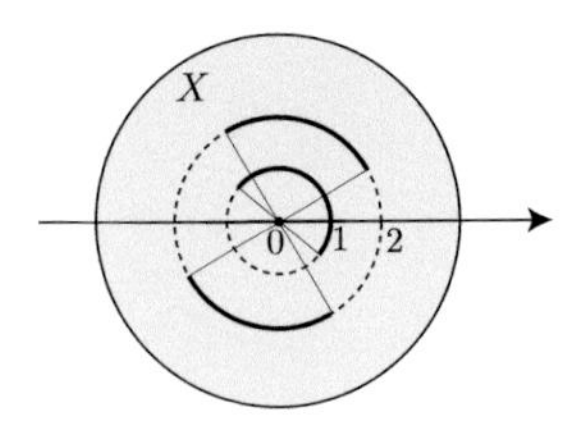

Fig. 5.19. Big points of an exponential of order 1 (a half-circle) and of an exponential of order 2 (two quarters of a circle)

Recall that, in X, the big points associated with an exponential of degree k are the closed arcs of length π/k bisected by the anti-Stokes directions of the exponential (directions of maximal decay) and lying on the circle of radius k in X. In X^k the big points are arches based on the previous arcs.

Suppose a series $\widetilde{f}(x)$ satisfies a linear differential equation or system to which are associated the two exponentials of figure 5.19 and these exponentials only. Then, $\widetilde{f}(x)$ is k-summable on any k-sector containing none of the big points drawn on figure 5.19. Moreover, it is (k_1, k_2)-summable on any (k_1, k_2)-sector containing none of the three big points. In particular, one can check easily that it is $(1,2)$-summable in almost all directions (here, all directions but the three anti-Stokes directions).

Chapter 6
Tangent-to-Identity Diffeomorphisms and the Birkhoff Normalization Theorem

Abstract Tangent-to-identity diffeomorphisms at infinity in $\mathbb{C}$ are studied in volume I as an illustration of the resurgent techniques. Here, we revisit the case of germs conjugated to the translation to show that the conjugation series is 1-summable. The result is obtained as a consequence of the Ramis-Sibuya theorem jointly with a normalization theorem by Birkhoff and Kimura which we prove.

6.1 Introduction

This chapter deals with the conjugation of tangent-to-identity germs of diffeomorphisms at 0. It aims at showing another example (not solution of a differential equation) where the Gevrey cohomological analysis is also efficient.

We consider the by-now classical case of a germ of "translation" (at the origin on the Riemann sphere)

$$g:\ x \longmapsto g(x) = \frac{x}{1+x}\cdot$$

As a homography, g is defined over the whole Riemann sphere $\overline{\mathbb{C}}$. In the chart of infinity, setting $z = 1/x$ and $G(z) = 1/g(x)$, the germ g reads

$$G : z \longmapsto G(z) = z+1$$

hence, the name of translation.

Convention. — As previously, we denote by x the coordinate about 0 and by $z = 1/x$ the coordinate about infinity. We denote by the same letter a given germ in the chart of 0 and in the chart at infinity, using a small letter at 0 and the corresponding capital one at infinity.

In this context, the formal and meromorphic gauge transformations of the classification of linear differential systems are replaced by formal and convergent tangent-

M. Loday-Richaud, *Divergent Series, Summability and Resurgence II*, Lecture Notes in Mathematics, DOI 10.1007/978-3-319-29075-1_6

to-identity diffeomorphisms $\widetilde{h}(x) = x + \sum_{n\geq 2} c_n x^n$ acting on g by conjugation, that is, by changing g into $\widetilde{h}^{-1} \circ g \circ \widetilde{h}$.

Definition 6.1.1 *A germ f is* formally conjugated (*or* analytically conjugated) *to g if there exists a formal (or a convergent) tangent-to-identity diffeomorphism*

$$\widetilde{h}(x) = x + \sum_{n\geq 2} c_n x^n$$

satisfying the conjugation equation

$$\boxed{\widetilde{h} \circ f = g \circ \widetilde{h}} \tag{6.1}$$

One can check that such an $\widetilde{h}$ exists if and only if f has the form

$$f(x) = x - x^2 + x^3 + \sum_{n\geq 4} a_n x^n.$$

The germ $f(x)$ is thus non-degenerate in the sense that $\mathrm{d}^2 f/\mathrm{d}x^2(0) \neq 0$. The conjugation germ

$$\widetilde{h}(x) = x + \sum_{p\geq 2} c_p x^p$$

is unique modulo the choice of c_2 and we can choose $c_2 = 0$. In the chart of infinity, setting $z = 1/x$ and $F(z) = 1/f(x)$, the condition reads

$$F(z) = z + 1 + \sum_{n\geq 2} \frac{A_n}{z^n} \quad \text{(observe } A_1 = 0)$$

and $\widetilde{H}(z) = 1/\widetilde{h}(1/z)$ is unique in the form

$$\widetilde{H}(z) = z + \sum_{p\geq 1} \frac{C_p}{z^p}.$$

The coefficient $-A_1$ is commonly called the *resiter* ρ of the germ $F(z)$; it is equal to 0 in the case here considered and it can be read in the coefficient of x^3 in the germ at the origin $f(x) = x - x^2 + (\rho + 1)x^3 + \ldots$. Notice that, in its formal class, g has the particularity of being best behaved with respect to iteration and, thus, plays the role of a *normal form.*

From now on, the diffeomorphisms $\widetilde{h}(x) = x + \sum_{p\geq 2} c_p x^p$ by which we conjugate are supposed to satisfy

$$\boxed{c_2 = 0}$$

We denote by

$$\widetilde{\mathbb{G}} = \Big\{ x + \sum_{n\geq 3} c_n x^n \in \mathbb{C}[[x]] \Big\}$$

the group of germs of formal tangent-to-identity diffeomorphisms of $\mathbb{C}$ at 0 satisfying $c_2 = 0$ and endowed with composition and by

$$\mathbb{G} = \Big\{x + \sum_{n\geq 3} c_n x^n \in \mathbb{C}\{x\}\Big\}.$$

the subgroup of convergent germs of $\widetilde{\mathbb{G}}$.

With this normalization, a conjugation map $\widetilde{h}$ when one exists is unique. It might be divergent although f and g are both convergent. One can prove, for instance, that a sufficient condition for $\widetilde{h}$ to be divergent is that f be an entire function.

Like for linear differential systems the analytic classification of the conjugation classes of diffeomorphisms is performed inside each formal class with a given normal form, here g. Our aim is not to extensively develop that classification but to give a proof of the main point in the given example of the translation g, that is to say, to prove that the conjugation maps $\widetilde{h}$ of g are 1-summable series.

A natural approach consists in analyzing the Borel transform of $\widetilde{h}(x)$ following so J. Écalle [Éca74]. This approach is thoroughly developed in volume I [MS16, Sect. 7]) including the meromorphic classification by means of the bridge equation. We choose to develop here the sectorial approach due to Kimura [Kim71, Thm. 6.1] (see also [Bir39, première partie, § 5]); the proof is based on the Ramis-Sibuya theorem (Thm. 5.2.1, p. 140) after constructing an adequate 1-quasi-sum.

We consider the following sheaves:

▷ $\mathscr{G} = \{f \in \mathscr{A}\,;\, Tf \in \widetilde{\mathbb{G}}\}$ the subsheaf of $\mathscr{A}$ consistng of (normalized) tangent-to-identity germs of diffeomorphisms (Recall that $\mathscr{A}$ is the sheaf over S^1 of germs of asymptotic functions at 0; cf. Sect. 2.1.5, p. 41). Equipped with the composition law, $\mathscr{G}$ is a sheaf of non commutative groups.

▷ $\mathscr{G}^{<0} = \{f \in \mathscr{G}\,;\, Tf = id\}$ the subsheaf of $\mathscr{G}$ consisting of its flat germs ("flat" in a multiplicative context means "asymptotic to identity"). $\mathscr{G}^{<0}$ is a subsheaf of groups of $\mathscr{G}$.

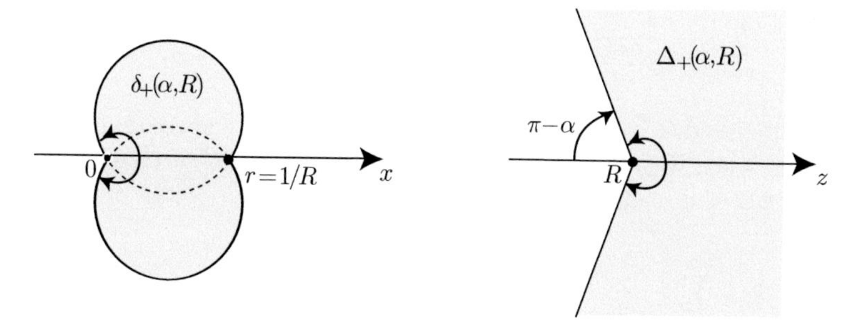

Fig. 6.1

Given $0 < \alpha < \pi$ we consider in the chart of infinity the sectors

$$\Delta_+(\alpha,R) = \{z;\, -\alpha < \arg(z-R) < \alpha\},$$
$$\Delta_-(\alpha,R) = \{z;\, \pi-\alpha < \arg(z+R) < \pi+\alpha\}.$$

$\Delta_+(\alpha,R)$ and $\Delta_+(\alpha,R)$ are symmetric to each other with respect $z=0$.

We denote by $\delta_+(\alpha,R)$ and $\delta_-(\alpha,R)$ their image in the coordinate $x = 1/z$.

Given g a germ of diffeomorphism we denote its pth power of composition by

$$g^p = \underbrace{g \circ g \circ \cdots \circ g}_{p \text{ times}}.$$

6.2 The Birkhoff-Kimura Sectorial Normalization theorem

Although we state the theorem in a chart of 0 (coordinate x) as we are use to do it, it is worth to perform the proof in the chart of infinity (coordinate z), taking thus benefit of the very simple expression of $G(z)$. Let us start with a technical lemma.

Lemma 6.2.1 *Let $0 < \alpha_0 < \pi$ and $R_0 > 1$ be given. For all $m \in \mathbb{N}^*$, there exists a constant $c > 0$ which depends on α_0 and m but not on R_0 such that,*

$$\sum_{p\geq 0} \frac{1}{|z+p|^{m+1}} \leq \frac{c}{|z|^m} \quad \textit{for all } z \in \Delta_+(\alpha_0,R_0).$$

Proof. The proof is elementary. We compare the sum to an integral as soon as possible, i.e., as soon as the general term of the series decreases and we estimate the extra terms.

Given $z \in \Delta_+(\alpha_0,R_0)$ let us denote by $p(z)+1 \geq 0$ the smallest integer such that $\Re(z+p(z)+1) > 0$. Notice that $p(z) \leq \max\big(0, |z|\cos(\pi-\alpha_0)\big)$ and $|z| \geq R_0 \sin\alpha_0$ for all z in $\Delta_+(\alpha_0,R_0)$. We split the series into

$$\sum_{p\geq 0} \frac{1}{|z+p|^{m+1}} = \sum_{p=0}^{p(z)+1} \frac{1}{|z+p|^{m+1}} + \sum_{p\geq p(z)+2} \frac{1}{|z+p|^{m+1}}.$$

We claim first that $|z+p| \geq |z|\sin\alpha_0 > 0$ for all $p \in \mathbb{N}$ and $z \in \Delta_+(\alpha_0,R_0)$; for, $|z+p| \geq |z|$ when $\Re(z) \geq 0$ and $|z+p| \geq \Im(z) = |z|\sin\theta \geq |z|\sin\alpha_0$ when $\Re(z) < 0$ since then $\pi/2 < \theta < \alpha_0$. It follows that

$$\sum_{p=0}^{p(z)+1} \frac{1}{|z+p|^{m+1}} \leq \frac{1}{|z|^{m+1}} + \frac{p(z)+1}{(|z|\sin\alpha_0)^{m+1}} \leq \frac{c_1}{|z|^m}$$

for a constant c_1 depending on α_0 and m but not on $R_0 \geq 1$.

Indeed, we have

$$\frac{1}{|z|}+\frac{p(z)+1}{|z|(\sin\alpha_0)^{m+1}}\leq\frac{1}{R_0\sin\alpha_0}+\frac{\cos(\pi-\alpha_0)}{(\sin\alpha_0)^{m+1}}+\frac{1}{R_0(\sin\alpha_0)^{m+2}}$$

and, since $R_0>1$, we can choose $c_1=3/(\sin\alpha_0)^{m+2}$.

Starting from $p=p(z)+1$ the function $p\mapsto\frac{1}{|z+p|^{m+1}}$ decreases and we have

$$\begin{aligned}\sum_{p\geq p(z)+2}\frac{1}{|z+p|^{m+1}}&\leq\int_{p(z)+1}^{+\infty}\frac{\mathrm{d}p}{|z+p|^{m+1}}=\int_0^{+\infty}\frac{\mathrm{d}q}{|z+p(z)+1+q|^{m+1}}\\&=\frac{1}{|z+p(z)+1|^m}\int_0^{+\infty}\frac{\mathrm{d}r}{(1+r)^{m+1}}\leq\frac{c_2}{|z|^m}\end{aligned}$$

for a constant $c_2=1/(\sin\alpha_0)^m$; indeed, $|z+p(z)+1|\geq|z|\sin\alpha_0$ and since $m\geq1$, we can write

$$\int_0^{+\infty}\frac{\mathrm{d}r}{(1+r)^{m+1}}\leq\int_0^{+\infty}\frac{\mathrm{d}r}{(1+r)^2}=1.$$

Hence, the result if one chooses the constant $c=c_1+c_2$. □

Theorem 6.2.2 (Birkhoff-Kimura Sectorial Normalization)
Let φ be a flat diffeomorphism over a proper sub-arc $I^+_{\alpha_0}=]-\alpha_0,+\alpha_0[$ of S^1 (that is, $0<\alpha_0<\pi$ and $\varphi\in H^0(I^+_{\alpha_0};\mathscr{G}^{<0})$).

Then, the diffeomorphism $g_1=\varphi\circ g$ belongs to $H^0(I^+_{\alpha_0};\mathscr{G})$ and is uniquely conjugated to g via a section of $\mathscr{G}^{<0}$: there exists a unique $\phi_+\in H^0(I^+_{\alpha_0};\mathscr{G}^{<0})$ such that

$$\phi_+\circ g_1=g\circ\phi_+\quad on\ I^+_{\alpha_0}.$$

Symmetrically, denote by $I^-_{\alpha_0}=[-\alpha_0+\pi,\alpha_0+\pi]$ the arc opposite to $I^+_{\alpha_0}$ on S^1 and suppose that $\varphi\in H^0(I^-_{\alpha_0};\mathscr{G}^{<0})$. Then, there exists a unique $\phi_-\in H^0(I^-_{\alpha_0};\mathscr{G}^{<0})$ such that

$$\phi_-\circ g_1=g\circ\phi_-\quad on\ I^-_{\alpha_0}.$$

Proof. We make the proof over $I^+_{\alpha_0}$. The proof on $I^-_{\alpha_0}$ is similar when applied to g^{-1} and g_1^{-1}. The fact that $g_1=\varphi\circ g$ be a diffeomorphism on $I^+_{\alpha_0}$ and have a Taylor expansion is clear since so do φ and g. Its Taylor expansion is equal to $Tg_1=T\varphi\circ Tg=\mathrm{id}\circ Tg=Tg$. Turn now to the variable z and denote by the corresponding capital letters the diffeomorphisms in the chart of infinity.

Given $\alpha<\alpha_0$ choose $\alpha_1\in]\alpha,\alpha_0[$ and $R_1>1$ so that $G_1(z)$ be well defined on $\Delta_+(\alpha_1,R_1)$. Denoting $K(z)=G_1(z)-G(z)$ and $\phi_+(z)=z+\psi_+(z)$ the condition $\phi_+\circ g_1=g\circ\phi_+$ becomes $K(z)+\psi_+\circ G_1(z)-\psi_+(z)=0$. A solution will be given by $\psi_+(z)=\sum_{p\geq0}K\circ G_1^p(z)$ if we prove that the series $\sum_{p\geq0}K\circ G_1^p(z)$ converges to a holomorphic function asymptotic to 0 at infinity.

▷ Given $R > R_1 + 2$, the function $K(z)$ being asymptotic to 0 on $\Delta_+(\alpha_1, R_1)$ it satisfies: for all $m \in \mathbb{N}$, there exists $a > 0$ such that

$$\big|K(z)\big| \leq \frac{a}{|z|^{m+1}} \quad \text{on} \quad \Delta_+(\alpha, R_1 + 1) \supset \Delta_+(\alpha, R - 1). \tag{6.2}$$

The constant a depends on m, α_1 and R_1 but not on R. Below, $R > R_1 + 2$ will be chosen conveniently large.

▷ Prove that, there exists a constant $A \geq a$ and independent of R such that,

$$\sup_{\substack{|z-z'|\leq \sin\alpha \\ z\in\Delta_+(\alpha,R)}} \big|K(z')\big| \leq \frac{A}{|z|^{m+1}}. \tag{6.3}$$

Indeed, the conditions $z \in \Delta_+(\alpha, R)$ and $|z - z'| \leq \sin\alpha$ imply $z' \in \Delta_+(\alpha, R-1)$.

We can then apply condition (6.2) to yield

$$\big|K(z')\big| \leq \frac{a}{|z'|^{m+1}} \leq \frac{a}{\big(|z| - \sin\alpha\big)^{m+1}} \leq \frac{A}{|z|^{m+1}}$$

with $A = a\big(\frac{R_1}{R_1-1}\big)^{m+1}$ since

$$\sup_{z\in\Delta_+(\alpha,R)} \frac{|z|}{|z| - \sin\alpha} \leq \sup_{|z|\geq R_1 \sin\alpha} \frac{|z|}{|z| - \sin\alpha} = \frac{R_1}{R_1 - 1}.$$

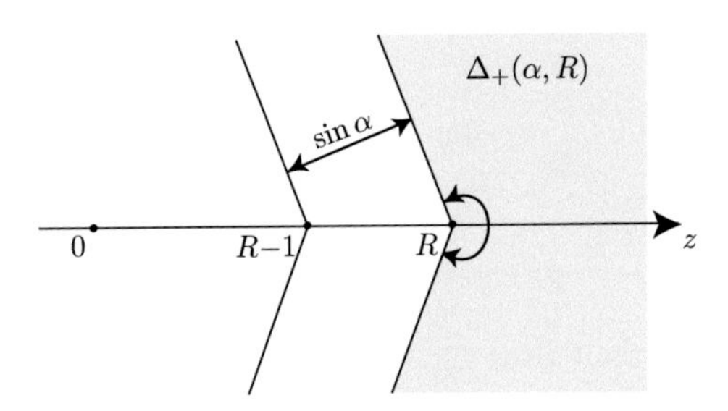

Fig. 6.2

▷ Choose R so large that, for all $m \geq 1$,

$$A \sum_{p\geq 0} \frac{1}{|z+p|^{m+1}} \leq \sin\alpha \quad \text{for all} \;\; z \in \Delta_+(\alpha, R). \tag{6.4}$$

Such a choice is possible. Indeed, from lemma 6.2.1, p. 188, applied to $\Delta_+(\alpha_1, R_1)$, we obtain

$$A \sum_{p\geq 0} \frac{1}{|z+p|^{m+1}} \leq \frac{Ac}{|z|^m}$$

and the constant Ac does not depend on R.

Now, we obtain $\max_{z\in\Delta_+(\alpha,R)} \frac{Ac}{|z|^m} = \frac{Ac}{(R\sin\alpha)^m}$. Assuming R large enough so that $R\sin\alpha > 1$ then, $\frac{Ac}{(R\sin\alpha)^m} \leq \frac{Ac}{R\sin\alpha}$ which is independent of m and can be made arbitrarily small by choosing R large.

▷ Prove by induction on p that, for all $p \geq 1$ and all $z \in \Delta_+(\alpha,R)$,

$$\left|G_1^p(z) - (z+p)\right| \leq A\sum_{q=0}^{p-1} \frac{1}{|z+q|^{m+1}} \tag{6.5}$$

(recall the notation $g^p = \underbrace{g\circ g\circ\cdots\circ g}_{p}$ times).

When $p=1$, the inequality reads $K(z) \leq A/|z|^{m+1}$ and follows from condition (6.2), p. 190, with $a < A$. Suppose condition (6.5) valid up to p. Then, from condition (6.4), we get $|G_1^p(z) - (z+p)| \leq \sin\alpha$.

This implies that:

(i) $\lim_{p\to\infty} G_1^p(z) = \infty$ for all $z \in \Delta_+(\alpha,R)$;

(ii) $G_1^p(z) \in \Delta_+(\alpha,R)$ since $z+p \in \Delta_+(\alpha,R+p) \subset \Delta_+(\alpha,R+1)$;

(iii) $\left|K\circ G_1^p(z)\right| \leq \frac{A}{|z+p|^{m+1}}$ (Estimate (6.3), p. 190, applied to $z' = G_1^p(z)$ and $z+p$ for z).

Since $G_1^p(z) \in \Delta_+(\alpha,R)$ it can be applied $G_1 = G+K$. We can then write

$$G_1^{p+1}(z) = G\big(G_1^p(z)\big) + K\big(G_1^p(z)\big) = G_1^p(z) + 1 + K\circ G_1^p(z)$$

from which we deduce $G_1^{p+1}(z) - (z+p+1) = G_1^p(z) - (z+p) + K\circ G_1^p(z)$. Applying the recurrence hypothesis and condition (iii) at rank p we obtain

$$\left|G_1^{p+1}(z) - (z+p+1)\right| \leq A\sum_{q=0}^{p-1} \frac{1}{|z+q|^{m+1}} + \frac{A}{|z+p|^{m+1}}$$

which is condition (6.5), p. 191, at rank $p+1$.

▷ Conclude on ψ_+. Condition (iii) for all p and $m \geq 1$ proves that the series

$$\sum_{p\geq 0} K\circ G_1^p(z)$$

converges uniformly on compact sets of $\Delta_+(\alpha,R)$. The functions $K\circ G_1^p$ being holomorphic, the sum $\Psi_+(z) = \sum_{p\geq 0} K\circ G_1^p(z)$ is holomorphic on $\Delta_+(\alpha,R)$. Moreover, for all $m \in \mathbb{N}^*$ and all $z \in \Delta_+(\alpha,R)$ (Recall that α and R do not depend on m) , there exist constants A and $c > 0$ such that

$$\begin{aligned} \left|\Psi_+(z)\right| &\leq \sum_{p\geq 0} \frac{A}{|z+p|^{m+1}} && \text{(Cond. (iii))} \\ &\leq \frac{Ac}{|z|^m} && \text{(Lem. 6.2.1, p. 188 for } \Delta_+(\alpha,R)) \end{aligned}$$

which shows that $\psi_+(z)$ is asymptotic to 0 at infinity.

▷ To prove the uniqueness of the solution it suffices to prove that the equation $\psi \circ G_1 - \psi = 0$ has a unique solution asymptotic to 0 on $\Delta_+(\alpha,R)$. And indeed, if we iterate the equation we obtain $\psi \circ G_1^p(z) - \psi(z) = 0$; letting p tend to infinity, we obtain $\psi(z) = \lim_{p\to+\infty} \psi \circ G_1^p(z) = \lim_{z'\to\infty} \psi(z')$ according to condition (i). Hence, $\psi(z) = 0$ for all $z \in \Delta_+(\alpha,R)$ and the proof is achieved. □

Actually, Birkhoff [Bir39] and Kimura [Kim71] stated the theorem in the following form (see also [Mal82] and [Éca74]).

Corollary 6.2.3 (Birkhoff-Kimura)
*Suppose f is a germ of diffeomorphism formally conjugated to $g(x)$ via the formal diffeomorphism $\widetilde{h}(x)$ (*cf.* Def. 6.1.1, p. 186):*

$$\widetilde{h} \circ f = g \circ \widetilde{h}. \tag{6.1}$$

With notations as before, there exist unique diffeomorphisms

$$h_+ \in H^0(I^+_{\alpha_0};\mathscr{G}) \quad \textit{and} \quad h_- \in H^0(I^-_{\alpha_0};\mathscr{G})$$

such that

$$h_+ \circ f = g \circ h_+ \quad \textit{and} \quad T_0 h_+ = \widetilde{h} \quad \textit{on } I^+_{\alpha_0}, \tag{6.6}$$

$$h_- \circ f = g \circ h_- \quad \textit{and} \quad T_0 h_- = \widetilde{h} \quad \textit{on } I^-_{\alpha_0}. \tag{6.7}$$

where $T_0 h_\pm$ stands for "Taylor expansion of $h_\pm$ at 0".

Proof. Again, we develop the proof over $I^+_{\alpha_0}$. We denote by $\delta_+(\alpha,R)$ the image in the chart of 0 of the domain $\Delta_+(\alpha,R)$ as built in the proof of theorem 6.2.2, p. 189. The Borel-Ritt theorem (Thm. 1.3.1 (i), p. 22) provides a function h holomorphic with Taylor expansion $Th(x) = \widetilde{f}(x)$ at 0 on $I^+_{\alpha_0}$. It results from the conjugation equation (6.1), p. 186 that the function $f_1 = h \circ f \circ h^{-1}$ is asymptotic to $\widetilde{h} \circ f \circ \widetilde{h}^{-1} = g$ on $I^+_{\alpha_0}$. Hence, $\varphi = f_1 \circ g^{-1}$ is flat and satisfies $f_1 = \varphi \circ g$ on $I^+_{\alpha_0}$.

Birkhoff-Kimura theorem 6.2.2, p. 189, applied to f_1 and g provides a ϕ_+ asymptotic to 0 and satisfying $\phi_+ \circ f_1 = g \circ \phi_+$ on $\delta_+(\alpha,R)$ for any $\alpha < \alpha_0$. Composing this relation by h to the right and denoting $h_+ = \phi_+ \circ h$ we obtain $h_+ \circ f = g \circ h_+$.

Uniqueness for h_+ is proved similarly as for ϕ_+ and is valid on $\delta(\alpha,R)$ for all $\alpha < \alpha_0$ (whereas R might depend on α) . Hence, condition (6.6) is satisfied.

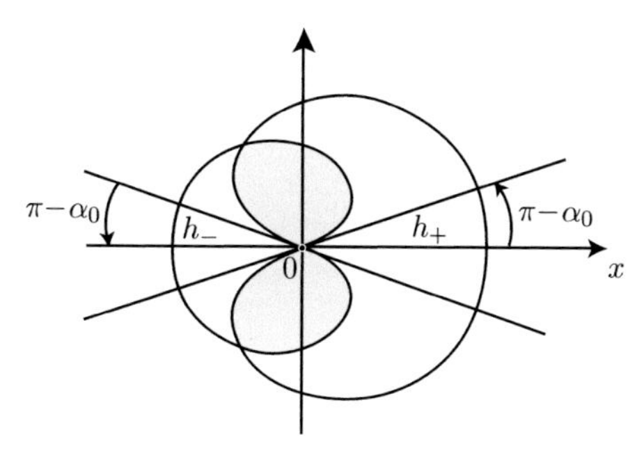

Fig. 6.3

Symmetrically, we prove the existence and uniqueness of h_- over $I^-_{\alpha_0}$ by the same method. □

When $\alpha_0 > \pi/2$ the domains of definition of h_+ and h_- overlap across the two imaginary directions as shown in Fig. 6.3.

6.3 The Invariance Equation of g

The invariance equation

$$u \circ g = g \circ u \tag{6.8}$$

of g is a particular case of the conjugation equation (6.1), p. 186. Hence, it admits the unique solution $u = \mathrm{Id}$ in $\widetilde{\mathbb{G}}$ and, given $0 < \alpha_0 < \pi$, it admits a unique solution u_+ section of the sheaf $\mathscr{G}$ over $I^+_{\alpha_0} =]-\alpha_0, +\alpha_0[$ and a unique solution u_- on $I^-_{\alpha_0} =]\pi - \alpha_0, \pi + \alpha_0[$, both asymptotic to Id. Since these solutions are unique and Id is a solution defined everywhere we can assert that u_+ and u_- are both equal to Id. The situation is different in a neighborhood of the imaginary axis where there might exist non trivial germs of solutions.

In this section we study the behavior of germs of flat solutions of the invariance equation (6.8) near the two imaginary half axis.

Proposition 6.3.1 *Let $\Delta_1 = \{|x| < r_1\,;\, \beta < \arg x < \pi - \beta\}$ with $0 < \beta < \pi/2$ be a sector with vertex 0 neighboring the positive imaginary axis. Any solution $u \in \mathscr{G}(\Delta_1)$ of the invariance equation (6.8) is exponentially flat of order 1 on Δ_1.*

The same result holds on a sector $\Delta_2 = \{|x| < r_2\,;\, \beta - \pi < \arg x < -\beta\}$ neighboring the negative imaginary axis.

Proof. It suffices to consider the case of Δ_2.

Again, it is more convenient to work in the chart of infinity. The sector $\mathcal{S}_2$ is changed into

$$\mathcal{S}_2 = \Big\{ |z| > R_2 = \frac{1}{r_2} \,;\, \beta < \arg z < \pi - \beta \Big\},$$

the solution u is changed into U and we set $U = \mathrm{Id}+V$. With these notations the invariance equation reads

$$V(z+1) = V(z) \tag{6.9}$$

whose solutions are the 1-periodic functions. Hence, to solutions $U \in \mathcal{G}(\mathcal{S}_2)$ there correspond functions V of the form $V(z) = \nu(\mathrm{e}^{2\pi \mathrm{i} z})$ that satisfy $\lim\limits_{z\to\infty} V(z) = 0$. Con-

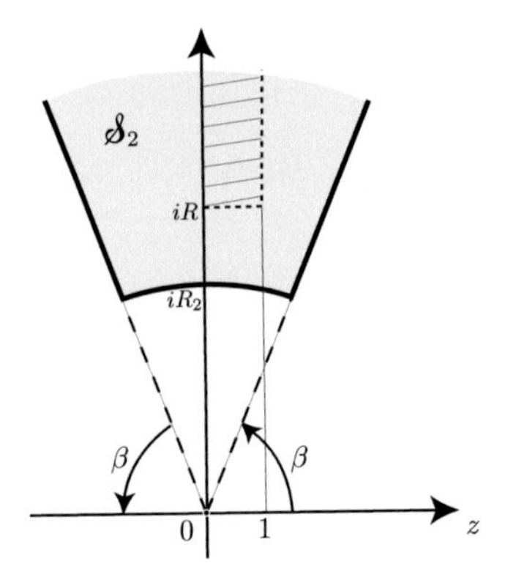

Fig. 6.4

sider, in $\mathcal{S}_2$, a vertical half-stripe $[iR, 1+iR[\times]iR, +i\infty[$ with width 1 (see Fig. 6.4, p. 194). It's easily checked that its image by the map $z \mapsto t = \mathrm{e}^{2\pi \mathrm{i} z}$ is a punctured disc Ω_2 centered at 0 in $\mathbb{C}$. Moreover, a fundamental system of neighborhoods of infinity in $\mathcal{S}_2$ is sent on a fundamental system of neighborhoods of 0. Hence, the condition $\lim_{\substack{z\to\infty \\ z\in\Sigma_1}} \nu(\mathrm{e}^{2\pi \mathrm{i} z}) = 0$ is equivalent to $\lim_{t\to 0} \nu(t) = 0$. Consequently, by the removable singularity theorem, ν can be continued into a holomorphic function at 0.

Now, suppose ν is not identically 0. Then, it has finite order, say k, at 0; we denote $\nu(t) = O(t^k)$. This implies that $V(z) = O(\mathrm{e}^{-2\pi k \Im(z)})$ as z tends to infinity in $\mathcal{S}_2$. However, on $\mathcal{S}_2$, one has $\Im(z) > |z| \sin\beta$ and consequently, $V(z) = O(\mathrm{e}^{-2\pi k \sin(\beta)|z|})$. Thus, V has (uniform) exponential decay of order one on $\mathcal{S}_2$ at infinity and so does the solution $u(x) = 1/\big(1/x + V(1/x)\big)$ at $x = 0$ on $\mathcal{S}_2$. □

6.4 1-Summability of the Conjugation Series $\widetilde{h}$

Recall that the conjugation equation

$$h \circ f = g \circ h \tag{6.1}$$

admits a unique formal solution $\widetilde{h}(x)$ in $\widetilde{\mathbb{G}}$. The 1-summability of $\widetilde{h}$ is now straightforward.

Theorem 6.4.1 *The series $\widetilde{h}(x)$ is 1-summable with singular directions the two imaginary half-axis.*

Proof. Let $\pi/2 < \alpha_0 < \pi$ and consider the two solutions $h_{\pm}(x) \in H^0(I^{\pm}_{\alpha_0}, \mathscr{G})$ of equation (6.1), both asymptotic to $\widetilde{h}(x)$ (cf. Cor. 6.2.3, p. 192). The non abelian 1-cocycle defined by

$$h_1 = h_-^{-1} \circ h_+ \text{ on }]\pi - \alpha_0, \alpha_0[\quad \text{and} \quad h_2 = h_+^{-1} \circ h_- \text{ on }]-\alpha_0, \alpha_0 - \pi[$$

satisfies the invariance equation (6.8), p. 193.
Denote $h_1 = \mathrm{Id} + u_1$ and $h_2 = \mathrm{Id} + u_2$. It follows from the previous section that u_1 and u_2 are exponentially flat of order one on $]\pi - \alpha_0, \alpha_0[$ and $]-\alpha_0, \alpha_0 - \pi[$ respectively.

To apply the Ramis-Sibuya theorem to $\widetilde{h}(x)$ on the covering $\mathscr{I} = (I^+_{\alpha_0}, I^-_{\alpha_0})$ of S^1 we must prove that the 1-cocycle equal to $h_+ - h_-$ on $]\pi - \alpha_0, \alpha_0[$ and to $h_- - h_+$ on $]-\alpha_0, \alpha_0 - \pi[$ is exponentially flat. To this end we observe from the form $h_1 = h_-^{-1} \circ h_+ = \mathrm{Id} + u_1$ and $h_2 = h_+^{-1} \circ h_- = \mathrm{Id} + u_2$ of h_1 and h_2 that

$$\begin{cases} h_+ - h_- = h_- \circ u_1 & \text{on }]\pi - \alpha_0, \alpha_0[, \\ h_- - h_+ = h_+ \circ u_2 & \text{on }]-\alpha_0, \alpha_0 - \pi[. \end{cases}$$

It follows that the abelian 1-cocycle is exponentially flat of order one since so are u_1 and u_2 whereas h_- and h_+ are asymptotic to the identity.

Since α_0 can be chosen arbitrarily close to π we can conclude that the series $\widetilde{h}(x)$ is 1-summable in all directions but the two imaginary half axis. □

One proves that these cocycles are not trivial in general and thus that $\widetilde{h}(x)$ is divergent; the non abelian 1-cocycle (h_1, h_2) classifies the analytic classes of diffeomorphisms $f(x)$ formally conjugated to $g(x)$. For classification problems we refer to volume I [MS16, Chap. 7].

Chapter 7
Six Equivalent Approaches to Multisummability

Abstract We consider the case of divergent series that are, in some way, relevant to several levels of summation together. We introduce the problems inherent to the situation with the example of the Ramis-Sibuya series and we show that this series is k-summable for no $k > 0$. We expound six different theories of multisummability which extend the theories of k-summability. In most of them we found useful to treat the case when the summability depends on only two levels $k_1 < k_2$ before to state general results depending on an arbitrary number of levels. We prove the equivalence of the relative Watson's lemma with the Tauberian theorem proved in Chapter 5. As an application we prove that any solution of a linear ordinary differential equation is multisummable for convenient levels $k_1 < k_2 < \cdots < k_\nu$ of summation.

7.1 Introduction and the Ramis-Sibuya Series

We can observe that the examples of series given in the previous chapters that are solution of linear differential equations are all k-summable for a convenient value of k. In theorem 5.2.5, p. 142, sufficient conditions are stated for the k-summability of solutions of linear differential equations (k-summability must be understood there in its global meaning, that is, k-summablility in almost all directions). Recall that the second Tauberian theorem of Martinet-Ramis (Cor. 5.3.16, p. 174), asserts that a series both k_1- and k_2-summable for two distinct values $k_1 \neq k_2$ of k is necessarily convergent. Though, such a result is no longer valid if one considers k_1- and k_2-summability in a given θ direction: as shown in example 5.3.17, p. 174, the Leroy series $\widetilde{\mathrm{L}}(x)$ is both 1- and 2-summable in all directions $\theta \in]-\pi/4, +\pi/4[\bmod \pi$.

A first natural question is to determine whether any series solution of a linear differential equation is k-summable for a convenient value of k. This question, known under the name of *Turrittin problem* although Turrittin after Trjitzinsky, Horn and al. formulated the question in different terms, received a negative answer by J.-P. Ramis and Y. Sibuya in 1984 (published later [RS89]) through a counter-example (cf. Exa. 7.1.1 below). A more intricate summation process called *multisummation* had become necessary.

M. Loday-Richaud, *Divergent Series, Summability and Resurgence II*, Lecture Notes in Mathematics, DOI 10.1007/978-3-319-29075-1_7

The counter-example given by J.-P. Ramis and Y. Sibuya with a proof of the fact that the Ramis-Sibuya series is k-summable for no $k>0$ is as follows.

Example 7.1.1 (Ramis-Sibuya Series)

The Ramis-Sibuya series is the series

$$\widetilde{\mathrm{RS}}(x)=\widetilde{\mathrm{E}}(x)+\widetilde{\mathrm{L}}(x)$$

sum of the Euler series $\widetilde{\mathrm{E}}(x)=\sum_{n\geq 0}(-1)^n n!x^{n+1}$ (cf. Exa. 1.1.4, p. 4) and of the Leroy series $\widetilde{\mathrm{L}}(x)=\sum_{n\geq 0}(-1)^n n!x^{2n+2}$ deduced from the Euler series by substituting x^2 for x (cf. Exa. 5.3.17, p. 174)). From the Euler equation $x^2y'+y=x$ and the Leroy equation $x^3y'+2y=2x^2$ one deduces that the Ramis-Sibuya series satisfies the Ramis-Sibuya equation

$$\mathrm{RS}(y)=4x+2x^2+10x^3-3x^4$$

where the operator RS reads

$$\mathrm{RS}=x^5(2-x)d^2/dx^2+x^2(4+5x^2-2x^3)d/dx+2(2-x+x^2).$$

It is worth to notice that the operator RS admits the following Newton polygon with the two slopes 1 and 2. We will see that this indicates that the series solution of the Ramis-Sibuya equation are all, at worst, (1,2)-summable as defined in the next sections.

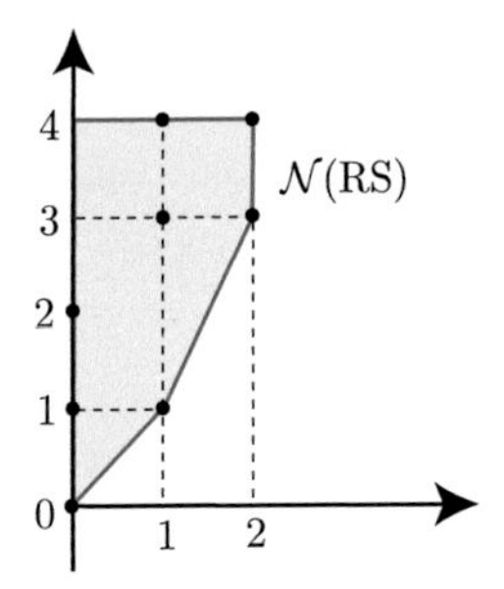

Fig. 7.1. Newton polygon of the Ramis-Sibuya operator

Check that the series $\widetilde{\mathrm{RS}}(x)$ *is k-summable for no $k>0$*. Indeed, as we saw earlier (cf. Com. 5.1.9, p. 138 and Sect. 1.1.2, p. 3), the Euler series $\widetilde{\mathrm{E}}(x)$ is 1-summable in all directions but the $\theta=\pi$ direction. As a consequence, the Leroy series $\widetilde{\mathrm{L}}(x)$ is 2-summable in all directions but the $\theta=\pm\pi/2$ directions. We saw in example 5.3.17, p. 174 that the Leroy series, and then also the Ramis-Sibuya series, is 1-summable in any θ direction satisfying $\theta\in]-\pi/4,+\pi/4[\bmod \pi$ and in these directions only. In particular, the Ramis-Sibuya series is not 1-summable. On another hand, the Euler series $\widetilde{\mathrm{E}}(x)$, and then also the Ramis-Sibuya series, is 2-summable in no direction since its 2-Borel transform does not converge. A fortiori, the Ramis-Sibuya series is not 2-summable. Consider now a direction $\theta\in]\pi/4,3\pi/4[\bmod \pi$ and show that $\widetilde{\mathrm{RS}}(x)$ is k-summable for no other value of $k>0$ in the θ direction. This is the case for $k>1$ (and in any direction) from the same argument as for $k=2$: the k-Borel transform of $\widetilde{\mathrm{RS}}(x)$ does not converge. Suppose there exists $k<1$ such that $\widetilde{\mathrm{RS}}(x)$ be k-summable in the θ direction. Then, since $\widetilde{\mathrm{RS}}(x)$ is a 1-Gevrey series, the Tauberian theorem (Thm. 5.3.15, p. 173) of Martinet-Ramis [MarR89] (taking $k_1=k<1$ and $k_2=1$) would imply that $\widetilde{\mathrm{RS}}(x)$ be 1-summable in the θ direction.

Hence, the contradiction and we can conclude that $\widetilde{\mathrm{RS}}(x)$ is k-summable for no $k>0$ since this is the case in each θ direction that satisfies $\theta\in]\pi/4,3\pi/4[\bmod \pi$.

A natural candidate for the sum of $\widetilde{\mathrm{RS}}(x)$ is the asymptotic function $\mathrm{E}(x)+\mathrm{L}(x)$ obtained by adding the 1-sum of $\widetilde{\mathrm{E}}(x)$ to the 2-sum of $\widetilde{\mathrm{L}}(x)$. Such a choice is not as trivial as it appears at first glance since there could exist several decompositions of $\widetilde{\mathrm{RS}}(x)$ in terms of 1- and 2-summable series.

The (1,2)-summability of $\widetilde{\mathrm{RS}}(x)$ following J. Écalle's approach is widely developed in [Lod90].

This example shows that the set of k-summable series for all $k>0$ is insufficient to embrace all series solutions of linear differential equations. Having in mind to sum solutions of linear differential equations another natural question is the following:

Is it possible to find
a (somewhat minimal) set of series equipped with a summation process
which is compatible with the various k-summation processes
and contains all solutions of linear differential equations?

From the example of the Ramis-Sibuya series we understand that any such set should contain the vector space $\sum_{0<k\leq+\infty}\mathbb{C}\{x\}_k$ of k-summable series for al $k>0$[1]. Observe, with the example of the 1-summable series $\widetilde{f}(x)=x/\widetilde{\mathrm{E}}(x)$, that not all k-summable series are solutions of linear differential equations. Since the derivative and the product of solutions of linear differential equations satisfy themselves linear differential equations such a set should also contain the differential algebra $\mathrm{Alg}_{\mathrm{k}>0}$ generated by the spaces $\mathbb{C}\{x\}_k$ for all $k>0$ including $k=+\infty$. And it results from the factorization theorem of solutions of linear differential systems [Ram85], [Lod94, Thm. III.2.5] that the algebra $\mathrm{Alg}_{\mathrm{k}>0}$ suffices. As a homomorphism of differential algebras a summation operator $\mathscr{S}$ on $\mathrm{Alg}_{\mathrm{k}>0}$, if one exists, is uniquely determined by its values on the spaces $\mathbb{C}\{x\}_k$ generating $\mathrm{Alg}_{\mathrm{k}>0}$. The problem lies in the *existence* of the operator $\mathscr{S}$. Indeed, the compatibility condition means that the restriction of $\mathscr{S}$ to each space $\mathbb{C}\{x\}_k$ has to be the k-summation operator. Hence, given an element of $\mathrm{Alg}_{\mathrm{k}>0}$ in the form of a sum of products of k-summable series its sum is obtained by replacing each factor by its k-sum. The point is that an element in $\mathrm{Alg}_{\mathrm{k}>0}$ may have several decompositions into sums of products of k-summable series and showing that these decompositions all provide the same sum is not obvious. There exists no direct proof of this fact. A solution is found in developing independently a theory of summation called *multisummation* which extends the k-summation processes with existence and uniqueness of sums and showing that the elements of $\mathrm{Alg}_{\mathrm{k}>0}$ are summable in that theory (cf. Prop. 7.2.14, p. 207).

The same question may be addressed in a given θ direction and the results are more precise. It was proved by W. Balser in [Bal92a] that, under a weak restrictive condition on the levels, any multisummable series in the θ direction lives in the vector space $\sum_{k>0}\mathbb{C}\{x\}_{k,\theta}$ spanned by all k-summable series in the θ direction for all $k>0$. The decomposition is essentially unique (cf. Prop. 7.5.1, p. 219) and

[1] One could also limit the choice to rational $k>0$ since all levels of linear differential equations are rational.

again the sum obtained by adding the k-sums of each term coincide with the multisum in any other usual sense whatever the decomposition. This property does not hold globally: when the series is multisummable (i.e., multisummable in almost all directions) the decomposition can be made in almost all directions but the decomposition depends on the chosen direction in general (cf. Sect. 7.5, p. 219). One could think of this approach as a good numerical tool based only on simple summation processes. Unfortunately, there is no algorithm to decompose a given series into a sum of products of k-summable series and the result remains indeed theoretical.

The aim of this chapter is to describe in a general setting various definitions of multisummability. Of course, we look forward to the same properties as those of k-summation, i.e., uniqueness, homomorphism of $\mathbb{C}$-differential algebras,.... We also compare these various approaches to prove their equivalence. Comparison being not evaluation, our aim is not to grade the different approaches. None approach can be considered as being the best, none as being the worst. But any of them might be better than another one depending on the question to answer.

All along the chapter we use repetedtly the Ramis-Sibuya series as our reference example.

7.2 The First Approach: Asymptotic Definition

In this section, we generalize the asymptotic approach of k-summability (cf. Sect. 5.1, p. 134) to the case of several levels $k_1 < k_2 < \cdots < k_\nu$. Watson's lemma has to be replaced by the so-called *relative Watson's lemma* although the relative Watson's lemma is not a parametric version of the classical Watson's lemma.

7.2.1 The Relative Watson's Lemma

The relative Watson's lemma is due to B. Malgrange and J.-P. Ramis [MalR92].

Theorem 7.2.1 (Relative Watson's Lemma) *Let $0 < k_1 < k_2$ be given and let I be a k_1-wide arc* (cf. *Def. 5.1.2, p. 135). Then,*

$$H^0\big(I; \mathscr{A}^{\le -k_1}/\mathscr{A}^{\le -k_2}\big) = 0.$$

When the length $|I|$ of I is smaller than 2π the arc I may be supposed to belong to S^1. Otherwise, it must be considered as an arc of the universal cover $\mathbb{R}$ of S^1. Recall that this latter case can be reduced to the first one by an adequate ramification of the variable x.

Compare corollary 5.1.4, p. 136 of Watson's lemma: here, instead of considering a k_1-exponentially flat function on a k_1-wide arc I one considers here a k_1-

exponentially flat 0-cochain with jumps (its 1-coboundary) small enough to be k_2-exponentially flat.

Roughly speaking, the theorem says:

A 0-cochain belonging to $H^0\big(I;\mathscr{A}^{\leq -k_1}/\mathscr{A}^{\leq -k_2}\big)$
has too small jumps on a too large arc I
to be not k_2-exponentially flat itself.

In [MalR92] the lemma is stated for closed k_1-wide arcs. It is equivalent to choose either closed or open k_1-wide arcs. Indeed, suppose I is closed; then an element of $H^0\big(I\,;\mathscr{A}^{\leq -k_1}/\mathscr{A}^{\leq -k_2}\big)$ is represented by a 0-cochain that lives on a larger open arc I'. If the lemma is true for open arcs then the cochain induces the null cohomology class in $H^0\big(I'\,;\mathscr{A}^{\leq -k_1}\,/\,\mathscr{A}^{\leq -k_2}\big)$ and also in $H^0\big(I\,;\mathscr{A}^{\leq -k_1}\,/\,\mathscr{A}^{\leq -k_2}\big)$. Conversely, suppose $H^0\big(I'\,;\mathscr{A}^{\leq -k_1}\,/\,\mathscr{A}^{\leq -k_2}\big)=0$ for any closed k_1-wide arc I'. Let I be an open k_1-wide arc and $f=(f_j)_{j\in J}$ a 0-cochain in $H^0\big(I\,;\mathscr{A}^{\leq -k_1}\,/\,\mathscr{A}^{\leq -k_2}\big)$ associated with a covering $\mathscr{I}=(I_j)_{j\in J}$ of I. Up to refining the covering $\mathscr{I}$ we can assume that it is indexed by $\mathbb{Z}$ and satisfies $I_j\cap I_\ell\neq\emptyset$ if $|j-\ell|=1$ and $I_j\cap I_\ell=\emptyset$ otherwise (and thus, in particular, it has no 3-by-3 intersection), since there exists arbitrarily fine such coverings of I. Write I as an increasing union of closed k_1-wide sub-arcs I'_ℓ. From the form of the covering $\mathscr{I}$ we deduce that any open arc I_j is contained in infinitely many closed arcs I'_ℓ; choose one of them denoted by I'_{ℓ_j}. Then, the restriction of the 0-cochain f to I'_{ℓ_j} induces 0 in $H^0\big(I'_{\ell_j}\,;\mathscr{A}^{\leq -k_1}\,/\,\mathscr{A}^{\leq -k_2}\big)$. This means, in particular, that f_j belongs to $\mathscr{A}^{\leq -k_2}(I_j)$. This being true for all $j\in\mathbb{Z}$ we can conclude that f induces 0 in $H^0\big(I\,;\mathscr{A}^{\leq -k_1}\,/\,\mathscr{A}^{\leq -k_2}\big)$.

The relative Watson's lemma can be reformulated as follows.

Corollary 7.2.2 *Under the conditions of the relative Watson's lemma the following natural map is injective:*

$$H^0\big(I;\mathscr{A}/\mathscr{A}^{\leq -k_2}\big)\longrightarrow H^0\big(I;\mathscr{A}/\mathscr{A}^{\leq -k_1}\big).$$

Proof. Consider the short exact sequence

$$0\longrightarrow \mathscr{A}^{\leq -k_1}/\mathscr{A}^{\leq -k_2}\longrightarrow \mathscr{A}/\mathscr{A}^{\leq -k_2}\longrightarrow \mathscr{A}/\mathscr{A}^{\leq -k_1}\longrightarrow 0\cdot$$

The associated long exact sequence of cohomology over I provides the exact sequence

$$0\longrightarrow H^0(I;\mathscr{A}^{\leq -k_1}/\mathscr{A}^{\leq -k_2})\longrightarrow H^0(I;\mathscr{A}/\mathscr{A}^{\leq -k_2})\longrightarrow H^0(I;\mathscr{A}/\mathscr{A}^{\leq -k_1})\cdot$$

Hence, the equivalence of the relative Watson's lemma 7.2.1, p. 200, and its corollary 7.2.2, p. 201. □

One can find in [MalR92] a direct proof of the relative Watson's lemma (see also [Mal95]). Instead of reproducing it we prefer to include a proof of the equivalence between the Tauberian theorem 5.3.15 and the relative Watson's lemma 7.2.1 [MalR92, Sect. 3 (ii)].

Lemma 7.2.3 (Malgrange-Ramis [MalR92, Lemme (2.5)])

Let $1/2 < k_1 < k_2$ and a closed k_1-wide arc I of S^1 be given.
To any h in $H^0\big(I;\mathscr{A}/\mathscr{A}^{\leq -k_2}\big)$ there exist sections

$$h' \in H^0(I;\mathscr{A}) \quad \text{and} \quad h'' \in H^0\big(S^1;\mathscr{A}/\mathscr{A}^{\leq -k_2}\big)$$

such that

$$h = \big(h' \bmod \mathscr{A}^{\leq -k_2}\big) + h''_{|I}.$$

The notation $h' \bmod \mathscr{A}^{\leq -k_2}$ stands for the element of $H^0\big(I;\mathscr{A}/\mathscr{A}^{\leq -k_2}\big)$ canonically induced by $h' \in H^0(I;\mathscr{A})$. It results from the condition $1/2 < k_1$ that the arc I is less than 2π long. Observe that h'', unlike h, exists all around S^1.

Roughly speaking the lemma says:

A section of $\mathscr{A}/\mathscr{A}^{\leq -k_2}$ over I can be continued into a section all over S^1 after "correction" by an adequate asymptotic function defined on I.

Proof. The section h can be represented as a finite 0-cochain $(h_j)_{j\in J}$ as follows. Let $J = \{j_1, j_2, \dots, j_p\}$. The components h_j are functions in $\overline{\mathscr{A}}(\Delta_j)$ for some open sectors $\Delta_j = I_j \times]0, r_j[$ with vertex 0; the 2-by-2 intersections of these sectors satisfy the conditions $\Delta_j \cap \Delta_{j+1} \neq \emptyset$ and $\Delta_j \cap \Delta_\ell = \emptyset$ when $|j-\ell| > 1$ and we assume that the global arc $I_1 \cup I_2 \cup \cdots \cup I_p$ is less than 2π wide, the union $I_2 \cup \cdots \cup I_{p-1}$ is included in I while I_1 and I_p are not. Moreover, the differences $-h_j + h_{j+1}$ belong to $\overline{\mathscr{A}}^{\leq -k_2}(\Delta_j \cap \Delta_{j+1})$ for all $j \in J$. Complete the family $(\Delta_j)_{j\in J}$ into a covering $\mathcal{S} = \{\Delta_j\}_{j\in J\cup K}$ of S^1 (denote also $\Delta_j = I_j \times]0, r_j[$ for $j \in K$) without 3-by-3 intersections and such that $(\cup_{j\in K} I_j) \cap I = \emptyset$. Consider the 1-cocycle $\dot{h} = (h_{j,j+1})_{j\in J\cup K}$ of $\mathcal{S}$ with values in $\mathscr{A}^{\leq -k_2}$ defined by

$$h_{j,j+1}(x) = \begin{cases} -h_j(x) + h_{j+1}(x) & \text{if } j \text{ and } j+1 \in J \\ 0 & \text{otherwise.} \end{cases}$$

From the Ramis-Sibuya theorem 5.2.1, p. 140, and shrinking the sectors Δ_j if necessary, there exist functions $g_j(x)$ belonging to $\overline{\mathscr{A}}_{1/k_2}(\Delta_j)$ (cf. Not. 1.2.8, p. 17) such that

$$h_{j,j+1} = -g_j + g_{j+1} \quad \text{for all } j \text{ and } j+1 \text{ in } J.$$

We obtain thus the equality $h_j(x) - g_j(x) = h_{j+1}(x) - g_{j+1}(x)$ on $\Delta_j \cap \Delta_{j+1}$ for all j and $j+1$ in J and the functions $h_j(x) - g_j(x)$ glue together into a section $h'(x)$ of $H^0(I;\mathscr{A})$. On another hand, by construction, the g_j's for $j \in J \cup K$ determine an element $h''(x)$ of $H^0\big(S^1;\mathscr{A}/\mathscr{A}^{\leq -k_2}\big)$ and we obtain

$$h' \bmod \mathscr{A}^{\leq -k_2} + h''_{|\Delta_j} = (h_j - g_j) + g_j = h_j \ \text{ on } \Delta_j \ \text{ for all } j \in J.$$

Hence, the result. □

Proposition 7.2.4 *The Tauberian theorem (Thm. 5.3.15, p. 173) and the relative Watson's lemma (Thm. 7.2.1, p. 200) are equivalent.*

Proof. Let $0 < k_1 < k_2$ and a closed k_1-wide arc I be given. By means of a convenient ramification we may assume that the arc I is less than 2π long (which implies $k_1 > 1/2$) and this allows us to work on S^1. As usually, we denote $s_1 = 1/k_1$ and $s_2 = 1/k_2$.

▷ *Show that the relative Watson's lemma implies the Tauberian theorem 5.3.15.* Let the series $\widetilde{f}(x)$ be both s_2-Gevrey and k_1-summable with sum $f(x)$ on a k_1-wide arc I. We must prove that $\widetilde{f}(x)$ is also k_2-summable on I.

As a s_2-Gevrey series and according to corollary 5.2.2, p. 140, the series $\widetilde{f}(x)$ can be identified to an element of $H^0\big(S^1;\mathscr{A}/\mathscr{A}^{\leq -k_2}\big)$, i.e., to a 0-cochain $g = (g_j)$ where g_j is asymptotic to $\widetilde{f}(x)$ for all j and $g_i - g_j$ takes its values in $\mathscr{A}^{\leq -k_2}$ for all i,j. It follows from the Ramis-Sibuya theorem 5.2.1 that $g_j(x)$ is actually s_2-Gevrey (and hence also, s_1-Gevrey) asymptotic to $\widetilde{f}(x)$. The 0-cochain g induces canonically an element of $H^0\big(S^1;\mathscr{A}/\mathscr{A}^{\leq -k_1}\big)$, thus characterizing $\widetilde{f}(x)$ as a s_1-Gevrey series. Since $f(x)$ is a k_1-sum of $\widetilde{f}(x)$ on I we deduce that $f = g \bmod \mathscr{A}^{\leq -k_1}$ on I (indeed, $f - g_{j_{|_I}}$ is s_1-Gevrey asymptotic to 0; cf. Prop. 1.2.17, p. 20). From Cor. 7.2.2, p. 201, of the relative Watson's lemma, it follows that $f = g \bmod \mathscr{A}^{\leq -k_2}$ on I, which proves that $\widetilde{f}(x)$ is k_2-summable on I with k_2-sum $f(x)$. Hence, the result.

▷ *Conversely, show that the Tauberian theorem 5.3.15 implies the relative Watson's lemma.*
Let $h(x)$ belong to $H^0\big(I;\mathscr{A}/\mathscr{A}^{\leq -k_2}\big)$. The section $h(x)$ admits a canonical image in $H^0\big(I;\mathscr{A}/\mathscr{A}^{\leq -k_1}\big)$: we specify $h(x) \bmod \mathscr{A}^{\leq -k_2}$ when $h(x)$ is seen as an element of $H^0\big(I;\mathscr{A}/\mathscr{A}^{\leq -k_2}\big)$ and $h(x) \bmod \mathscr{A}^{\leq -k_1}$ to denote its canonical image in $H^0\big(I;\mathscr{A}/\mathscr{A}^{\leq -k_1}\big)$. We must prove (Cor. 7.2.2, p. 201) that $h(x) = 0 \bmod \mathscr{A}^{\leq -k_1}$ on I implies $h(x) = 0 \bmod \mathscr{A}^{\leq -k_2}$ on I.

From lemma 7.2.3, p. 201, there exists $h' \in H^0(I;\mathscr{A})$ and $h'' \in H^0\big(S^1;\mathscr{A}/\mathscr{A}^{\leq -k_2}\big)$ such that $h = \big(h' \bmod \mathscr{A}^{\leq -k_2}\big) + h''_{|_I}$. By definition, $h''(x)$ can be seen as a 0-cochain $h'' = (h''_j)$ where the various components h''_j are asymptotic to a same Taylor series $\widetilde{h}''(x)$ and $h_i - h_j$ takes its values in $\mathscr{A}^{\leq -k_2}$ for all i,j. And we know, from the Ramis-Sibuya theorem 5.2.1, p. 140 that the $h''_j(x)$'s are actually s_2-asymptotic to $\widetilde{h}''(x)$.

The assumption $h(x) = 0 \bmod \mathscr{A}^{\leq -k_1}$ on I implies $h''(x) = -h'(x) \bmod \mathscr{A}^{\leq -k_1}$ on I and the series $\widetilde{h}''(x)$ is k_1-summable on I with k_1-sum $-h'(x)$ (h' is a true function; see Def.5.2.4, p. 141). On another hand, by definition, $\widetilde{h}''(x)$ is a s_2-Gevrey series. By the Tauberian theorem 5.3.15, p. 173 the series $\widetilde{h}''(x)$ is then k_2-summable on I with the same sum $-h'(x)$. This implies the equality $h''(x) = -h'(x) \bmod \mathscr{A}^{\leq -k_2}$. Hence, $h(x) = 0 \bmod \mathscr{A}^{\leq -k_2}$ as followed. □

7.2.2 An Asymptotic Definition of Multisummablilty

Towards the generalization of the asymptotic definition (cf. Def. 5.1.6, p. 137) of k-summability we proceed as follows. Begin with the case of two levels $k_1 < k_2$, that is, the case of (k_1,k_2)-summability.

Suppose we are given a series $\widetilde{f}(x)$ and two arcs $I_1 \supseteq I_2$ of S^1 respectively k_1- and k_2-wide. Set as usually, $s_1 = 1/k_1$.

Definition 7.2.5 ((k_1, k_2)-Summability)
The series $\widetilde{f}(x)$ is said to be (k_1,k_2)-summable on (I_1,I_2) with sum (f_1,f_2) *if*
(i) $f_1(x)$ *belongs to* $H^0\big(I_1;\mathscr{A}_{s_1}/\mathscr{A}^{\leq -k_2}\big)$*;*
(ii) $f_2(x)$ *belongs to* $H^0\big(I_2;\mathscr{A}_{s_1}\big)$ *(thus, is a true asymptotic function on* I_2*);*
(iii) f_1 *and* f_2 *agree on* I_2*,* i.e., $f_{1|_{I_2}} = f_2 \mod \mathscr{A}^{\leq -k_2}$*;*
(iv) f_1 *and* f_2 *are* s_1*-Gevrey asymptotic to* $\widetilde{f}(x)$ *on* I_1 *and* I_2 *respectively:*

$$T_{s_1,I_1} f_1(x) = T_{s_1,I_2} f_2(x) = \widetilde{f}(x).$$

To be more precise the sum (f_1,f_2) is also called *multisum* or (k_1,k_2)*-sum of* $\widetilde{f}(x)$ *on* (I_1,I_2). Sometimes and especially when I_1 and I_2 have a same bisecting θ direction, one talks of f_2 as a (k_1,k_2)-sum of $\widetilde{f}(x)$ on (I_1,I_2) letting f_1 understood. In the latter case, one also says that f_2 is a (k_1,k_2)*-sum of* $\widetilde{f}(x)$ *in the* θ *direction.*

Remark 7.2.6 Suppose I_1 and I_2 are closed arcs. Sections over I_1 or I_2 live then on larger open arcs. From condition (ii), one can represent f_1 by a 0-cochain containing f_2 as a component. One can also choose a 0-cochain over an open covering of I_1 by arcs with no 3-by-3 intersection and no intersection 2-by-2 on I_2. Thus, definition 7.2.5, p. 204, can be reformulated as follows:

The series $\widetilde{f}(x)$ is (k_1,k_2)-summable on (I_1,I_2) if there exists a 0-cochain f, s_1-Gevrey asymptotic to $\widetilde{f}(x)$ on I_1, that has no jump on I_2 and only k_2-exponentially flat jumps on $I_1 \setminus I_2$.
The couple (f_1,f_2) where f_1 is the natural image of f in $H^0(I_1;\mathscr{A}_{s_1}/\mathscr{A}^{\leq -k_2})$ and f_2, its restriction to I_2, is a (k_1,k_2)-sum of $\widetilde{f}(x)$ on (I_1,I_2).

Recall that, in general, a 0-cochain which is s_1-Gevrey asymptotic to a given series may have jumps (its coboundary) as large as k_1-exponentially flat (Prop. 1.2.17, p. 20). The condition that the jumps are at least k_2-exponentially flat is strong and guaranties the uniqueness of the (k_1,k_2)-sum of $\widetilde{f}(x)$ on (I_1,I_2) as we show below.

Watson's lemma and the relative Watson's lemma imply uniqueness of (k_1,k_2)-sums providing thus a well defined notion of (k_1,k_2)-summability.

Proposition 7.2.7 (Uniqueness of (k_1, k_2)-Sums)
The multisum (f_1,f_2) of $\widetilde{f}(x)$ on (I_1,I_2), when it exists, is unique.

Proof. Suppose we are given two (k_1,k_2)-sums (f_1,f_2) and (f_1',f_2') of $\widetilde{f}(x)$ on (I_1,I_2). By Prop. 1.2.17, p. 20, the difference $f_1 - f_1'$ belongs to $H^0\big(I_1;\mathscr{A}^{\leq -k_1}/\mathscr{A}^{\leq -k_2}\big)$ and the relative Watson's lemma (Thm. 7.2.1, p. 200) implies that $f_1 - f_1' = 0$. This, in turn, implies that $f_2 = f_2'$ mod $\mathscr{A}^{\leq -k_2}$ and, from the classical Watson's lemma (Thm. 5.1.3, p. 136), that $f_2 = f_2'$ since I_2 is k_2-wide. □

Example 7.2.8

The Ramis-Sibuya series $\widetilde{\mathrm{RS}}(x)$ (Exa. 7.1.1, p. 198) is $(1,2)$-summable on (I_1,I_2) if and only if I_1 does not contain the Stokes arc of the Euler series $[\pi/2,3\pi/2]$ and I_2 contains none of the two Stokes arcs of the Leroy series $[\pi/4,3\pi/4]$ mod π.

Let us make explicit the $(1,2)$-summability of $\widetilde{\mathrm{RS}}(x)$ according to Def. 7.2.5, p. 204, say, for,

$$I_1 = [0,\pi] \quad \text{and} \quad I_2 = [0,\pi/2] \subset I_1.$$

Choose $0<\varepsilon<\pi/4$ and consider the open covering $\mathscr{I}$ of I_1 by the arcs

$$I_1' =]\pi/2,\pi+\varepsilon[\quad \text{and} \quad I_2' =]-\varepsilon,\pi/2+\varepsilon[.$$

Notice that $I_1' \cap I_2 = \emptyset$ so that the covering has no intersection 2-by-2 on I_2 as it was pointed out in remark 7.2.6, p. 204. Denote temporarily by $\mathrm{E}(x)$ the determination of the Euler function defined on the arc $-\varepsilon < \arg(x) < \pi+\varepsilon$. Denote by E_1' and E_2' the restrictions of E to I_1' and I_2' respectively. Clearly, the 0-cochain $(\mathrm{E}_1',\mathrm{E}_2')$ of $\mathscr{I}$ takes its values in $\mathscr{A}_1$ and has a trivial coboundary $\mathrm{E}_1' - \mathrm{E}_2' \equiv 0$. Denote by L_1' and L_2' the 2-sums of $\mathrm{L}(x)$ on I_1' and I_2' respectively. The 0-cochain $(\mathrm{L}_1',\mathrm{L}_2')$ of $\mathscr{I}$ takes its values in $\mathscr{A}_{1/2}$ hence also in $\mathscr{A}_1$ and its coboundary $\mathrm{L}_1' - \mathrm{L}_2'$ belongs to $H^0\big(I_1' \cap I_2';\mathscr{A}^{\leq -2}\big)$. It follows that the 0-cochain $(\mathrm{E}_1'+\mathrm{L}_1',\mathrm{E}_2'+\mathrm{L}_2')$ determines an element f_1 of $H^0\big(I_1;\mathscr{A}_1/\mathscr{A}^{\leq -2}\big)$. Denote by f_2 the element of $H^0(I_2;\mathscr{A}_1)$ defined by $\mathrm{E}_2'+\mathrm{L}_2'$. Both f_1 and f_2 are 1-asymptotic to the Ramis-Sibuya series $\widetilde{\mathrm{RS}}(x)$. The couple (f_1,f_2) is the $(1,2)$-sum of $\widetilde{\mathrm{RS}}(x)$ on (I_1,I_2).

Conversely, if I_1 contained the Stokes arc $[\pi/2,3\pi/2]$ of the Euler series $\widetilde{\mathrm{E}}(x)$ then we would have to use two different determinations of E on I_1 generating a non trivial coboundary with values in $\mathscr{A}^{\leq -1}$ and not in $\mathscr{A}^{\leq -2}$ and, thus, condition (i) of definition 7.2.5, p. 204 would fail. If I_2 contained a Stokes arc $[\pi/4,3\pi/4]$ or $[-3\pi/4,-\pi/4]$ of the Leroy series $\widetilde{\mathrm{L}}(x)$ then we would have to split the 2-sum L_2' of $\widetilde{\mathrm{L}}(x)$ into a 0-cochain with non trivial coboundary on I_2 and condition (ii) of definition 7.2.5 would fail.

Let us now state the general case of an arbitrary number of levels.

Definition 7.2.9 (Multi-Level $\underline{k}$ and $\underline{k}$-Multi-Arc)

▷ *We call* multi-level, *and we denote by* $\underline{k} = (k_1,k_2,\ldots,k_\nu)$, *any finite sequence of numbers* $k_1,k_2,\ldots,k_\nu$ *satisfying the conditions*

$$0 < k_1 < k_2 < \cdots < k_\nu.$$

▷ *We call* $\underline{k}$-multi-arc, *and we denote by* $\underline{I} = (I_1,I_2,\ldots,I_\nu)$, *any sequence of arcs satisfying the conditions*

$$\begin{cases} I_1 \supseteq I_2 \supseteq \cdots \supseteq I_\nu \\ \textit{for } j=1,\ldots,\nu, \textit{ the arc } I_j \textit{ is } k_j\textit{-wide (cf. Def. 5.1.2, p. 135).} \end{cases}$$

Recall that we always order the levels in increasing order: $k_1 < k_2 < \cdots < k_\nu$ whereas some authors order them in decreasing order.

From now, suppose we are given

- a series $\widetilde{f}(x)$,
- a multi-level $\underline{k} = (k_1, k_2, \ldots, k_\nu)$,
- a $\underline{k}$-multi-arc $\underline{I} = (I_1, I_2, \ldots, I_\nu)$.

As usually, we set $s_1 = 1/k_1$.

Definition 7.2.10 (Multisummability)

A series $\widetilde{f}(x)$ is said to be $\underline{k}$-summable on $\underline{I}$ with $\underline{k}$-sum $\underline{f} = (f_1, f_2, \ldots, f_\nu)$ *if*

▷ *f_j belongs to $H^0\big(I_j; \mathscr{A}_{s_1}/\mathscr{A}^{\leq -k_{j+1}}\big)$ for all $j = 1, 2, \ldots, \nu - 1$;*
▷ *f_ν belongs to $H^0\big(I_j; \mathscr{A}_{s_1}\big)$;*
▷ *the f_j's are compatible:*

$$f_{j|_{I_{j+1}}} = f_{j+1} \bmod \mathscr{A}^{\leq -k_{j+1}} \textit{ for all } j = 1, 2, \ldots, \nu - 1;$$

▷ *for all j, the section f_j is s_1-Gevrey asymptotic to the series $\widetilde{f}(x)$ on I_j:*

$$T_{s_1, I_j} f_j(x) = \widetilde{f}(x) \textit{ for all } j = 1, 2, \ldots, \nu.$$

In the case when the arcs $I_1, I_2, \ldots, I_\nu$ are all bisected by a same θ direction then $\underline{f}$ is called a *$\underline{k}$-sum of $\widetilde{f}(x)$ in the θ direction.* By abuse of language, one sometimes talks of f_ν as a $\underline{k}$-sum of $\widetilde{f}(x)$ on $\underline{I}$, the components $f_1, f_2, \ldots, f_{\nu-1}$ being understood.

Remark 7.2.11 Remark 7.2.6, p. 204, can be generalized to an arbitrary number of levels as follows. Suppose $\underline{I}$ is a closed $\underline{k}$-multi-arc. Definition 7.2.10, p. 206 is equivalent to saying:

The series $\widetilde{f}(x)$ is $\underline{k}$-summable on $\underline{I}$ with $\underline{k}$-sum $\underline{f} = (f_1, f_2, \ldots, f_\nu)$ if there exists a 0-cochain f on I_1 which is s_1-Gevrey asymptotic to $\widetilde{f}(x)$, which has no jump on I_ν and otherwise, has jumps that are at least k_ν-exponentially flat on $I_{\nu-1}$, $k_{\nu-1}$-exponentially flat on $I_{\nu-2}, \ldots$ and k_2-exponentially flat on I_1.

If so, the $\underline{k}$-sum $\underline{f}$ is defined by taking as f_j, for $j = 1, 2, \ldots, \nu - 1$, the natural image of f in $H^0\big(I_j; \mathscr{A}_{s_1}/\mathscr{A}^{\leq -k_{j+1}}\big)$ and taking as f_ν the restriction of f to I_ν.

Proposition 7.2.12 (Uniqueness) *The $\underline{k}$-sum $\underline{f}(x)$, when it exists, is unique.*

Proof. The proof proceeds as in the case of two levels (cf. Prop. 7.2.7, p. 204). Suppose $(f_1, f_2, \ldots, f_\nu)$ and $(f'_1, f'_2, \ldots, f'_\nu)$ are $\underline{k}$-sums of $\widetilde{f}(x)$ on $\underline{I}$. By proposition 1.2.17, p. 20, the difference $f_1 - f'_1$ belongs to $H^0\big(I_1; \mathscr{A}^{\leq -k_1}/\mathscr{A}^{\leq -k_2}\big)$ and, I_1 being k_1-wide, the relative Watson's lemma (cf. Thm. 7.2.1, p. 200) implies that $f_1 - f'_1 = 0$. This, in turn, implies that $f_2 - f'_2$ belongs to $H^0\big(I_2; \mathscr{A}^{\leq -k_2}/\mathscr{A}^{\leq -k_3}\big)$. Again, the relative Watson's lemma implies that $f_2 - f'_2 = 0$ and that $f_3 - f'_3$ belongs to $H^0\big(I_3; \mathscr{A}^{\leq -k_3}/\mathscr{A}^{\leq -k_4}\big)$. And so on, until the νth step where $f_\nu - f'_\nu$ belongs to $H^0\big(I_\nu; \mathscr{A}^{\leq -k_\nu}\big)$ and we conclude by the classical Watson's lemma (Thm. 5.1.3, p. 136) that $f_\nu = f'_\nu$ since I_ν is k_ν-wide. □

Notation 7.2.13 (Multisummable Series)

Given a multi-level $\underline{k}$ and a $\underline{k}$-multi-arc $\underline{I}$ we denote by

▷ $\mathbb{C}\{x\}_{\{\underline{k},\underline{I}\}}$ the set of $\underline{k}$-summable series on $\underline{I}$;

▷ $\mathbb{C}\{x\}_{\{\underline{k},\theta\}}$ the set of $\underline{k}$-summable series in the θ direction.

▷ $\mathrm{Sum}_{\{\underline{k},\underline{I}\}}$ the subset of $\prod_{j=1}^{\nu-1} H^0\big(I_j;\mathscr{A}_{s_1}/\mathscr{A}^{\leq -k_{j+1}}\big)\times H^0\big(I_\nu;\mathscr{A}_{s_1}\big)$ consisting of the elements satisfying the compatibility condition of definition 7.2.10 (iii), p. 206.

▷ $\mathscr{S}_{\{\underline{k},\underline{I}\}} : \mathbb{C}\{x\}_{\{\underline{k},\underline{I}\}} \longrightarrow \mathrm{Sum}_{\{\underline{k},\underline{I}\}}$ the $\underline{k}$–summation operator on $\underline{I}$ which to any $\underline{k}$-summable series on $\underline{I}$ associates its unique $\underline{k}$-sum on $\underline{I}$ according to proposition 7.2.12, p. 206.

▷ $\mathrm{Sum}_{\{\underline{k},\theta\}}$ and $\mathscr{S}_{\{\underline{k},\theta\}}$ instead of $\mathrm{Sum}_{\{\underline{k},\underline{I}\}}$ and $\mathscr{S}_{\{\underline{k},\underline{I}\}}$ in the θ direction.

We leave as an exercise the proof of the following proposition generalizing proposition 5.1.10, p. 138.

Proposition 7.2.14 *Let $\underline{k}=(k_1,k_2,\dots,k_\nu)$ and $\underline{I}=(I_1,I_2,\dots,I_\nu)$ be a multi-level and a $\underline{k}$-multi-arc* (cf. *Def. 7.2.9, p. 205).*

(i) *The set $\mathbb{C}\{x\}_{\{\underline{k},\underline{I}\}}$ is a differential subalgebra of the differential $\mathbb{C}$-algebra $\mathbb{C}[[x]]_{s_1}$ of s_1-Gevrey series.*

(ii) *Let $\underline{k}'$ be a multi-level extracted from $\underline{k}$ and $\underline{I}'$ the corresponding $\underline{k}'$-multi-arc extracted from $\underline{I}$.*

Then, $\mathbb{C}\{x\}_{\{\underline{k}',\underline{I}'\}}$ is a differential subalgebra of $\mathbb{C}\{x\}_{\{\underline{k},\underline{I}\}}$.

In particular, the differential algebras $\mathbb{C}\{x\}_{\{k_j,I_j\}}$ of k_j-summable series on I_j for $j=1,2,\dots,\nu$ are differential subalgebras of $\mathbb{C}\{x\}_{\{\underline{k},\underline{I}\}}$.

(iii) *The Taylor map*

$$T_{s_1,\underline{I}} : \mathrm{Sum}_{\{\underline{k},\underline{I}\}} \longrightarrow \mathbb{C}\{x\}_{\{\underline{k},\underline{I}\}}$$

is an isomorphism of differential $\mathbb{C}$-algebras with inverse the $\underline{k}$-summation operator $\mathscr{S}_{\underline{k},\underline{I}}$ on $\underline{I}$.

Remark 7.2.15 The previous proposition asserts that $\mathbb{C}\{x\}_{\{\underline{k},\underline{I}\}}$ contains the differential algebra generated by the algebras $\mathbb{C}\{x\}_{\{k_j,I_j\}}$, $j=1,2,\dots,\nu$ of k_j-summable series on I_j. It will be shown in section 7.5, p. 219, that for $k_1>1/2$, the two algebras are actually equal.

Although we do not provide an extensive proof of proposition 7.2.14 let us observe how a k_j-summable series may be regarded as a $\underline{k}$-summable series. Consider the example of $j=\nu$, all cases being similar. Suppose $\widetilde{f}(x)$ is k_ν-summable on I_ν. This means that there exists a function (its k_ν-sum) $f_\nu \in H^0\big(I_\nu;\mathscr{A}_{s_\nu}\big)$ satisfying $T_{s_\nu,I_\nu} f_\nu(x)=\widetilde{f}(x)$. This implies also that the series $\widetilde{f}(x)$ is s_ν-Gevrey and the Borel-Ritt theorem (Cor. 1.3.4, p. 25) allows to complete the sum f_ν into a 0-cochain (f'_ν) over I_1 whose components are all s_ν-Gevrey asymptotic (hence, s_1-Gevrey asymptotic) to $\widetilde{f}(x)$ and its coboundary has values in $\mathscr{A}^{\leq -k_\nu}$. Recall that the sheaves $\mathscr{A}^{\leq -k}$ satisfies the inclusions $\mathscr{A}^{\leq -k_\nu}\subset \mathscr{A}^{\leq -k_{\nu-1}}\subset\cdots\subset\mathscr{A}^{\leq -k_1}$.

Thus, the 0-cochain (f'_ν) induces canonically elements $f_1 \in H^0\big(I_1;\mathscr{A}_{s_1}/\mathscr{A}^{\leq -k_1}\big)$, $f_2 \in H^0\big(I_2;\mathscr{A}_{s_1}/\mathscr{A}^{\leq -k_2}\big)$, etc..., and $f_\nu \in H^0\big(I_\nu;\mathscr{A}_{s_1}\big)$ defined on $I_1, I_2, \ldots, I_\nu$ respectively and satisfying Def. 7.2.10, p. 206. Thus, $(f_1, f_2, \ldots, f_\nu)$ defines the $\underline{k}$-sum of $\widetilde{f}(x)$ on $\underline{I}$.

To end this section let us mention the fact that Prop. 5.1.11, p. 138, and its corollary 5.1.12, p. 139, remain valid if one replaces k-summability by multisummability.

Recall the notations of sections 1.2.2 and 5.1: given a series $\widetilde{g}(t)$ we denote by $\widetilde{g}_j$ its r-rank reduced series defined for $j = 0, 1, \ldots, r-1$, by $\widetilde{g}(t) = \sum_{j=0}^{r-1} t^j \widetilde{g}_j(t^r)$; given an arc $I = (\alpha, \beta)$we denote by $I^\ell_{/r}$ the arc $\big((\alpha + 2\ell\pi)/r, (\beta + 2\ell\pi)/r\big)$.

We can state:

Proposition 7.2.16 *Let $r > 1$ be an integer.*

(i) Extension of the variable.

A series $\widetilde{f}(x)$ is $\underline{k}$-summable on $\underline{I} = (I_1, I_2, \ldots, I_\nu)$ if and only if the series $\widetilde{g}(x) = \widetilde{f}(x^r)$ is $r\underline{k}$-summable on $\underline{I}_{/r} = (I_{1/r}, I_{2/r}, \ldots, I_{\nu/r})$.

(ii) Rank reduction.

The series $\widetilde{g}$ is $r\underline{k}$-summable on the arcs $I^\ell_{/r}$ for all $\ell = 0, 1, \ldots, r-1$ if and only if the series $\widetilde{g}_j$ for $j = 0, 1, \ldots, r-1$, are $\underline{k}$-summable on $\underline{I}$.

Proof. (i) Let $\underline{f}(x) = (f_1(x), f_2(x), \ldots, f_\nu(x))$ be the $\underline{k}$-sum of $\widetilde{f}(x)$ on $\mathscr{I}$. Then, $\underline{g}(x) = (g_1(x) = f_1(x^r), g_2(x) = f_2(x^r), \ldots, g_\nu(x) = f_\nu(x^r))$ is the $r\underline{k}$-sum of $\widetilde{g}(x)$ on $\mathscr{I}_{/r}$.

(ii) Suppose the series $\widetilde{g}_j(x)$ are all $\underline{k}$-summable on $\underline{I}$. By definition, these series satisfy $\widetilde{g}(t) = \sum_{j=0}^{r-1} t^j \widetilde{g}_j(t^r)$. From (i) and Prop. 7.2.14, p. 207, it follows that $\widetilde{g}(t)$ is $r\underline{k}$-summable on the arcs corresponding to the various determinations of $t = x^{1/r}$. Conversely, use formula $rt^j \widetilde{g}_j(t^r) = \sum_{\ell=0}^{r-1} \omega^{\ell(r-j)} \widetilde{g}(\omega^\ell t)$ where $\omega = e^{2\pi i/r}$ to conclude that the series $\widetilde{g}_j(t^r)$ are $r\underline{k}$-summable on $I^0_{/r}$ and then, using (i) again, the series $\widetilde{g}_j(x)$ are $\underline{k}$-summable on $\underline{I}$. □

Proposition 7.2.16, p. 208 allows us to assume k large or small at convenience. In what follows, the assumption $k > 1/2$ is quite often convenient and this allows us to consider sectors of opening π/k in $\mathbb{C}$.

7.3 The Second Approach: Malgrange-Ramis Definition

For convenience, we assume any $k > 1/2$ so that arcs of length π/k are proper sub-arcs of the circle S^1. This is always made possible by a convenient change of variable $x = t^r$ according to proposition 7.2.16, p. 208.

Like in the Ramis-Sibuya approach for k-summability, the aim is now to get rid of Gevrey asymptotics.

7.3.1 Definition

Suppose we are given an s_1-Gevrey series $\widetilde{f}(x)$, a multi-level $\underline{k} = (k_1, k_2, \ldots, k_\nu)$ and a $\underline{k}$-multi-arc $\underline{I} = (I_1, I_2, \ldots, I_\nu)$. Denote by $\varphi_0 \in H^0\big(S^1; \mathscr{A}/\mathscr{A}^{\leq -k_1}\big)$ the k_1-quasi sum of $\widetilde{f}(x)$. Multisummability can be given a definition as follows:

Definition 7.3.1 (Malgrange-Ramis Multisummability)

*The series $\widetilde{f}(x)$ is said to be $\underline{k}$-*summable on $\underline{I}$ with $\underline{k}$-sum $\underline{f} = (f_1, f_2, \ldots, f_\nu)$ *if the following conditions are satisfied:*

- ▷ *f_j belongs to $H^0\big(I_j; \mathscr{A}/\mathscr{A}^{\leq -k_{j+1}}\big)$ for all $j = 1, 2, \ldots, \nu-1$;*
- ▷ *f_ν belongs to $H^0\big(I_\nu; \mathscr{A}\big)$;*
- ▷ *the f_j's are compatible:*

$$f_{j|_{I_{j+1}}} = f_{j+1} \bmod \mathscr{A}^{\leq -k_{j+1}} \text{ for all } j = 1, 2, \ldots, \nu-1;$$

- ▷ *the f_j's are compatible with the k_1-quasi-sum φ_0 of $\widetilde{f}(x)$:*

$$\varphi_{0|_{I_j}} = f_j \bmod \mathscr{A}^{\leq -k_1} \quad \text{for all} \quad j = 1, 2, \ldots, \nu.$$

Proposition 7.3.2 *Definitions 7.2.10, p. 206, and 7.3.1, p. 209, of multisummability are equivalent (with same sums). In particular, Malgrange-Ramis $\underline{k}$-sum, when it exists, is unique.*

Proof. Suppose $\underline{f} = (f_1, f_2, \ldots, f_\nu)$ is a $\underline{k} = (k_1, k_2, \ldots, k_\nu)$-sum of $\widetilde{f}(x)$ on $\underline{I}$ in the sense of Def. 7.3.1, p. 209. By the Ramis-Sibuya theorem 5.2.1, p. 140 the k_1-quasi-sum φ_0 of $\widetilde{f}(x)$ belongs to $H^0\big(S^1; \mathscr{A}_{s_1}/\mathscr{A}^{\leq -k_1}\big)$ and is asymptotic to $\widetilde{f}(x)$. The compatibility condition (iv) implies that the same is true for $\underline{f}$; hence, definition 7.2.10, p. 206, is satisfied.

Conversely, suppose that $\underline{f}$ is a $\underline{k}$-sum of $\widetilde{f}(x)$ on $\underline{I}$ following definition 7.2.10, p. 206. By the Borel-Ritt theorem 1.3.1, p. 22, a 0-cochain representing f_1 can be completed into a 0-cochain over S^1 representing φ_0; hence, satisfying definition 7.3.1. □

Example 7.3.3

Let us go back to example 7.2.8, p. 205, in view to make explicit Malgrange-Ramis definition for the Ramis-Sibuya series $\widetilde{\mathrm{RS}}(x)$ and prove its (1,2)-summability on the (1,2)-multi-arc $\underline{I} = (I_1, I_2)$ where $I_1 = [0, \pi]$ and $I_2 = [0, \pi/2] \subset I_1$. We want to represent the 1-quasi-sum of $\widetilde{\mathrm{RS}}(x)$ by a 0-cochain with no jump on I_2, flat jumps of exponential order at most 2 on I_1 and flat jumps of order at most 1 outside of I_1. To this end, choosing again $0 < \varepsilon < \pi/4$, we consider the open covering $\mathscr{I}'$ of S^1 by

$$I_1' =]\pi/2, 3\pi/2[, \quad I_2' =]-\varepsilon, \pi/2+\varepsilon[\quad \text{and} \quad I_3' =]-\pi/2-\varepsilon, 0[.$$

Denote by E^+ the determination of the Euler function $\mathrm{E}(x)$ defined on $]-\pi/2, 3\pi/2[$ and by E^- its determination on $]-3\pi/2, \pi/2[$.

Denote by L^+ the 2-sum of the Leroy series $\widetilde{\mathrm{L}}(x)$ on $]-3\pi/4, 3\pi/4[$ and by L^- its 2-sum on $]\pi/4, 7\pi/4[$.

The 1-quasi-sum φ_0 of $\widetilde{\mathrm{RS}}(x)$ is represented by the 0-cochain of $\mathscr{I}'$ defined as follows:

$$\begin{cases} \mathrm{RS}_1(x) &= \mathrm{E}^+(x) + \mathrm{L}^-(x) \text{ on } I'_1 \\ \mathrm{RS}_2(x) &= \mathrm{E}^+(x) + \mathrm{L}^+(x) \text{ on } I'_2 \\ \mathrm{RS}_3(x) &= \mathrm{E}^-(x) + \mathrm{L}^+(x) \text{ on } I'_3. \end{cases}$$

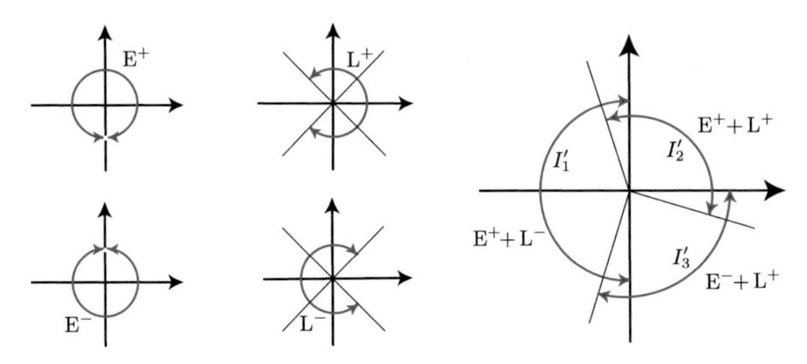

Fig. 7.2

We observe that $\mathrm{RS}_1 - \mathrm{RS}_2 = \mathrm{L}^- - \mathrm{L}^+ = c\,e^{1/x^2}$ (c is the corresponding Stokes multiplier) is exponentially flat of order 2 on $I'_1 \cap I'_2$ while $\mathrm{RS}_2 - \mathrm{RS}_3$ and $\mathrm{RS}_3 - \mathrm{RS}_1$ are exponentially flat of order 1 on $I'_2 \cap I'_3$ and $I'_3 \cap I'_1$ respectively. The (1,2)-sum (f_1, f_2) of $\widetilde{\mathrm{RS}}(x)$ on (I_1, I_2) is given by the restriction of φ_0 to I_1 and I_2 respectively, i.e., , for f_1 by the 0-cochain $(\mathrm{RS}_1, \mathrm{RS}_2)$ and for f_2 by $f_2 = \mathrm{RS}_2$.

7.3.2 Application to Differential Equations

In this section, we extend theorem 5.2.5, p. 142, to the case of several levels.

We consider a linear differential equation (or system) $Dy = 0$ with meromorphic coefficients at 0 and we suppose that the equation $Dy = 0$ has a series solution $\widetilde{f}(x)$ and multi-level $\underline{k} = (k_1, k_2, \ldots, k_\nu)$ (cf. Def. 3.3.4 (i), p. 77). Since $\widetilde{f}(x)$ stands for $\widetilde{f}(x)\mathrm{e}^{q_j}$ with $q_j = 0$ then, $\underline{k}$ is also the multi-level of $\widetilde{f}(x)$ (cf. Def. 3.3.4 (ii)). Recall that levels are always given in increasing order: they satisfy

$$0 < k_1 < k_2 < \cdots < k_\nu.$$

Definition 7.3.4 *A* $\underline{k}$*-multi-arc* $\underline{I} = (I_1, I_2, \ldots, I_\nu)$ *is said to be* $\underline{k}$-generic (*or simply,* generic) for $\widetilde{f}(x)$ *if, for all* j*, the arc* I_j *contains no Stokes arc of level* $\leq k_j$ *of* $\widetilde{f}(x)$.

Compare Def. 3.3.9, p. 79.

Theorem 7.3.5 *A series* $\widetilde{f}(x)$ *solution of a linear differential equation with multi-level* $\underline{k} = (k_1, k_2, \ldots, k_\nu)$*, is* $\underline{k}$*-summable on any* $\underline{k}$*-generic multi-arc* $\underline{I}$.

Proof. Recall that, with no loss of generality, we assume that $k_1 > 1/2$. In section 4.2, p. 123, we defined the sheaves $\mathscr{V}^{\leq -k}$ as being the sheaves of germs of solutions with exponential decay of order at least k.

They satisfy

$$\mathscr{V}^{\leq -k_\nu} \subset \mathscr{V}^{\leq -k_{\nu-1}} \subset \cdots \mathscr{V}^{\leq -k_1} = \mathscr{V}^{<0}.$$

To set the theorem we proceed by recurrence on the levels and we argue like we already did in the proof of the main asymptotic existence theorem (Cor. 3.4.2, p. 87). Since the smallest level of the equation is k_1 the series $\widetilde{f}(x)$ is at most Gevrey of order $1/k_1$. Let $f(x)$ be a k_1-quasi sum of $\widetilde{f}(x)$ (cf. Def. 5.2.3, p. 141). We have to prove that $f(x)$ can be chosen so that it has only exponentially small jumps of order at most k_{j+1} on I_j for $j = 1, \ldots, \nu - 1$ and no jump on I_ν.
From the short exact sequence $0 \to \mathscr{V}^{\leq -k_1}/\mathscr{V}^{\leq -k_2} \to \mathscr{V}/\mathscr{V}^{\leq -k_2} \to \mathscr{V}/\mathscr{V}^{\leq -k_1} \to 0$ we deduce from the long exact sequence of cohomology the exact sequence

$$H^0\big(I_1\,;\mathscr{V}/\mathscr{V}^{\leq -k_2}\big) \longrightarrow H^0\big(I_1\,;\mathscr{V}/\mathscr{V}^{\leq -k_1}\big) \longrightarrow H^1\big(I_1\,;\mathscr{V}^{\leq -k_1}/\mathscr{V}^{\leq -k_2}\big)$$

However, $H^1\big(I_1\,;\mathscr{V}^{\leq -k_1}/\mathscr{V}^{\leq -k_2}\big) = 0$ (observe that $\mathscr{V}^{\leq -k_1}/\mathscr{V}^{\leq -k_2} \simeq \mathscr{V}^{-k_1}$) since the arc I_1 contains no Stokes arc of level k_1 (cf. Lem. 3.4.3, p. 88). It follows that the map $H^0\big(I_1\,;\mathscr{V}/\mathscr{V}^{\leq -k_2}\big) \longrightarrow H^0\big(I_1\,;\mathscr{V}/\mathscr{V}^{\leq -k_1}\big)$ is surjective and we can choose the k_1-quasi-sum $f(x)$ so that it has only jumps of k_2-exponentially small size on I_1.

Then, to prove the result on the arc I_2 we consider the short exact sequence $0 \to \mathscr{V}^{\leq -k_2}/\mathscr{V}^{\leq -k_3} \to \mathscr{V}/\mathscr{V}^{\leq -k_3} \to \mathscr{V}/\mathscr{V}^{\leq -k_2} \to 0$. We apply now lemma 3.4.3, p. 88, to $\mathscr{V}^{-k_2} \simeq \mathscr{V}^{\leq -k_2}/\mathscr{V}^{\leq -k_3}$ and so we obtain $H^1\big(I_2\,;\mathscr{V}^{\leq -k_2}/\mathscr{V}^{\leq -k_3}\big) = 0$. We can again deduce from the long exact sequence of cohomology that the natural map $H^0\big(I_2\,;\mathscr{V}/\mathscr{V}^{\leq -k_3}\big) \longrightarrow H^0\big(I_2\,;\mathscr{V}/\mathscr{V}^{\leq -k_2}\big)$ is surjective. Hence, we can choose the k_1-quasi-sum $f(x)$ so that it had only jumps of k_3-exponentially small size on I_2.

We iterate the process until I_ν where we can choose the k_1-quasi-sum $f(x)$ with no jump at all. Hence $\widetilde{f}(x)$ is $\underline{k}$-summable on $\underline{I}$. Notice that $f(x)$ may have (and has in general) jumps of exponentially small size of order k_1 outside of I_1. □

7.4 The Third Approach: Iterated Laplace Integrals

The method is due to W. Balser [Bal92b]. It proceeds by recursion and is based on the fact that a convenient Borel transform of the series is itself summable with conditions that we make explicit below. Among the known ones this approach is probably the best from a numerical view point to numerically evaluate multisums.

As previously, we develop first the case of two levels (k_1, k_2). Suppose we are given $0 < k_1 < k_2$ and a (k_1, k_2)-multi-arc (I_1, I_2) (cf. Def. 7.2.9, p. 205) and set

$$1/\kappa_1 = 1/k_1 - 1/k_2$$

Denote by θ_1, θ_2 the bisecting directions of I_1, I_2 and by $\widehat{I_1}, \widehat{I_2}$ arcs centered at θ_1, θ_2 with length $|\widehat{I_1}| = |I_1| - \pi/k_2 \geq \pi/\kappa_1$ and $|\widehat{I_2}| = |I_2| - \pi/k_2$ respectively. We assume that I_1 and $\widehat{I_1}$, resp. I_2 and $\widehat{I_2}$, are simultaneously open or closed[2].

Definition 7.4.1 ((k_1, k_2)-Li-Summability) *A series $\widetilde{f}(x)$ is said to be (k_1,k_2)-summable by Laplace iteration on (I_1,I_2) (in short, (k_1,k_2)-Li-summable on (I_1,I_2)) if its k_2-Borel transform $\widehat{g}(\xi) = \mathscr{B}_{k_2}(\widetilde{f})(\xi)$ satisfies the following two conditions:*

▷ *$\widehat{g}(\xi)$ is κ_1-summable on $\widehat{I_1}$*

▷ *its κ_1-sum $g(\xi)$ can be analytically continued to an unlimited open sector Σ containing $\widehat{I_2}\times]0,+\infty[$ with exponential growth of order k_2 at infinity.*

The (k_1,k_2)-Li-sum $f(x)$ of $\widetilde{f}(x)$ on (I_1,I_2) is defined as

$$\boxed{f^{\mathrm{Li}}(x) = \mathscr{L}_{k_2,\theta}(g)(x)}$$

for each $\theta \in \Sigma$ direction and corresponding x.

It follows from the definition, that the (k_1,k_2)-Li-sum of $\widetilde{f}(x)$ on (I_1,I_2), when it exists, is unique.

Example 7.4.2 (Even Part of the Ramis-Sibuya Series)

Again, we illustrate the definition on the example of the Ramis-Sibuya series $\widetilde{\mathrm{RS}}(x)$ (Exa. 7.1.1, p. 198) for which $\underline{k} = (k_1,k_2) = (1,2)$ and, like in example 7.2.8, p. 205, we choose $\underline{I} = (I_1,I_2)$ with $I_1 = [0,\pi]$ and $I_2 = [0,\pi/2] \subset I_1$. Then, $\widehat{I_1} = [\pi/4, 3\pi/4]$, $\widehat{I_2} = \{\pi/4\}$ and $\kappa_1 = 2$ so that we must apply to $\widetilde{\mathrm{RS}}(x)$ a 2-Borel transform. In order to make the calculations simpler we choose to use rank reduction of order two, that is, to perform the calculations separately on the even and the odd part of the series. Treat now the case of the even part and thus, consider the series $\widetilde{\mathrm{M}}(x) = \widetilde{\mathrm{E}}^0(x) + \widetilde{\mathrm{L}}(x)$ where $\widetilde{\mathrm{E}}^0(x) = \sum_{n\geq 1} -(2n-1)!x^{2n}$ is the even part of the Euler series $\widetilde{\mathrm{E}}(x)$.

▷ Look first at what happens to $\widetilde{\mathrm{E}}^0(x)$ after a 2-Borel transform.

The series $\widetilde{\mathrm{E}}^0(x)$ satisfies the equation

$$Dy \equiv x^4y'' + 2x^3y' - y = x^2. \tag{7.1}$$

The homogeneous equation $Dy = 0$ admits the two linearly independent solutions $e^{1/x}$ and $e^{-1/x}$. It follows that the anti-Stokes directions for $\widetilde{\mathrm{E}}^0(x)$ are the two real half-axis. To apply a 2-Borel transform one has to apply a ramification $x = t^{1/2}$ followed by a 1-Borel transform and the inverse ramification. So, set $x = t^{1/2}$. The series $\widetilde{\mathrm{E}}^0(t^{1/2})$ is a series in integer powers of t (this is why we separated the even and the odd part of $\widetilde{\mathrm{RS}}(x)$) and it satisfies the equation

$$4t^3Y'' + 6t^2Y' - Y = t.$$

Its 1-Borel transform $\widehat{Y}(\tau)$ satisfies the equation

$$4\tau^2\widehat{Y}' + (6\tau - 1)\widehat{Y} = 1$$

[2] In case I_2 is closed of size $|I_2| = \pi/k_2$ then $\widehat{I_2}$ reduces to one point. Recall that, handling presheaves or sheaves, asymptotic conditions of sections over a closed set are valid on a convenient larger open set.

(Substitute τ for $t^2 d/dt$ and $d/d\tau$ for $1/t$. If we had not restricted the study to an even series, terms in $t^{1/2}$ would appear and this Borel equation would be a much more complicated convolution equation). With the inverse ramification $\tau = \xi^2$, the formal 2-Borel transform $\widehat{U}(\xi) = \widehat{Y}(\tau)$, of $\widetilde{\mathrm{E}}^0(x)$, satisfies the equation

$$\Delta \widehat{U} \equiv 2\xi^3 \widehat{U}' + (6\xi^2 - 1)\widehat{U} = 1. \tag{7.2}$$

The Newton polygon of Δ has a unique slope, equal to 2, at 0 and a null slope at infinity.

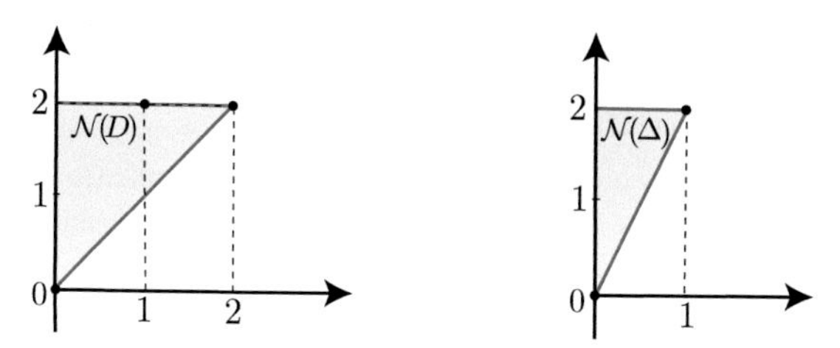

Fig. 7.3. Newton polygons of D (slope 1) and Δ (slope 2)

The 1-dimensional space of solutions of the homogeneous equation $\Delta \widehat{U} = 0$ is generated by $\widehat{u}(\xi) = e^{-1/(4\xi^2)}/\xi^3$. The anti-Stokes directions of equation (7.2), p. 213, are then the two real half-axis and, in all other θ direction, the series $\widehat{U}(\xi)$ is 2-summable (cf. Thm. 5.2.5, p. 142). For the choice $\theta = \pi/2$ this means that $\widehat{U}(\xi)$ is 2-summable on $\widehat{I}_1$. Hence, the first condition of Definition 7.4.1, p. 212, is satisfied. On another hand, the 2-sum $U(\xi)$ of $\widehat{U}(\xi)$ satisfies equation (7.2) which admits 0 and infinity as unique singular points. It can then be analytically continued up to infinity in the $\theta = \pi/4$ direction and the neighboring ones, i.e., to an unlimited open sector Σ containing $\widehat{I}_2 \times]0, +\infty[$. From the Newton polygon of Δ we know that all solutions of (7.2) have moderate growth at infinity and the second condition of definition 7.4.1, p. 212, is satisfied. It follows that the 2-Laplace transforms $\mathrm{E}^0_\theta(x)$ of $U(\xi)$ are defined in any θ direction provided $\theta \in \Sigma$. The functions $\mathrm{E}^0_\theta(x)$ are analytic continuation from each others since $U(\xi)$ admits no singular point in these directions and therefore, they define the $(1,2)$-Li-sum $\mathrm{E}^0(x)$ of $\widetilde{\mathrm{E}}^0(x)$ on (I_1, I_2).

To summarize:

$$\mathrm{E}^0_\theta(x) = \mathscr{L}_{2,\theta} \circ \mathscr{L}_{2,\theta} \circ \mathscr{B}_{2,\theta} \circ \mathscr{B}_{2,\theta}\big(\widetilde{\mathrm{E}}^0(x)\big)$$

where the Borel series are replaced by their sum together with analytic continuation when necessary. Notice that the function $\mathrm{E}^0_\theta(x)$, although asymptotic to $\widetilde{\mathrm{E}}^0(x)$, is not 1/2-Gevrey asymptotic since, otherwise, the series $\widetilde{\mathrm{E}}^0(x)$ would be a 1/2-Gevrey series (cf. 1.2.10, p. 17), which is not.

▷ Look now at what happens to the series $\widetilde{\mathrm{L}}(x)$ in this procedure.
The 2-Borel transform of $\widetilde{\mathrm{L}}(x)$ produces the convergent series

$$V(\xi) = \sum_{n \geq 0} (-1)^n \xi^{2n} = 1/(1+\xi^2).$$

It is then, 2-summable in all directions, and especially on $\widehat{I}_1$. It can be continued up to infinity with moderate growth but in the $\theta = \pm\pi/2$ directions. It can then be applied a 2-Laplace transform in any direction of an unlimited open sector Σ containing $\widehat{I}_2 \times]0, +\infty[$ to define the (1,2)-Li-sum $\mathrm{L}(x)$ of $\widetilde{\mathrm{L}}(x)$ on (I_1, I_2).

We conclude that the series $\widetilde{\mathrm{M}}(x)$ is (1,2)-Li-summable on (I_1, I_2).

▷ Compare (1,2)-sum and (1,2)-Li-sum.

Denote by $M^{\mathrm{Li}}(x)$ the $(1,2)$-sum of $\widetilde{\mathrm{M}}(x)$ on (I_1, I_2). The (1,2)-Li-sum $\mathrm{E}^0(x)$ of $\widetilde{\mathrm{E}}^0(x)$ can be continued all over I_1 by applying 2-Laplace transforms in the directions θ from $\pi/4 - \varepsilon$

to $3\pi/4+\varepsilon$ (indeed, the unique anti-Stokes directions for $\widehat{U}(\xi)$ are $\theta=0$ and $\theta=\pi$; therefore, the 2-sum $U(x)$ of $\widehat{U}(\xi)$ exists. It can be continued with moderate growth at infinity on the unlimited sector $-\pi/4<\arg(\xi)<5\pi/4$). Taking 2-Laplace transforms in the θ directions belonging to $]-3\pi/4,-\pi/4[$ allows to complete $\mathrm{E}^0(x)$ into an element $\mathscr{E}^0$ of $H^0(S^1;\mathscr{A}/\mathscr{A}^{\leq -1})$. Similarly, the section L over I_2 can be completed into an element $\mathscr{L}\in H^0(S^1;\mathscr{A}/\mathscr{A}^{\leq -k_2})$ and the sum $\mathscr{E}^0+\mathscr{L}$ provides a k_1-quasi-sum of $\widetilde{\mathrm{M}}(x)$ in the form of a section over S^1 with no jump on I_2 and flat jumps of exponential order 2 on I_1 and of order 1 outside of I_1, so that, Definition 7.3.1 is satisfied. By restriction to I_1 and I_2, this k_1-quasi-sum determines the $(1,2)$-sum (M_1,M_2) of $\widetilde{\mathrm{M}}(x)$ on (I_1,I_2) in the sense of definition 7.3.1. It follows that $M_2(x)=M^{\mathrm{Li}}(x)$ on I_2. This fact is general as it is proved below (Balser-Tougeron theorem).

Let us finally observe that the previous procedure can as well be applied to the odd part of $\widetilde{\mathrm{RS}}(x)$ after factoring x. This shows that the procedure applies to the Ramis-Sibuya series $\widetilde{\mathrm{RS}}(x)$ itself giving rise to the same $(1,2)$-sum as before.

The main result is as follows.

Theorem 7.4.3 (Balser-Tougeron: The Case of Two Levels)
(k_1,k_2)-Li-summability on (I_1,I_2) and (k_1,k_2)-summability on (I_1,I_2) are equivalent with "same" sum.

Precisely, if f^{Li} denotes the (k_1,k_2)-Li-sum of a series $\widetilde{f}(x)$ and (f_1,f_2) denotes its (k_1,k_2)-sum on (I_1,I_2) then, $f_2(x)=f^{\mathrm{Li}}(x)$ on I_2 and thus,

$$\boxed{f_2(x)=\mathscr{L}_{k_2,\theta_2'}\circ\mathscr{L}_{\kappa_1,\theta_1'}\circ\mathscr{B}_{\kappa_1}\circ\mathscr{B}_{k_2}\circ\widetilde{f}(x)}$$

when the formula makes sense and especially, for directions θ_1' and θ_2' close enough to the bisecting direction θ_2 of I_2.

The latter formula explains the denomination "by Laplace iteration".

Proof. For simplicity of language assume that I_1 and I_2 are closed arcs.

$\triangleright$ *Prove that (k_1,k_2)-summability implies (k_1,k_2)-Li-summability.*

We use the notations of Definition 7.4.1 and above. In particular, κ_1 is given by the formula $1/\kappa_1=1/k_1-1/k_2$. By hypothesis, there exists a k_1-quasi-sum of $\widetilde{f}(x)$ which induces the function $f_2(x)$ on I_2 and the 0-cochain $f_1(x)$ on I_1 (using the same notation for the 0-cochain and the element of $H^0(I_1;\mathscr{A}/\mathscr{A}^{\leq -k_2})$ it defines); the coboundary of f_1 has values in $\mathscr{A}^{\leq -k_2}$ (no jump allowed on I_2 and exponentially flat jumps of order at most k_2 on $I_1\setminus I_2$; cf. Def. 7.3.1). Denote again $\widehat{g}(\xi)=\mathscr{B}_{k_2}(\widetilde{f})(\xi)$.

For simplicity, we assume that $f_1(x)$ has only one jump, the case of more jumps being treated similarly. Thus, assume that, in restriction to I_1 (i.e., to a neighborhood of I_1), the k_1-quasi-sum reduces to two components: $f(x)=f_2(x)$ over an open arc I containing I_2 and $f^*(x)$ over an open arc I^* which we can assume to satisfy $I^*\cap I_2=\emptyset$ jointly with $I\cup I^*\supset I_1$.

The proof of theorem 5.3.14 (part (i) $\Longrightarrow$ (ii)) remains valid for $f(x)$ and $I\supset I_2$ although I is shorter than π/k_1. Like in theorem 5.3.14, denote by $\widehat{I}$ the (open since I is open) arc deduced from I by shortening it of $\pi/(2k_2)$ on both sides. It follows that there exists an unlimited sector Σ containing $\widehat{I}\times]0,+\infty[$ on which there is an analytic function $g(\xi)$ both κ_1-Gevrey asymptotic to $\widehat{g}(\xi)$ at 0 and having exponential growth of order at most k_2 at infinity. As I is smaller than I_1 the sector Σ has opening smaller than $|\widehat{I_1}| = \pi/\kappa_1$ but it contains the θ_2 direction. The question is to analytically continue $g(\xi)$ into a κ_1-Gevrey asymptotic function on $\widehat{I_1}$. To this end, let us use again the variables $Z = 1/x^{k_2}$, $\zeta = \xi^{k_2}$ and notations as in the proof of theorem 5.3.14 choosing $\theta_2 = 0$ by means of a rotation. In particular, the series $\widetilde{f}(x)$ reads now $\widetilde{F}(Z) = \sum_{n\geq k_0} a_n/Z^{n/k_2}$. We are led to the following situation (cf. Fig. 7.4):

- a function $F(Z)$ satisfying the asymptotic condition (5.11) at infinity on a sector $\mathscr{S} = [-\omega_1, +\omega_2] \times [R_0, +\infty[$
  which contains the right half-plane $\Re(Z) > 0$ but the disc $|Z| < R_0$,
- a function $F^*(Z)$ satisfying the asymptotic condition (5.11) at infinity on a sector $\mathscr{S}^* = [\omega_1^*, \omega_2^*] \times [R_0, +\infty[$ where, say, $\pi/2 < \omega_1^* < \omega_2 < \omega_2^*$,
- the difference $F(Z) - F^*(Z)$ being exponentially flat of order 1 on the intersection
  $\mathscr{S} \cap \mathscr{S}^* = [\omega_1^*, \omega_2] \times [R_0, +\infty[$.

Recall condition (5.11) for $F(Z)$ on $\mathscr{S}$: there exist $A, C > 0$ such that

$$\left| F(Z) - \sum_{n=k_0}^{N-1} \frac{a_n}{Z^{n/k_2}} \right| \leq C N^{N/k_1} \frac{A^N}{|Z|^{N/k_2}} \quad \text{for all } N \text{ and all } Z \in \mathscr{S}.$$

We also assume that $-\pi < -\omega_1$ and $\omega_1^* < \omega_2 < \omega_2^* < \pi$. Otherwise, we would proceed in several steps like in the proof of theorem 5.3.14.

Defining $G(\zeta)$ by $G(\zeta) = 1/(2\pi \mathrm{i}) \int_\gamma F(U) e^{\zeta U} dU$ where $\gamma = \gamma_1 \cup \gamma_2 \cup \gamma_3$ (see Fig 7.4) is the boundary of $\mathscr{S}$ provides a function satisfying condition (ii) of theorem 5.3.14 on the sector $\Sigma = \{-\omega_2 + \pi/2 < \arg(\zeta) < +\omega_1 - \pi/2\}$ (condition translated in the variable $\zeta = \xi^{k_2}$). Observe that Σ contains the $\theta_2 = 0$ direction.

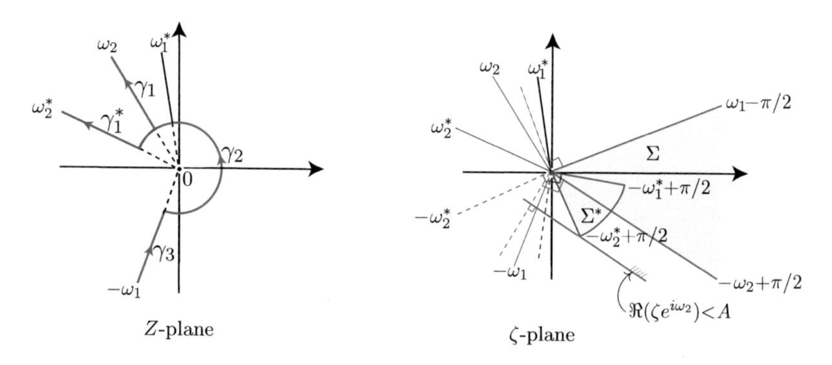

Fig. 7.4

Extend now the domain of definition of $G(\zeta)$ by moving γ_1 towards γ_1^* in the direction ω_2^*. To this end, set

$$G^*(\zeta) = \frac{1}{2\pi i}\Big(\int_{\gamma_1^*} F^*(U)e^{\zeta U}\,dU + \int_{\gamma_2\cup\gamma_3} F(U)e^{\zeta U}\,dU\Big).$$

By Cauchy's theorem the difference $\Delta(\zeta) = G(\zeta) - G^*(\zeta)$ is given by

$$\begin{aligned}\Delta(\zeta) &= \frac{1}{2\pi i}\int_{\gamma_1}(F(U) - F^*(U))e^{\zeta U}\,dU \\ &= \frac{e^{i\omega_2}}{2\pi i}\int_{R_0}^{+\infty}\big(F(Ve^{i\omega_2}) - F^*(Ve^{i\omega_2})\big)e^{\zeta Ve^{i\omega_2}}\,dV.\end{aligned}$$

However, by hypothesis, there exists $A > 0$ such that $|F(U) - F^*(U)| \le e^{-A|U|}$ (the coboundary in the variable x is exponentially flat of order k_2) so that

$$|\big(F(Ve^{i\omega_2}) - F^*(Ve^{i\omega_2})\big)e^{\zeta Ve^{i\omega_2}}| \le e^{\big(-A+\Re(\zeta e^{i\omega_2})\big)V}$$

and the function $\Delta(\zeta)$ is defined and holomorphic on the half-plane $\Re(\zeta e^{i\omega_2}) < A$. The function $G^*(\zeta) + \Delta(\zeta)$ provides the analytic continuation of $G(\zeta)$ to a (limited) sector Σ^* based on $\widehat{I^*} = [-\omega_2^* + \pi/2, -\omega_1^* + \pi/2]$ near 0. From now on, denote by $G(\zeta)$ this analytic continuation of the initial $G(\zeta)$ to $\Sigma^* \cup \Sigma$. Observe that the arguments above are unable to prove that the analytic continuation on Σ^* can be pushed up to infinity. This might however be possible, for instance, by theorem 5.3.14, when the series $\widetilde{f}(x)$ is not only (k_1,k_2)-summable on (I_1,I_2) but k_1-summable on I_1.

Recall condition (5.10) for $G(\zeta)$ on Σ: there exist $A', C' > 0$ such that

$$\Big|G(\zeta) - \sum_{n=k_0}^{N-1}\frac{a_n}{\Gamma(n/k_2)}\zeta^{n/k_2-1}\Big| \le C'N^{N/\kappa_1}A'^N|\zeta|^{N/k_2-1} \quad \text{for all } N \text{ and all } \zeta\in\Sigma.$$

Like the initial one the new function $G(\zeta)$ satisfy an asymptotic condition of the same type as condition (5.10) at 0 on $\Sigma\cup\Sigma^*$ since this is the case for $G(\zeta)$ on Σ and for $G^*(\zeta)$ and $\Delta(\zeta)$ on Σ^*. We can thus conclude that $g(\xi) = G(\xi^{k_2})$ is κ_1-asymptotic to $\widetilde{g}(\xi)$ on $\widehat{I_1}$. This achieves the proof of the fact that $\widetilde{f}(x)$ is (k_1,k_2)-Li-summable on (I_1,I_2).

▷ *Prove that (k_1,k_2)-Li-summability implies (k_1,k_2)-summability.*

Conversely, suppose $G(\zeta)$ is defined on $\Sigma\cup\Sigma^*$, has exponential growth at infinity on Σ and satisfies at 0 the asymptotic condition (5.10) on $\Sigma\cup\Sigma^*$.

From the proof of theorem 5.3.14 we know that the Laplace transforms in the directions belonging to Σ provide a function $F(Z)$ which satisfies condition (5.11) on $[-\omega_1, \omega_2]$.

Let $b = |b|e^{i\beta} \in \Sigma\cap\Sigma^*$ with, say, $\beta = -\omega_2 + \pi/2$, and let d_β be the half-line from b in the β direction. Consider the truncated Laplace transform defined in formula (1.16) by $F^b(Z)=\int_0^b G(\zeta)e^{-Z\zeta}d\zeta$. We prove, like for P_1 in the proof of theorem 5.3.14, that $F^b(Z)$ satisfies, on some half-plane $\Re(Ze^{i\beta})>R_0$, an asymptotic estimate similar to (5.11) (with new constants). Since $G(\zeta)$ has exponential growth $|G(\zeta)|\le C''e^{B|\zeta|}$ on d_β the difference $F(Z)-F^b(Z)=\int_{d_b}G(\zeta)e^{-Z\zeta}d\zeta$ satisfies

$$|F(Z)-F^b(Z)|\le C''\int_{|b|}^{+\infty}e^{(B-\Re(Ze^{i\beta}))\tau}d\tau$$

and it has then exponential decay on the half-plane $\Re(Ze^{i\beta})>B$ based on the arc $]\omega_2-\pi,\,\omega_2[$ (i.e., bisected by $-\beta=\omega_2-\pi/2$).

Consider now $b^*=|b|e^{i\beta^*}$ with $\beta^*=-\omega_2^*+\pi/2$ in Σ^*.

The function $F^{b^*}(Z)=\int_0^{b^*}G(\zeta)e^{-Z\zeta}d\zeta$ satisfies an estimate of type (51) on the half-plane $\Re(Ze^{i\beta^*})>0$ and the difference $F^b(Z)-F^{b^*}(Z)=\int_{\widehat{\beta^*\beta}}G(\zeta)e^{-Z\zeta}d\zeta$ is exponentially small of order one on the intersection of the half-planes $\Re(Ze^{i\beta})>0$ and $\Re(Ze^{i\beta^*})>0$ (since $G(\zeta)$ is bounded on the arc $\widehat{\beta^*\beta}$). Turning back to the variable x this provides functions $f(x)=F(x^{k_2})$ and $f^{b^*}(x)=F^{b^*}(x^{k_2})$ that are k_1-Gevrey asymptotic to $\widetilde{f}(x)$ on $I\supset I_2$ and $I^*\supset I_1\setminus I_2$ respectively, the difference $f(x)-f^*(x)$ being exponentially flat of order k_2 on $I\cap I^*$. In other words, the couple $(f^{b^*}(x)=F^{b^*}(x^{k_2}),f(x)=F(x^{k_2}))$ defines a (k_1,k_2)-sum of $\widetilde{f}(x)$ on (I_1,I_2).

$\triangleright$ *Prove the formula.* We have proved that the (k_1,k_2)-Li-sum and the final (k_1,k_2)-sum are both equal to $f(x)$. By construction, $F(Z)$ is the Laplace transform of $G(\zeta)$ in the directions belonging to Σ, that is, $f(x)$ is the k_2-Laplace transform of $g(\xi)$ in the θ_2' directions close to the bisecting line θ_2 of I_2. The κ_1-sum $g(\xi)$ of $\widehat{g}(\xi)$ reads $g(\xi)=\mathscr{L}_{\kappa_1,\theta_1'}\circ\mathscr{B}_{\kappa_1}(\widehat{g})(\xi)$ for $\theta_1'\in\widehat{I_1}$ and $\widehat{g}(\xi)=\mathscr{B}_{k_2}(\widetilde{f})(\xi)$ by definition. Hence, the result

$$f(x)=\mathscr{L}_{k_2,\theta_2'}\circ\mathscr{L}_{\kappa_1,\theta_1'}\circ\mathscr{B}_{\kappa_1}\circ\mathscr{B}_{k_2}\circ\widetilde{f}(x)$$

for all compatible choices of θ_1',θ_2' and x. This ends the proof. $\square$

Summability by Laplace iteration can be generalized by induction to the case of any multi-level $\underline{k}=(k_1,k_2,\dots,k_\nu)$ as follows.

Let $\underline{I}=(I_1,I_2,\cdots,I_\nu)$ be a $\underline{k}$-wide multi-arc. Recall (cf. Def. 7.2.9) that this means that $0<k_1<k_2<\cdots<k_\nu$, the arcs $I_1,I_2,\dots,I_\nu$ are respectively k_1-, k_2-,..., k_ν-wide and they satisfy $I_1\supseteq I_2\supseteq,\cdots\supseteq I_\nu$.

Denote by $\widehat{I_1},\widehat{I_2},\dots,\widehat{I_\nu}$ the arcs deduced from $I_1,I_2,\dots,I_\nu$ by truncating an arc of length $\pi/(2k_\nu)$ on both sides of each arc and set $\widehat{\underline{I}}=(\widehat{I_1},\widehat{I_2},\dots,\widehat{I}_{\nu-1})$.

Define $\widehat{\underline{k}}=(\widehat{k}_1,\widehat{k}_2,\dots,\widehat{k}_{\nu-1})$ by setting

$$1/\widehat{k}_j=1/k_j-1/k_\nu \quad \text{for all } j=1,2,\dots,\nu-1. \tag{7.3}$$

Definition 7.4.4 (Summability by Laplace Iteration: The General Case)
A series $\widetilde{f}(x)$ *is said to be* $\underline{k}$-summable by Laplace iteration on $\underline{I}$ (*in short,* $\underline{k}$-Li-summable on $\underline{I}$) *if its* k_ν*-Borel transform* $\widehat{g}(\xi)=\mathscr{B}_{k_\nu}(\widetilde{f})(\xi)$ *satisfies the following two conditions:*

▷ $\widehat{g}(\xi)$ *is* $\widehat{\underline{k}}$*-Li-summable on* $\widehat{\underline{I}}$*,*

▷ *its* $\widehat{\underline{k}}$*-Li-sum* $g(\xi)$ *can be analytically continued to an unlimited open sector* Σ *containing* $\widehat{I}_\nu\times]0,+\infty[$ *with exponential growth of order* k_ν *at infinity.*

The $\underline{k}$*-Li-sum* $f(x)$ *of* $\widetilde{f}(x)$ *on* $\underline{I}$ *is defined as* $f^{\mathrm{Li}}(x)=\mathscr{L}_{k_\nu,\theta}(g)(x)$ *for all* θ *direction in* Σ *and corresponding x.*

From the definition, the $\underline{k}$-Li-sum f^{Li} when it exists is unique.

Denote by $\kappa_1,\kappa_2,\ldots,\kappa_\nu$ the numbers given by

$$1/\kappa_j = 1/k_j - 1/k_{j+1} \quad \text{for } j=1,2,\ldots,\nu \text{ setting } 1/k_{\nu+1}=0 \tag{7.4}$$

or equivalently, by

$$\begin{cases} \dfrac{1}{k_\nu} = \dfrac{1}{\kappa_\nu} \\ \dfrac{1}{k_{\nu-1}} = \dfrac{1}{\kappa_\nu} + \dfrac{1}{\kappa_{\nu-1}} \\ \vdots \\ \dfrac{1}{k_1} = \dfrac{1}{\kappa_\nu} + \dfrac{1}{\kappa_{\nu-1}} + \cdots + \dfrac{1}{\kappa_1} \end{cases}$$

Theorem 7.4.3 can be generalized as follows.

Theorem 7.4.5 (Balser-Tougeron: The General Case)
$\underline{k}$*-Li-summability on* $\underline{I}$ *and* $\underline{k}$*-summability on* $\underline{I}$ *are equivalent with "same" sum.*

Precisely, the $\underline{k}$*-Li-sum of a series* $\widetilde{f}(x)$ *with* $\underline{k}$*-sum* $(f_1,f_2,\ldots,f_\nu)$ *on* $\underline{I}$ *is equal to* f_ν *and consequently,* f_ν *reads*

$$\boxed{f_\nu(x)=\mathscr{L}_{\kappa_\nu,\theta'_\nu}\circ\cdots\circ\mathscr{L}_{\kappa_2,\theta'_2}\circ\mathscr{L}_{\kappa_1,\theta'_1}\circ\mathscr{B}_{\kappa_1}\circ\cdots\circ\mathscr{B}_{\kappa_{\nu-1}}\circ\mathscr{B}_{\kappa_\nu}\circ\widetilde{f}(x)}$$

when the formula makes sense and, especially, for directions $\theta'_1,\ldots,\theta'_\nu$ *close to the bisecting direction* θ_ν *of* I_ν *and corresponding x.*

Proof. The theorem can be proved by recurrence as follows. It is trivially true for $\nu=1$ (and proved for $\nu=2$ in theorem 7.4.3). Suppose it is true for $\nu-1$ and prove it for ν. The fact that $\widetilde{f}(x)$ be $\underline{k}$-Li-summable on $\underline{I}$ is now equivalent to the fact $\widehat{g}(\xi)$ be $\widehat{\underline{k}}$-summable on $\widehat{\underline{I}}$ with $\underline{k}$-sum $(g_1,g_2,\ldots,g_{\nu-1})$; and that moreover, $g_{\nu-1}$ be defined with exponential growth of order k_ν at infinity on Σ. The proof that this is equivalent to saying that $\widetilde{f}(x)$ is $\underline{k}$-summable on $\underline{I}$ with sum $(f_1,f_2,\ldots,f_\nu)$ satisfying $f_\nu=\mathscr{L}_{k_\nu}(g_{\nu-1})$ is similar to the proof of theorem 7.4.3 but the fact that the 1-cochain f_1 has to be replaced by $(f_1,f_2,\ldots,f_{\nu-1})$ with jumps of order k_ν

on $I_{\nu-1} \setminus I_\nu$, of order $k_{\nu-1}$ on $I_{\nu-2} \setminus I_{\nu-1}, \ldots$, of order k_2 on $I_1 \setminus I_2$. We leave the details to the reader.

Suppose by recurrence that the sum $g_{\nu-1}(\xi)$ of $\widehat{g}(\xi)$ satisfies the formula of the theorem computed with values $(k_1, k_2, \ldots, k_\nu)$ replaced by $(\widehat{k}_1, \widehat{k}_2, \ldots \widehat{k}_{\nu-1})$. Then, the associated values $(\kappa_1, \kappa_2, \ldots, \kappa_{\nu-1})$ remain unchanged. Moreover, f_ν is given by $\mathscr{L}_{\kappa_\nu, \theta'_\nu}(g_{\nu-1})$ (observe that $k_\nu = \kappa_\nu$) and the formula follows. This ends the proof. □

7.5 The Fourth Approach: Balser's Decomposition into Sums

Suppose again that we are given a multi-level $\underline{k} = (k_1, k_2, \ldots, k_\nu)$ (cf. definition 7.2.9) and a $\underline{k}$-multi-arc $\underline{I} = (I_1, I_2, \ldots, I_\nu)$.

We saw in proposition 7.2.14 (ii) that a sum $\sum_{j=1}^{\nu} \widetilde{f}_j(x)$ of k_j-summable series on I_j is a $\underline{k}$-summable series on $\underline{I}$. We address now the converse question:

Do such splittings characterize $\underline{k}$-summable series on $\underline{I}$?

The answer is yes when $k_1 > 1/2$. Otherwise, one might have to introduce series in a fractional power of x. The condition $k_1 > 1/2$ is weakened in theorem 7.6.7.

7.5.1 The Case $k_1 > 1/2$

Look first at the relations between the various splittings of a given series.

Proposition 7.5.1 *Splittings are essentially unique.*

Precisely, suppose the series $\widetilde{f}(x)$ admits two splittings

$$\widetilde{f}(x) = \widetilde{f}_1(x) + \widetilde{f}_2(x) + \cdots + \widetilde{f}_\nu(x) = \widetilde{f}'_1(x) + \widetilde{f}'_2(x) + \cdots + \widetilde{f}'_\nu(x)$$

where, for $j = 1, 2, \ldots, \nu$, the series $\widetilde{f}_j(x)$ and $\widetilde{f}'_j(x)$ are k_j-summable on I_j.
Then, there exist series $\widetilde{u}_j(x)$ such that, for $j = 1, 2, \ldots, \nu$,

$$\widetilde{f}'_j(x) = \widetilde{u}_j(x) + \widetilde{f}_j(x) - \widetilde{u}_{j+1}(x)$$

where $\widetilde{u}_1 = \widetilde{u}_{\nu+1} = 0$ and, for $j = 2, \ldots, \nu$, the series $\widetilde{u}_j$ is k_j-summable on I_{j-1}.

Moreover, the k_j-sums f_j of the $\widetilde{f}_j$'s and f'_j of the $\widetilde{f}'_j$'s satisfy

$$f_1(x) + f_2(x) + \cdots + f_\nu(x) = f'_1(x) + f'_2(x) + \cdots + f'_\nu(x) \quad \text{on} \quad I_\nu$$

Notice that since $\widetilde{u}_j(x)$ is k_j-summable not only on the k_j-wide arc I_j but on the k_{j-1}-wide arc I_{j-1} it is also k_{j-1}-summable on I_{j-1}.

Proof. The series $\widetilde{u}_\nu(x) = \widetilde{f}'_\nu(x) - \widetilde{f}_\nu(x)$ is k_ν-summable on I_ν and, in particular, is s_ν-Gevrey. Being equal to $\big(\widetilde{f}_1(x)+\cdots+\widetilde{f}_{\nu-1}(x)\big) - \big(\widetilde{f}'_1(x)+\cdots+\widetilde{f}'_{\nu-1}(x)\big)$ it is also $(k_1,\ldots,k_{\nu-1})$-summable on $(I_1,\ldots,I_{\nu-1})$. From the Tauberian theorem 7.7.5 we deduce that $\widetilde{u}_\nu(x)$ is k_ν-summable on $I_{\nu-1}$ and, a fortiori, $k_{\nu-1}$-summable on $I_{\nu-1}$. Applying the same argument to the series $\widetilde{f}(x) - \widetilde{f}_\nu(x)$ and its two splittings $\widetilde{f}_1(x)+\cdots+\widetilde{f}_{\nu-1}(x)$ and $\widetilde{f}'_1(x)+\cdots+\widetilde{f}'_{\nu-2}(x)+\big(\widetilde{f}'_{\nu-1}(x)+\widetilde{u}_\nu(x)\big)$ proves the existence of $\widetilde{u}_{\nu-1}(x)$ and we conclude to the existence of all $\widetilde{u}_j$'s by decreasing recurrence. The equality of the sums follows directly. □

We can now give a new definition of multisummability as follows.

Definition 7.5.2 *Assume* $k_1 > 1/2$.

- ▷ *A series* $\widetilde{f}(x)$ *is said to be* $\underline{k}$-split-summable on $\underline{I}$ *if, for* $j = 1,2\ldots,\nu$, *there exist* k_j*-summable series* $\widetilde{f}_j(x)$ *on* I_j *such that*

$$\widetilde{f}(x) = \widetilde{f}_1(x) + \widetilde{f}_2(x) + \cdots + \widetilde{f}_\nu(x).$$

- ▷ *The* $\underline{k}$-split-sum of $\widetilde{f}(x)$ on $\underline{I}$ *is the function* $f(x)$ *uniquely defined on* I_ν *from any splitting of* $\widetilde{f}(x)$ *by*

$$f(x) = f_1(x) + f_2(x) + \cdots + f_\nu(x),$$

where, for $j = 1,2,\ldots,\nu$, $f_j(x)$ *denotes the* k_j*-sum of* $\widetilde{f}_j(x)$ *on* I_j.

Theorem 7.5.3 (Balser [Bal92a])
Assume $k_1 > 1/2$.
A series $\widetilde{f}(x)$ *is* $\underline{k}$*-split-summable on* $\underline{I}$ *if and only if it is* $\underline{k}$*-summable on* $\underline{I}$.
Moreover, the $\underline{k}$*-split-sum and the* $\underline{k}$*-sum agree.*

Proof. The "only if" part was considered in remark 7.2.15 and above. Prove the converse assertion: if $\widetilde{f}(x)$ is $\underline{k}$-summable on $\underline{I}$ then it is $\underline{k}$-split-summable on $\underline{I}$.

Treat first the case when $\nu = 2$. Set $s_1 = 1/k_1$ and $s_2 = 1/k_2$ as usually.
The series $\widetilde{f}(x)$ being s_1-Gevrey has a k_1-quasi-sum $f_0(x) \in H^0\big(S^1;\mathscr{A}/\mathscr{A}^{\le -k_1}\big)$ (cf. Def. 5.2.3) and, by hypothesis (cf. Def. 7.3.1), there exists a 0-cochain $f_1(x)$ in $H^0\big(I_1;\mathscr{A}/\mathscr{A}^{\le -k_2}\big)$ such that $f_1 \bmod \mathscr{A}^{\le -k_1} = f_0$ on I_1 and there exists $f_2(x)$ in $H^0\big(I_2;\mathscr{A}\big)$ such that $f_2 \bmod \mathscr{A}^{\le -k_2} = f_1$ on I_2. In other words, f_0 can be represented by a 0-cochain φ_0 with values in $\mathscr{A}/\mathscr{A}^{\le -k_1}$ and satisfying the following properties: its restriction $\varphi_1 = \varphi_{0|_{I_1}}$ to I_1 represents f_1 and has values in $\mathscr{A}/\mathscr{A}^{\le -k_2}$; its restriction to I_2 is the asymptotic function $\varphi_{0|_{I_2}} = f_2$. From lemma 7.2.3 applied to $f_1(x)$ on I_1 we are given $f'_1(x) \in H^0\big(I_1;\mathscr{A}\big)$ and $f''_1(x) \in H^0\big(S^1;\mathscr{A}/\mathscr{A}^{\le -k_2}\big)$ such that

$$f_1 = f'_1 \bmod \mathscr{A}^{\le -k_2} + f''_1 \quad \text{on} \quad I_1.$$

There exists then a 0-cochain φ_1'' with values in $\mathscr{A}/\mathscr{A}^{\leq -k_2}$ representing f_1'' which satisfies $\varphi_1'' = \varphi_1 - f_1'$ in restriction to I_1. From corollary 5.2.2, $f_1''(x)$ can be identified to an s_2-Gevrey series $\widetilde{f}_1''(x)$ of which $f_1''(x)$ is a k_2-quasi-sum. In restriction to I_2 the 0-cochain φ_1'' belongs to $H^0(I_2;\mathscr{A})$ since $\varphi''_{1|_{I_2}} = f_2 - f'_{1|_{I_2}}$. Therefore, according to definition 5.2.4, the series $\widetilde{f}_1''(x)$ is k_2-summable on I_2 with k_2-sum $f_2(x) - f'_{1|_{I_2}}(x)$. Consider now the 0-cochain $\varphi_1' = \varphi_0 - \varphi_1''$ which belongs to $H^0\big(S^1;\mathscr{A}/\mathscr{A}^{\leq -k_1}\big)$ and denote by $\widetilde{f}_1(x)$ the s_1-Gevrey series it defines (see Cor. 5.2.2). The 0-cochain φ_1' has no jump on I_1 since

$$\varphi_{0|_{I_1}} - \varphi''_{1|_{I_1}} = \varphi_1 - (\varphi_1 - f_1') = f_1'.$$

And this, again by definition 5.2.4, means that $\widetilde{f}_1(x)$ is k_1-summable on I_1. We have thus proved that $\widetilde{f}(x) = \widetilde{f}_1(x) + \widetilde{f}_2(x)$ where $\widetilde{f}_1$ is k_1-summable on I_1 and $\widetilde{f}_2 = \widetilde{f}_1''$ is k_2-summable on I_2.

To prove the general case one proceeds by recurrence. It suffices to prove that when $\widetilde{f}(x)$ is $\underline{k}$-summable on $\underline{I}$ there exist a k_ν-summable series $\widetilde{f}_\nu(x)$ on I_ν and a $\underline{k}'$-summable series $\widetilde{g}(x)$ on $\underline{I}'$ (where $\underline{k}' = (k_1,k_2,\ldots,k_{\nu-1})$ and $\underline{I}' = (I_1,I_2,\ldots,I_{\nu-1})$) such that $\widetilde{f}(x) = \widetilde{g}(x) + \widetilde{f}_\nu(x)$. Indeed, let $(f_1,f_2,\ldots,f_\nu)$ denote the $\underline{k}$-sum of $\widetilde{f}(x)$ on $\underline{I}$. The k_1-quasi-sum $f_0(x)$ is now represented by a 0-cochain φ_0 with values in $\mathscr{A}/\mathscr{A}^{\leq -k_1}$ with the following properties: for $j = 1,2,\ldots,\nu$ its restriction φ_j to I_j represents f_j on I_j; for $j = 1,2,\ldots,\nu-1$ the restriction to I_j has values in $\mathscr{A}/\mathscr{A}^{-k_{j+1}}$ and for $j=\nu$ the restriction to I_ν is $\varphi_{0|_{I_\nu}} = f_\nu$. Apply lemma 7.2.3 to $f_{\nu-1}$ on $I_{\nu-1}$ to get $f'(x) \in H^0(I_{\nu-1};\mathscr{A})$ and $f''(x) \in H^0\big(S^1;\mathscr{A}/\mathscr{A}^{\leq -k_\nu}\big)$ such that

$$f_{\nu-1} = f' \bmod \mathscr{A}^{\leq -k_\nu} + f'' \text{ on } I_{\nu-1}.$$

Like for f_1'' above, the section f'' determines a series $\widetilde{f}_\nu(x)$ which is k_ν-summable on I_ν. There exists a 0-cochain φ'' with values in $\mathscr{A}/\mathscr{A}^{\leq -k_\nu}$ which represents f'' and satisfies the condition $\varphi'' = \varphi_{\nu-1} - f'$ on $I_{\nu-1}$. The 0-cochain $\varphi_0 - \varphi''$ shows that the series $\widetilde{g}(x) = \widetilde{f}(x) - \widetilde{f}_\nu(x)$ is $\underline{k}'$-summable on $\underline{I}'$. Hence, the result. □

Remark 7.5.4 One must be aware of the fact that the splitting strongly depends on the choice of the multi-arc of summation (on the direction of summation if all arcs are bisected by the same direction). It would be interesting to know which series admit a global splitting, that is, the same splitting in almost all directions.

7.5.2 The Case $k_1 \leq 1/2$

Choose $r \in \mathbb{N}$ such that $rk_1 > 1/2$. We know from proposition 7.2.16 (i) that the series $\widetilde{f}(x)$ is $\underline{k}$-summable on $\underline{I}$ if and only if the series $\widetilde{g}(x) = \widetilde{f}(x^r)$ is $r\underline{k}$-summable on $\underline{I}_{/r}$. We can then apply Balser's theorem 7.5.3 to $\widetilde{g}(x)$ to write $\widetilde{g}(x) = \sum_{j=1}^{\nu} \widetilde{g}_j(x)$ where the series $\widetilde{g}_j(x)$ are rk_j-summable on $I_{j/r}$. This way, we obtain a splitting $\widetilde{f}(x)=\sum_{j=1}^{\nu}\widetilde{f}_j(x)$ of $\widetilde{f}(x)$ by setting $\widetilde{f}_j(x)=\widetilde{g}_j(x^{1/r})$ for all j. However, in general, the series $\widetilde{f}_j(x)$ thus obtained contains roots of x (they are series in the variable $x^{1/r}$) and the splitting does not fit the statement of theorem 7.5.3. Actually, as shown by the example below (cf. proof of Prop. 7.5.7) in the case when $k_1=1/2$ there might exist no splitting in integer powers of x and we are driven to set the following definition.

Definition 7.5.5 *Suppose $k_1\le 1/2$.*
*A series $\widetilde{f}(x)$ is said to be $\underline{k}$-*split-summable on $\underline{I}$ *if, given $r\in\mathbb{N}$ such that $rk_1>1/2$, the series $\widetilde{g}(x)=\widetilde{f}(x^r)$ satisfies definition 7.5.2.*

With this definition and proposition 7.2.16 (i) we can assert in all cases the equivalence of $\underline{k}$-summability and $\underline{k}$-split-summability on $\underline{I}$ with "same" sum.
Uniqueness holds as follows:
Let r and r' be such that rk_1 and $r'k_1>1/2$. Set $\widetilde{g}(x)=\widetilde{f}(x^r)$ and $\widetilde{g}'(x)=\widetilde{f}(x^{r'})$ with splittings $\widetilde{g}(x)=\widetilde{g}_1(x)+\cdots+\widetilde{g}_\nu(x)$ and $\widetilde{g}'(x)=\widetilde{g}'_1(x)+\cdots+\widetilde{g}'_\nu(x)$ respectively. Denote by $R=\rho r=\rho' r'$ the l.c.m. of r and r'. Then, $\widetilde{g}_1(x^\rho)+\cdots+\widetilde{g}_\nu(x^\rho)$ and $\widetilde{g}'_1(x^{\rho'})+\cdots+\widetilde{g}'_\nu(x^{\rho'})$ are two splittings of $\widetilde{f}(x^R)$ into Rk_1-, $\ldots$, Rk_ν-summable series. Henceforth, they are essentially equal (cf. Prop. 7.5.1).

Show now that there might exist no splitting into integer power series.

To this end, consider the case of a multi-level $\underline{k}=(k_1,k_2)$ satisfying

$$1/\kappa_1 := 1/k_1 - 1/k_2 \ge 2.$$

Let (I_1,I_2) be a $\underline{k}$-multi-arc. Assume, for instance, that I_1 and I_2 are closed with same middle point θ_0 and, by means of a rotation, that $\theta_0=0$. For simplicity, assume that $1/\kappa_1<4$ and that $|I_2|=\pi/k_2$. Thus, with notations of sections 5.3.3 and 7.4, the closed arc $\widehat{I}_1$ centered at 0 with length $|\widehat{I}_1|=|I_1|-\pi/k_2$ overlaps just once (since $2\pi\le|\widehat{I}_1|<4\pi$) and $\widehat{I}_2$ reduces to $\theta=0$.

Consider a series $\widetilde{f}(x)=\widetilde{f}_1(x)+\widetilde{f}_2(x)$ which is the sum of a k_1-summable series $\widetilde{f}_1(x)$ on I_1 and of a k_2-summable series $\widetilde{f}_2(x)$ on I_2 (series in integer powers of x). Denote by $\widehat{g}=\mathscr{B}_{k_2}(\widetilde{f})$, $\widehat{g}_1=\mathscr{B}_{k_2}(\widetilde{f}_1)$ and $\widehat{g}_2=\mathscr{B}_{k_2}(\widetilde{f}_2)$ the series deduced from $\widetilde{f},\widetilde{f}_1$ and $\widetilde{f}_2$ by a k_2-Borel transform. We know from the pre-Tauberian theorem 5.3.14 that $\widehat{g}_1(\xi)$ is κ_1-summable on $\widehat{I}_1$ with κ_1-sum $g_1(\xi)$. The theorem asserts also that $g_1(\xi)$ has an analytic continuation to an unlimited open sector $\Sigma=\widehat{J}_1\times]0,+\infty[$ containing $\widehat{I}_1\times]0,+\infty[$ with exponential growth of order k_2 at infinity. Since Σ is wider than 2π it has a self-intersection $\dot{\Sigma}$ that contains the negative real axis. Narrowing it, if necessary, we can assume that Σ overlaps just once like $\widehat{I}_1$ does. Denote by $\dot{g}_1(\xi)=g_1(\xi)-g_1(\xi e^{2\pi i})$ the difference of the two determinations of g_1 on $\dot{\Sigma}$. On the other hand, $\widehat{g}_2(\xi)$ is convergent with sum $g_2(\xi)$. Hence, the difference of two determinations $\dot{g}_2(\xi)=g_2(\xi)-g_2(\xi e^{2\pi i})$ is identically 0 near 0 and can then be continued all over $\dot{\Sigma}$ by 0. Set $\dot{g}(\xi)=\dot{g}_1(\xi)+\dot{g}_2(\xi)$. We can thus state:

Lemma 7.5.6 *The germ $\dot{g}(\xi)$ can be analytically continued all over $\dot{\Sigma}$.*

Proposition 7.5.7 *With notations and conditions as before (and especially, with the condition $1/\kappa_1 := 1/k_1 - 1/k_2 \geq 2$) there exists series that are (k_1,k_2)-summable on (I_1,I_2) but cannot be split into the sum of a k_1-summable series on I_1 and a k_2-summable series on I_2 if one restricts the splitting to series in integer powers of x.*

Proof. To exhibit a counter-example to the splitting of (k_1,k_2)-summable series suppose that $\dot{g}(-1) \neq 0$ and consider the series $\widehat{G}(\xi) = \widehat{g}(\xi).\sum_{n\geq 0}(-1)^n\xi^n$.
Set then $G(\xi) = g(\xi)/(1+\xi)$ and $\dot{G}(\xi) = \dot{g}(\xi)/(1+\xi)$ and denote the k_2-Laplace transform of the series $\widehat{G}(\xi)$ by $\widetilde{F}(x) = \mathscr{L}_{k_2}(\widehat{G})(x)$. The function $G(\xi)$ is κ_1-asymptotic to $\widehat{G}(\xi)$ on $\widehat{I_1}$ (cf. Prop.1.2.12) and it can be analytically continued with exponential growth of order k_2 at infinity to an unlimited open sector σ containing $\widehat{I_2}\times]0,+\infty[$. Indeed, the function $1/(1+\xi)$ is bounded at infinity and has a pole at -1. The function $g(\xi)$ is analytic with exponential growth of order k_2 at infinity on an unlimited open sector σ'. In case σ' does not contain the negative real axis one can take $\sigma = \sigma'$; otherwise, set $\sigma = \sigma' \cap \Re(\xi) > 0$ for instance. According to definition 7.4.1 and Balser-Tougeron theorem 7.4.3 this shows that the series $\widetilde{F}(x)$ is (k_1,k_2)-summable on (I_1,I_2). However, $\dot{G}(\xi) = \dot{g}(\xi)/(1+\xi)$ has a pole at $\xi = -1$ which belongs to $\dot{\Sigma}$ and, thus, $\dot{G}(\xi)$ cannot be continued up to infinity over $\dot{\Sigma}$. From the lemma we conclude that the series $\widetilde{F}(x)$, however (k_1,k_2)-summable on (I_1,I_2), is not the sum of a k_1-summable series on I_1 and of a k_2-summable series on I_2 if one requires series in integer powers of x. □

7.6 The Fifth Approach: Écalle's Acceleration

Historically, this approach called *accelero-summation* was first to solve the problem of summation in the case of several levels. Introduced by J. Écalle in a very general setting applying to series solutions of non-linear equations and more general functional equations, it was adapted by J. Martinet and J.-P. Ramis to the case of solutions of linear differential equations in [MarR91]. The method proceeds by recursion on increasing levels whereas the iterated Laplace method proceeds with decreasing levels. Each step is performed by means of special integral operators called *accelerators* which involve the successive levels taken two-by-two. J. Écalle defined more general accelerators to treat the case of solutions of a variety of equations. We do not consider them here.

In this section, for simplicity, we work in a given direction θ_0, that is, we consider only multi-arcs $\underline{I} = (I_1,I_2,\ldots,I_\nu)$ with common middle point θ_0.

We begin with the study of the example of the Ramis-Sibuya series $\widetilde{\mathrm{RS}}(x)$ (Exa. 7.1.1) which is our reference example in multisummability.

Example 7.6.1 (Accelero-Summation of $\widetilde{\mathrm{RS}}$)

We saw in example 7.1.1 that the series $\widetilde{\mathrm{RS}}(x)$ is k-summable for no $k>0$ in the θ directions satisfying $\theta \in [\pi/4, 3\pi/4] \bmod \pi$ and therefore, no k-Borel-Laplace process applies in these directions. In the case of $\widetilde{\mathrm{RS}}(x)$ and more generally of a $(1,2)$-summable series the method consists, in some way, in applying simultaneously a 1- and a 2-Borel-Laplace process as shown below.

Fix a non anti-Stokes direction θ belonging to $]\pi/4, 3\pi/4[\bmod \pi$ and, when no confusion is possible, denote simply by $\mathscr{B}_1$ and $\mathscr{B}_2$ instead of $\mathscr{B}_{1,\theta}$ and $\mathscr{B}_{2,\theta}$ the 1- and the 2-Borel transforms in the θ direction and by $\mathscr{L}_1$ and $\mathscr{L}_2$ instead of $\mathscr{L}_{1,\theta}$ and $\mathscr{L}_{2,\theta}$ the 1- and the 2-Laplace integrals in the θ direction (cf. Def. 5.3.5). Contrary to the 2-Borel transform the (formal) 1-Borel transform applied to $\widetilde{\mathrm{E}}(x)$ and $\widetilde{\mathrm{L}}(x)$ provides convergent series. This invites us to begin with the 1-Borel-Laplace process followed by the 2-Borel-Laplace process. The 1-Borel transform of $\mathrm{E}(x)$ can be continued to infinity in the θ direction (and the neighboring directions) with exponential growth of order one and can then be applied a Laplace operator $\mathscr{L}_1$. On the contrary, the 1-Borel transform of $\mathrm{L}(x)$ can only be continued with exponential growth of order two (cf. Exa. 5.3.17). Hence, the Laplace operator $\mathscr{L}_1$ does not apply to $\mathscr{B}_1\big(\widetilde{\mathrm{RS}}(x)\big)(\xi)$.

A solution to this problem consists in merging the next two arrows of the process as indicated in the diagram:

$$\widetilde{\mathrm{RS}}(x) \xrightarrow{\ \mathcal{B}_1\ } \varphi(\xi) \overset{\mathcal{L}_1}{\dashrightarrow} \bullet \overset{\mathcal{B}_2}{\dashrightarrow} \psi(\zeta) \xrightarrow{\ \mathcal{L}_2\ } \mathrm{RS}(x).$$
$$\underbrace{\qquad\qquad\qquad}_{\mathbf{A}_{2,1}}$$

Formally, we can write

$$\begin{aligned}
\mathbf{A}_{2,1}(\varphi)(\zeta) &= \mathscr{B}_2\big(\mathscr{L}_1(\varphi)\big)(\zeta)\\
&= \frac{1}{2\pi \mathrm{i}}\int_{\gamma_{2\theta}} \Big(e^{\zeta^2/t}\int_0^{e^{\mathrm{i}\theta}\infty} \varphi(\xi)e^{-\xi/t^{1/2}}d\xi\Big)\frac{dt}{t^2}\\
&= \frac{1}{2\pi \mathrm{i}}\int_0^{e^{\mathrm{i}\theta}\infty} \varphi(\xi)\int_{\gamma_{2\theta}} \exp\Big(\frac{\zeta^2}{t}-\frac{\xi}{t^{1/2}}\Big)\frac{dt}{t^2}\,d\xi \quad \text{(commuting the integrals)}\\
&= \frac{1}{2\pi \mathrm{i}}\int_0^{e^{\mathrm{i}\theta}\infty} \varphi(\xi)\int_{\mathscr{H}} \exp\Big(u-\frac{\xi}{\zeta}u^{1/2}\Big)\frac{du}{\zeta^2}\,d\xi \quad \text{(setting } u=\zeta^2/t)
\end{aligned}$$

where $\mathscr{H}$ denotes a Hankel contour around the negative real axis. Setting $u^{1/2}=iv$ in the integral kernel $\mathscr{C}_2(\tau)=\frac{1}{2\pi \mathrm{i}}\int_{\mathscr{H}} e^{u-\tau u^{1/2}}du$ we recognize the derivative of the Fourier transform of the Gauss function e^{-v^2} and we obtain $\mathscr{C}_2(\tau)=\tau e^{-\tau^2/4}/(2\sqrt{\pi})$. This kernel is (for $\Re\tau>0$) exponentially small of order two and can then be applied to the 1-Borel transform $\varphi(\xi)$ of the Ramis-Sibuya series $\widetilde{\mathrm{RS}}(x)$. The operator $\mathbf{A}_{2,1}$ defined by

$$\mathbf{A}_{2,1}\left(\varphi(\xi)\right)(\zeta)=\frac{1}{\zeta^2}\int_0^{e^{i\theta}\infty}\varphi(\xi)\mathscr{C}_2(\xi/\zeta)d\xi$$

is called $(2,1)$-*accelerator*.

Now, the function $\mathbf{A}_{2,1}\left(\varphi(\xi)\right)(\zeta)$ satisfies the same equation as $\mathscr{B}_2(\widetilde{\mathrm{RS}}(x))(\zeta)$ and can thus be applied a 2-Laplace transform (see Exa. 7.4.2). Finally, we obtain an asymptotic function on a quadrant bisected by θ. This function has the Ramis-Sibuya series $\widetilde{\mathrm{RS}}(x)$ as Taylor series since, formally, the followed process is the identity.

The formal calculation made above to define the accelerator $\mathbf{A}_{2,1}$ from the formal composite $\mathscr{B}_2\circ\mathscr{L}_1$ can be made with $\mathscr{B}_{k_2}\circ\mathscr{L}_{k_1}$ for any pair of levels $k_1<k_2$. We

obtain

$$\boxed{\mathbf{A}_{k_2,k_1}\big(\varphi(\xi)\big)(\zeta)=\frac{1}{\zeta^{k_2}}\int_{\xi=0}^{e^{i\theta}\infty}\varphi(\xi)\,\mathscr{C}_{k_2/k_1}\big((\xi/\zeta)^{k_1}\big)d(\xi^{k_1})} \tag{7.5}$$

where the kernel $\mathscr{C}_{k_2/k_1}$ is defined by

$$\boxed{\mathscr{C}_\alpha(\tau)=\frac{1}{2\pi\mathrm{i}}\int_{\mathscr{H}} e^{w-\tau w^{1/\alpha}}dw, \quad \alpha>1} \tag{7.6}$$

with $\mathscr{H}$ denoting again a Hankel contour around the negative real axis oriented positively.

For simplicity, we chose to denote the accelerators without making explicit the θ direction in which these operators are considered. In case it should be useful to make it explicit we would denote for instance by $\mathbf{A}^\theta_{k_2,k_1}$ the accelerator of levels $k_1<k_2$ in the θ direction.

Proposition 7.6.2 *Given $\alpha>1$ let $\beta>1$ denote its conjugate number:*

$$1/\alpha+1/\beta=1,$$

and let Δ_β denote the sector $|\arg(\tau)|<\pi/(2\beta)$.

The kernel $\mathscr{C}_\alpha(\tau)$ is flat of exponential order β at infinity on the sector Δ_β, i.e., *for all $\delta>0$ there exist constants $c_1,c_2>0$ such that*

$$\boxed{|\mathscr{C}_\alpha(\tau)|\le c_1\exp(-c_2|\tau|^\beta)\quad on\quad |\arg(\tau)|\le\pi/(2\beta)-\delta}$$

Proof. The proof was already given in the part "(i) implies (ii) point 3" of the proof of the pre-Tauberian theorem 5.3.14, p. 166.

To fit the notations in the pre-Tauberian theorem, perform the change of variable $U=\tau^\alpha w$ in the integral defining $C_\alpha(\tau)$. Set $F(U)=\exp(-U^{1/\alpha})$, $\zeta=1/\tau^\alpha$ and $G(\zeta)=C_\alpha(\zeta^{1/\alpha})/\zeta$. We obtain $G(\zeta)=1/(2\pi\mathrm{i})\int_{\mathscr{H}}F(U)e^{\zeta U}dU$. Then, set $k_1=1$, $k_2=\alpha$ and $\kappa_1=\beta$. Set $a_n=0$ for all n in formula (5.11), p. 168 (the exponential $F(U)$ is flat on $\arg(U)<\alpha\pi/2$).

We conclude from estimate (5.10), p. 168 and proposition 1.2.17, p. 20. □

Corollary 7.6.3 *Set, as before, $1/\kappa_1=1/k_1-1/k_2$.*
The accelerator $\mathbf{A}_{k_2,k_1}$ applies to any function φ with exponential growth of order κ_1 at infinity in the θ direction.

With this result the accelerator $\mathbf{A}_{k_2,k_1}$ appears like similar to a κ_1-Laplace operator and has similar properties. In the case when φ has exponential growth not only of order κ_1 but of order $k_1<\kappa_1$ the function φ can be applied the Laplace operator $\mathscr{L}_{k_1}$ followed by the Borel operator $\mathscr{B}_{k_2}$ and we have thus $\mathbf{A}_{k_2,k_1}(\varphi)=\mathscr{B}_{k_2}\big(\mathscr{L}_{k_1}(\varphi)\big)$.

Let us now state the following result generalizing theorem 5.3.14 “(ii) $\Longrightarrow$ (i)”, p. 170 (cf. [Bal94, Thm 5.2.1]).

Lemma 7.6.4 *Let $\widehat{k}, k_1, k_2 > 0$ be given. Assume $k_1 < k_2$ and define κ and k by*

$$1/\kappa = 1/k_1 - 1/k_2 \quad \text{and} \quad 1/k = 1/\kappa + 1/\widehat{k}.$$

Let $\widehat{\sigma} = \widehat{I} \times]0, +\infty[$ be an unlimited sector and denote by I the arc with same middle point θ_0 as $\widehat{I}$ and length $|\widehat{I}| + \pi/\kappa$.

Suppose that $g(\xi)$ is analytic on $\widehat{\sigma}$, belongs to $\mathscr{A}_{1/\widehat{k}}(\widehat{I})$ at 0 and has exponential growth of order κ at infinity on $\widehat{\sigma}$.

Then, the function $h(\xi) = \mathbf{A}_{k_2,k_1}(g)(\xi)$ belongs to $\mathscr{A}_{1/k}(I)$.
Moreover, if $\widehat{g}(\xi) = \sum c_n \xi^{n-k_1} / \Gamma(n/k_1)$ is the Taylor expansion of $g(\xi)$ on $\widehat{I}$ then,

$$\widehat{h}(\xi) = \mathscr{B}_{k_2} \circ \mathscr{L}_{k_1}(\widehat{g})(\xi) = \mathscr{B}_{k_2}\left(\sum c_n \xi^n\right) = \sum c_n x^{n-k_2} / \Gamma(n/k_2)$$

is the Taylor expansion of $h(\xi)$ on I.

With no loss of generality, we took $\widehat{g}(\xi)$ in the form of a k_1-Borel transformed series because this is the form in which it appears in the accelero-summation process.

Definition 7.6.5 *Let $\underline{k} = (k_1, k_2, \dots, k_\nu)$ be a multi-level and θ_0 a direction.*
*A series $\widetilde{f}(x)$ is said to be $\underline{k}$-*accelero-summable (*or,* accelero-summable) in the θ_0 direction *if it can be applied the following sequence of operators in the $\theta = \theta_0$ direction and the neighboring ones resulting in the* accelero-sum $f(x)$*:*

$$\widetilde{f}(x) \xrightarrow{\mathscr{B}_{k_1}} \bullet \xrightarrow{\mathbf{A}_{k_2,k_1}} \bullet \xrightarrow{\mathbf{A}_{k_3,k_2}} \bullet \cdots \bullet \xrightarrow{\mathbf{A}_{k_\nu,k_{\nu-1}}} \bullet \xrightarrow{\mathscr{L}_{k_\nu}} f(x).$$

The schematic diagram above requires some explanation: typically we start with a divergent series $\widetilde{f}(x)$ although the process could be applied to a convergent series as well. If we apply to $\widetilde{f}(x)$ a formal Borel transform of level k_1 we get a formal series $\widehat{f}_1(\xi)$ and, in view to apply an accelerator, we have to “change” $\widehat{f}_1(\xi)$ into an analytic function defined up to infinity in the θ_0 direction and the neighboring directions. In the process, we suppose that $\widehat{f}_1(\xi)$ is convergent (i.e., $\widetilde{f}(x)$ is k_1-Gevrey) and that its sum can be analytically continued up to infinity in the θ_0 direction and the close enough directions. We suppose moreover that the continuation has exponential growth of order κ_1 at infinity (cf. formulæ (7.4), p. 218, for the definition of the κ_j's). We then apply the accelerator $\mathbf{A}_{k_2,k_1}$ and we assume that the resulting function can be analytically continued up to infinity in the θ_0 direction and the neighboring directions with exponential growth of order κ_2, and so on... until we apply the Laplace transform of order $\kappa_\nu = k_\nu$ to get the accelero-sum $f(x)$ of $\widetilde{f}(x)$.

The term accelero-summation is commonly used for a larger class of operators associated with various kernels depending on the type of problem one wants to solve. Here, we refer always to the definition given above.

Theorem 7.6.6 *$\underline{k}$-multisummability and $\underline{k}$-accelero-summability in a given θ_0 direction are equivalent with "same" sum.*

Precisely, if $(f_1, f_2, \dots, f_\nu)$ is the $\underline{k}$-multisum of a series $\widetilde{f}(x)$ in direction θ_0 then, $f = f_\nu$ is its $\underline{k}$-accelero-sum in the θ_0 direction.

Proof. We sketch the case of two levels $\underline{k} = (k_1, k_2)$ letting the reader perform the general case by iteration. Since the theorem holds true for polynomials we can assume that the given series have valuation greater than k_2 so that their k_2-Borel series contain only positive powers of ξ.

▷ *multisummability implies accelero-summability.*

Without loss of generality we can assume that $k_1 > 1/2$. From theorem 7.5.3, p. 220, it is then sufficient to prove that k_1- and k_2-summable series in the θ_0 direction are accelero-summable in that direction.

Suppose first that $\widetilde{f}(x)$ is k_1-summable in the θ_0 direction with k_1-sum $f_1(x)$. Then, its Borel transform $\mathscr{B}_{k_1}(\widetilde{f})(\xi)$ is convergent at 0 and its sum $\varphi(\xi)$ can be analytically continued with exponential growth of order k_1 at infinity in the θ_0 direction and the neighboring ones. Since $\kappa_1 > k_1$ one can apply the accelerator $\mathbf{A}_{k_2,k_1}$ to $\varphi(\xi)$ in the θ_0 direction and the neighboring ones. The resulting function can be analytically continued to infinity with moderate growth; then, it can be applied a k_2-Laplace transform to produce a (k_1, k_2)-sum $f(x)$. Actually in that case, we have $\mathbf{A}_{k_2,k_1}(\varphi) = \mathscr{B}_{k_2} \circ \mathscr{L}_{k_1}(\varphi)$ (the integrals exist). It follows that $f(x) = \mathscr{L}_{k_2} \circ \mathscr{B}_{k_2}(f_1)(x)$ and $f(x)$ is the restriction of f_1 to an open sector with opening larger than π/k_2 (but possibly smaller than π/k_1) centered at θ_0.

Suppose now that $\widetilde{f}(x)$ is k_2-summable in the θ_0 direction with k_2-sum $f(x)$ and let κ_1 be defined by $1/\kappa_1 = 1/k_1 - 1/k_2$. Since the theorem holds true for polynomials we can assume that $k_0 \geq \max(k_2, \kappa_1)$, hence also $k_0 > k_1$. Since the series $\widetilde{f}(x)$ is k_2-summable it is $1/k_2$-Gevrey and there exist constants $b, c > 0$ such that $|a_n| \leq c\Gamma(n/k_2) b^n$ for all n. This implies that $\widehat{g}_1(\xi) = \mathscr{B}_{k_1}(\widetilde{f})(\xi)$ converges for all $\xi \in \mathbb{C}$. Its sum $g_1(\xi)$ is then an entire function and we have to prove that it has exponential growth of order κ_1 at infinity.

To this end, we write $g_1(\xi) = \sum_{N \geq k_0 - k_1} a_{N+k_1} \xi^N / \Gamma(1 + N/k_1)$ and, since the coefficients a_n satisfy $|a_{N+k_1}| \leq c\Gamma\big((N+k_1)/k_2\big) b^{N+k_1} \leq c\Gamma(1+N/k_2) b^{N+k_1}$ we obtain

$$|g_1(\xi)| \leq cb^{k_1} \sum_{N \geq k_0 - k_1} \frac{\Gamma(1+N/k_2)}{\Gamma(1+N/k_1)} (b|\xi|)^N.$$

It follows from the Stirling formula that there exist constants $b', c' > 0$ such that

$$\begin{aligned}\Gamma(1+N/k_1)/\Gamma(1+N/k_2) &= \sqrt{k_2/k_1}\, N^{N/\kappa_1} \mathrm{e}^{-N/\kappa_1} \big(k_2^{1/k_2} / k_1^{1/k_1}\big)^N (1 + o(1)) \\ &\geq c'\Gamma(1+N/\kappa_1) b'^{-N}.\end{aligned}$$

Henceforth, there exists a constant $C > 0$ such that

$$|g_1(\xi)| \leq C \sum_{N \geq 0} \frac{(bb'|\xi|)^N}{\Gamma(1+N/\kappa_1)} = C\,\mathrm{E}_{\kappa_1}(bb'|\xi|)$$

where $\mathrm{E}_{\kappa_1}(t) = \sum_{N\geq 0} \frac{t^N}{\Gamma(1+N/\kappa_1)}$ is the Mittag-Leffler function of index κ_1. It is well known (cf. [Val42, Sect. 241]) that this Mittag-Leffler function is an entire function with exponential growth of order κ_1 at infinity. Hence the same result holds for $g_1(\xi)$ which can thus be applied the accelerator $\mathbf{A}_{k_2,k_1}$ in the θ_0 direction.

Lemma 7.6.4, p. 226, applied to $g_1(\xi)$ with $1/\widehat{k} = 0$ shows that $\mathbf{A}_{k_2,k_1}(g_1)(\xi)$ is κ_1-Gevrey asymptotic to $\widehat{g}_2(\xi) = \mathscr{B}_{k_2}(\widetilde{f})(\xi)$ on a sector of opening larger than π/κ_1 bisected by θ_0. Since $\widetilde{f}(x)$ is k_2-summable in the θ_0 direction its k_2-Borel series $\widehat{g}_2(\xi)$ is convergent and $\mathbf{A}_{k_2,k_1}(g_1)(\xi)$, being $1/\kappa$-Gevrey asymptotic to $\widehat{g}_2(\xi)$ on an arc of length more than π/κ, coincide with the sum $g_2(\xi)$ of the series $\widehat{g}_2(\xi)$. A k_2-Laplace transform provides then the k_2-sum $f(x)$ of $\widetilde{f}(x)$ in the θ_0 direction.

▷ *Accelero-summability implies multisummability.*

Suppose $\widetilde{f}(x)$ is (k_1,k_2)-accelero-summable in the θ_0 direction. This means that its k_1-Borel transform $\widehat{g}(\xi) = \sum_{p>k_2} c_p x^p$ converges on a disc $\widehat{D}_\rho = \{|\xi| < \rho\}$ and that the sum $g(\xi)$ of $\widehat{g}(\xi)$ can be analytically continued to an open sector $\widehat{\sigma}$ neighboring the θ_0 direction, with exponential growth of order κ_1 at infinity (recall that $1/\kappa_1 = 1/k_1 - 1/k_2$). We choose $\widehat{\sigma}$ so narrow about θ_0 that $\widetilde{f}(x)$ is (k_1,k_2)-accelero-summable in all θ directions belonging to $\widehat{\sigma}$. Without loss of generality, we assume that $1/\kappa_1 < 2$. Let us show that under this condition the series $\widetilde{f}$ splits into a sum $\widetilde{f}(x) = \widetilde{f}_1(x) + \widetilde{f}_2(x)$ where $\widetilde{f}_1(x)$ and $\widetilde{f}_2(x)$ are respectively k_1- and k_2-summable in the θ_0 direction (cf. [Bal92a, Lem. 1 and 2]).

Show first that it suffices to consider the case when $\widetilde{f}(x)$ is (k_1,k_2)-accelero-summable (i.e., (k_1,k_2)-accelero-summable in almost all directions). Let $0 < r < \rho$. The circle γ centered at 0 with radius r belongs to $\widehat{D}_\rho$. Denote by γ_1 the arc of γ oriented positively outside the sector $\widehat{\sigma}$ and by γ_2 the arc oriented positively inside $\widehat{\sigma}$. From Cauchy's integral formula we know that, on the interior $\widehat{D}_r$ of γ, the function $g(\xi)$ satisfies

$$g(\xi) = \frac{1}{2\pi i} \int_\gamma \frac{g(\eta)}{\eta - \xi} d\eta$$

and $g(\xi) = g_1(\xi) + g_2(\xi)$ there if one sets

$$g_1(\xi) = \frac{1}{2\pi i} \int_{\gamma_1} \frac{g(\eta)}{\eta - \xi} d\eta \quad \text{and} \quad g_2(\xi) = \frac{1}{2\pi i} \int_{\gamma_2} \frac{g(\eta)}{\eta - \xi} d\eta.$$

The function $g_1(\xi)$ has an analytic continuation to $\widehat{D}_r \cup \widehat{\sigma}$ which is bounded at infinity. The k_1-Laplace transform $\widetilde{f}_1(x)$ of the Taylor series $\widehat{g}_1(\xi)$ of $g_1(\xi)$ is therefore a k_1-summable series in any direction belonging to $\hat{\sigma}$. Denote by $c\ell(\widehat{\sigma})$ the closure of $\widehat{\sigma}$ and set $\widehat{\sigma}' = \mathbb{C} \setminus c\ell(\widehat{\sigma})$. Thus, $\mathbb{C} = \widehat{\sigma} \cup \widehat{\sigma}' \cup d' \cup d''$ where d' and d'' are the two half-lines limiting $\widehat{\sigma}$ and $\widehat{\sigma}'$. The function $g_2(\xi)$ has an analytic continuation to $\widehat{D}_r \cup \widehat{\sigma}'$ which is bounded at infinity. The k_1-Laplace transform $\widetilde{f}_2(x)$ of its Taylor series $\widetilde{g}_2(\xi)$ is thus k_1-summable, hence (k_1,k_2)-accelero-summable, in all directions of $\widehat{\sigma}'$. On another hand, $\widetilde{f}_2(x) = \widetilde{f}(x) - \widetilde{f}_1(x)$, is (k_1,k_2)-accelero-summable in all directions of $\widehat{\sigma}$. We conclude that $\widetilde{f}_2(x)$ is (k_1,k_2)-accelero-summable in all directions but, maybe, the singular directions d' and d'', that is, (k_1,k_2)-accelero-summable. Since $\widetilde{f}_1(x)$ is k_1-summable –hence also (k_1,k_2)-summable– in the θ_0 direction it suffices to prove the theorem with $\widetilde{f}_2(x)$ in place of $\widetilde{f}(x)$. To continue the proof we can then assume that $\widetilde{f}(x)$ is (k_1,k_2)-accelero-summable (in almost all directions).

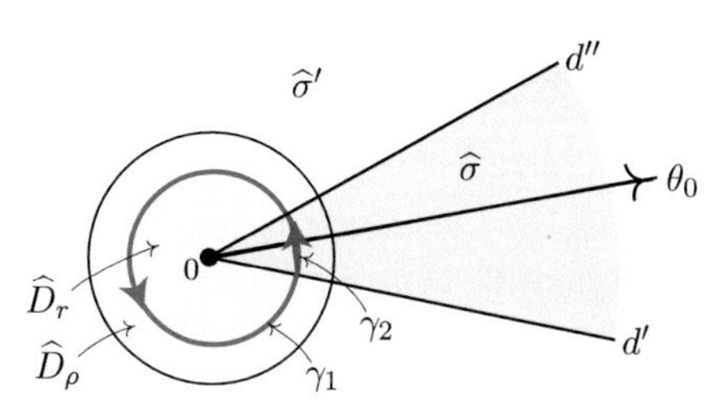

Fig. 7.5

Prove now the followed splitting $\widetilde{f}(x) = \widetilde{f}_1(x) + \widetilde{f}_2(x)$ where $\widetilde{f}_1(x)$ and $\widetilde{f}_2(x)$ are respectively k_1- and k_2-summable in the θ_0 direction.

To begin we build a 0-cochain with good properties as follows:

It results from lemma 7.6.4, p. 226, that the functions $h^\theta(\xi) = \mathbf{A}^\theta_{k_2,k_1}(g)(\xi)$ are κ_1-Gevrey asymptotic to the series $\widehat{h}(\xi) = \mathscr{B}_{k_2}(\widetilde{f})(\xi)$ on a sector of size π/κ_1 in all non-singular θ direction. One proves like in Prop. 5.3.7, p. 152, with the functions $f_\theta(x)$ that these functions are analytic continuations of each other as long as the direction θ does not pass a singular direction. With $\widetilde{f}(x)$ we can thus associate a 0-cochain $(\varphi_j(\xi))_{j\in J}$ with $\varphi_j \in \overline{\mathscr{A}}_{1/\kappa_1}(U_j)$ all asymptotic to the same series $\widehat{h}(\xi)$, the sectors U_j having opening $|U_j| > \pi/\kappa_1$ and making a good covering $\mathscr{U} = (U_j)_{j\in\mathbb{Z}/p\mathbb{Z}}$ of a punctured neighborhood of 0 in $\mathbb{C}$ (cf. Def. 2.2.9, p. 53). We can choose the covering $\mathscr{U}$ such that U_0 is bisected by θ_0. Such a covering exists because of the condition $1/\kappa_1 < 2$. Denoting by θ_j the direction bisecting U_j observe that there might be zero, one or several singular directions between θ_j and θ_{j+1} (we do not need one singular direction between each pair of consecutive θ_j). Denote, as previously, by $\dot{U}_j = U_j \cap U_{j+1}$ the nerve of $\mathscr{U}$. From Prop. 1.2.17, p. 20, we know that the functions $\dot{\varphi}_j = \varphi_j - \varphi_{j+1}$ have values in $\overline{\mathscr{A}}^{\leq -\kappa_1}(\dot{U}_j)$. For all $j \in \mathbb{Z}/p\mathbb{Z}$ choose $a_j \in \dot{U}_j$ and apply the Cauchy-Heine theorem (Thm. 1.4.2, p. 27) to build a new 0-cochain with associated 1-cocycle $(\dot{\varphi}_j = \varphi_j - \varphi_{j+1})$. The construction is as follows. Decompose the 1-cocycle $(\dot{\varphi}_j)$ into the sum of the elementary 1-cocycles $\varphi_j' = \dot{\varphi}_j$ on $\dot{U}_j$ and $\varphi_j' = 0$ on $\dot{U}_\ell$ when $\ell \neq j$. Set $r' = \min(|a_j|)$ and $\dot{U}_j' = \dot{U}_j \cap \{|\xi| < r'\}$. Denote by U_j' the sector with self intersection $\dot{U}_j'$.

The Cauchy-Heine theorem (Thm. 1.4.2, p. 27) says that the function

$$\psi'_j(\xi) = \frac{1}{2\pi i}\int_0^{a_j} \frac{\varphi'_j(t)}{t-\xi}\,dt$$

can be analytically continued to U'_j with 1-cocycle $\varphi'_j(\xi)$ and $\psi'_j(\xi)$ is κ_1-Gevrey asymptotic to the series $\sum c_m \xi^m$ where $c_m = 1/(2\pi i)\int_0^{a_j} \dot{\varphi}_j(t)/t^{m+1}dt$. It has also an analytic continuation to $\mathbb{C}$ deprived of the half-line $d_j = [0, a_j]$, and it tends to 0 at infinity. Define the analytic function $\psi_j(\xi)$ on U_j by setting (choose the branches of ψ'_ℓ that are analytic on all of U_j):

$$\psi_j(\xi) = \sum_{\ell\in\mathbb{Z}/p\mathbb{Z}} \psi'_\ell(\xi), \quad \xi \in U_j.$$

We can now define the series $\widetilde{f}_1(x)$ being k_1-summable in the θ_0 direction. Denote by α_0 and α_{p-1} the arguments of a_0 and a_{p-1} respectively and suppose that a_0 and a_{p-1} are chosen so that the angle $|\alpha_0 - \alpha_{p-1}|$ is $> \pi/\kappa_1$ and bisected by θ_0. This is possible since the opening of U_0 is larger than $\pi/(2\kappa_1)$ on both sides of θ_0. Denote by V_0 the unlimited open sector $]\alpha_{p-1}, \alpha_0[\times]0, +\infty[$ and by $\Psi_0(\xi)$ the analytic continuation of $\psi_0(\xi)$ to V_0. The sector V_0 is κ_1-wide; the function $\Psi_0(\xi)$ has a κ_1-asymptotic expansion $\widehat{\Psi}_0(\xi)$ at 0 and it has an exponential growth of order less than k_2 at infinity on V_0 since it tends to 0. Denote by $\widetilde{f}_1(x)$ the k_2-Laplace transform of the series $\widehat{\Psi}_0(\xi)$. From theorem 5.3.14 (ii)$\Longrightarrow$(i), p. 166, we can conclude that the series $\widetilde{f}_1(x)$ is k_1-summable in the θ_0 direction.

The series $\widetilde{f}_2(x)$ is obtained as follows. Consider $\varphi_0(\xi)$ and its asymptotic series $\widehat{h}(\xi)$. By hypothesis, one can apply a $\mathscr{L}_{k_2}$-Laplace transform to φ_0 in the θ_0 direction and the neighboring ones. This means that φ_0 has an analytic continuation to an unlimited sector V'_0 containing the θ_0 direction with exponential growth of order k_2. The 0-cochains $(\varphi_j(\xi))_{j\in\mathbb{Z}/p\mathbb{Z}}$ and $(\psi_j(\xi))_{j\in\mathbb{Z}/p\mathbb{Z}}$ induce the same 1-cocycle on $(\dot{U}'_j)_{j\in\mathbb{Z}/p\mathbb{Z}}$. This means that they satisfy the relation

$$\varphi_j(\xi) - \varphi_{j+1}(\xi) = \psi_j(\xi) - \psi_{j+1}(\xi) \quad \text{for all } j$$

and we deduce that the functions $\varphi_j(\xi) - \psi_j(\xi)$ glue together into an analytic function $\Phi(\xi)$ on the disc $D' = \{|\zeta| < r'\}$. The function $\Phi(\xi)$ is the sum of the series $\widehat{h}(\xi) - \widehat{\Psi}_0(\xi)$. Denote by $\widetilde{f}_2(x)$ the k_2-Laplace transform of that series. It follows that the function $\varphi_0(\xi) - \psi_0(\xi)$ can be continued into an analytic function on $V'_0 \cup D'$ with Taylor series $\mathscr{B}_{k_2}(\widetilde{f}_2)(\xi)$ and it has exponential growth of order k_2 at infinity. This means that the series $\widetilde{f}_2(x)$ is k_2-summable in the θ_0 direction. Moreover, $\widetilde{f}(x) = \widetilde{f}_1(x) + \widetilde{f}_2(x)$ and the result follows. □

The second part of the proof of theorem 7.6.6, p. 227, provides the following improvement of theorem 7.5.3, p. 220:

Theorem 7.6.7 (Balser [Bal93]) *Let $\underline{k} = (k_1, k_2)$ be a multi-level and let $\underline{I} = (I_1, I_2)$ be a $\underline{k}$-multi-arc.*
Denote $1/\kappa_1 = 1/k_1 - 1/k_2$. Under the condition

$$\kappa_1 > 1/2$$

then, $\underline{k}$-split-summability and $\underline{k}$-summability on $\underline{I}$ are equivalent with same sum.

The property extends to the case of any multi-level $\underline{k} = (k_1, \ldots, k_\nu)$ and any multi-arc $\underline{I} = (I_1, \ldots, I_\nu)$ under the conditions

$$\kappa_j > 1/2 \ \text{ where } \ 1/\kappa_j = 1/k_j - 1/k_{j+1} \ \text{ for } \ j = 1, \ldots, \nu - 1.$$

Observe that the counter-example in Prop. 7.5.7, p. 223, corresponds to $\nu = 2$ and $\kappa_1 = 1/2$.

7.7 The Sixth Approach: Wild Analytic Continuation

7.7.1 $\underline{k}$-Wild-Summability

Like the Ramis-Sibuya definition of k-summability was translated in terms of analytic continuation in the infinitesimal neighborhood X^k of 0 equipped with the sheaf $\mathscr{F}^k$ (cf. Sect. 5.4.1, p. 180) the Malgrange-Ramis definition of $\underline{k}$-multisummability can be translated in terms of analytic continuation in the infinitesimal neighborhood $X^{\underline{k}}$ of 0 equipped with the sheaf $\mathscr{F}^{\underline{k}}$ (cf. Sect. 3.6.3, p. 119).

The definition is as follows:

Definition 7.7.1 ($\underline{k}$-Wild-Summability)

Let $\underline{k} = (k_1, \ldots, k_\nu)$ be a multi-level and
$\underline{I} = (I_1, \ldots, I_\nu)$ be a $\underline{k}$-multi-arc (Def. 7.2.9, p. 205).
Set $\{k_{\nu+1}, 0\} = +\infty$ (cf. notation in Sect. 3.6.2, p. 116).

▷ *A series $\widetilde{f}(x) = \sum_{n \geq 0} a_n x^n$ is said to be $\underline{k}$-*wild-summable *on $\underline{I}$ if it can be wild analytically continued in the infinitesimal neighborhood $(X^{\underline{k}}, \mathscr{F}^{\underline{k}})$ of 0 to a domain containing the closed disc $\overline{D}(0, \{k_1, 0\})$ and the sectors $I_j \times]0, \{k_{j+1}, 0\}]$ for all $j = 1, \ldots \nu$.*

▷ *Its* sum *is the germ of analytic function defined on I_ν by this wild analytic continuation. It is said to be $\underline{k}$-*wild-summable in the *θ* direction *if all arcs I_j* are *bisected by the direction θ*.

▷ *The series is said to be $\underline{k}$-*wild-summable *if it is $\underline{k}$-wild-summable in almost all directions, that is, all directions but finitely many called* singular directions.

It follows from the relative Watson's lemma 7.2.1, p. 200, and Watson's lemma 5.1.3, p. 136, that the continuation, hence the sum in the sense of wild-summation, when it exists, is unique.

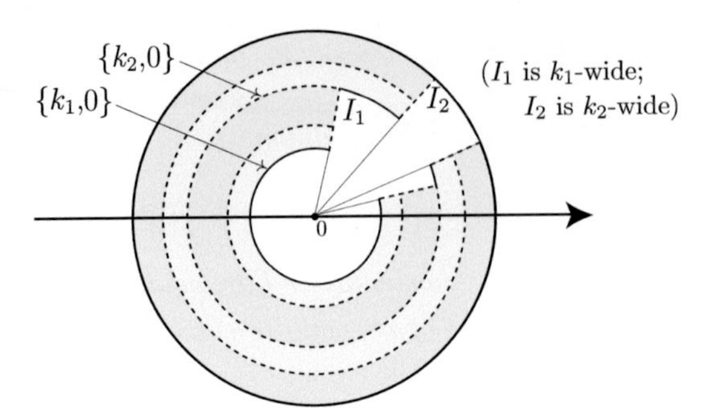

Fig. 7.6. Domain for a (k_1,k_2)-sum in X^{k_1,k_2} (in white)

Since definition 7.7.1, p. 231 exactly translates the Malgrange-Ramis definition of multisummability (Def. 7.3.1, p. 209) we can state:

Proposition 7.7.2 *$\underline{k}$-wild-summability is equivalent to $\underline{k}$-summability in any of the previous sense with same sum.*

Definition 7.7.3 *Let $\underline{I}$ be a $\underline{k}$-multi-arc.*

$\triangleright$ *A sector built on $\underline{I}$ like in definition 7.7.1, p. 231* (cf. *Fig. 7.6) is called a $\underline{k}$-*sector in $X^{\underline{k}}$.

$\triangleright$ *A k_j-arc I_j such that $\widetilde{f}(x)$ can be wild analytically continued to the open sector $I_j\times]0,\{k_j,0\}[$ but not to the closed sector $I_j\times]0,\{k_j,\infty\}]$ in $X^{\underline{k}}$ is said to be* a singular arc of level k_j (for $\widetilde{f}(x)$)*; otherwise it is said to be* non singular.

$\triangleright$ *A direction θ bisecting one or several singular arcs of $\widetilde{f}(x)$ is said to be a* singular direction for $\widetilde{f}(x)$*; otherwise it is said to be* non singular.

From the point of view of wild analytic continuation the following results are straightforward.

Proposition 7.7.4 *Let $\underline{k}=(k_1,k_2,\ldots,k_\nu)$ be a multi-level.*

$\triangleright$ *A series is $\underline{k}$-summable if and only if it admits finitely many singular arcs in $X^{\underline{k}}$.*

$\triangleright$ *Let $\underline{k}'$ be a multi-level containing all levels $k_1,k_2,\ldots,k_\nu$ of $\underline{k}$. A series which is $\underline{k}$-summable in a direction is also $\underline{k}'$-summable in that direction. In other words, $\underline{k}$-summability is stronger than $\underline{k}'$-summability.*

Proof. $\triangleright$ If there is finitely many singular arcs then the series is $\underline{k}$-summable in all directions but the finitely many bisecting directions of the singular arcs. Conversely, suppose that the series has infinitely many singular arcs. For the series to be $\underline{k}$-summable the levels of its singular arcs must belong to $\underline{k}$. Since $\underline{k}$ contains finitely many levels, at least one of them supports infinitely many singular arcs. The infinitely many bisecting directions of these latter singular arcs are singular directions for the $\underline{k}$-summation of the series. Hence, the series is not $\underline{k}$-summable.

$\triangleright$ From the point of view of wild analytic continuation the domain to which one has to continue the series towards its $\underline{k}$-summability in the θ direction contains the domain to which one has to continue it towards its $\underline{k}'$-summability in the same direction. $\square$

7.7.2 Application to Tauberian Theorems

The Tauberian theorems 5.3.15, p. 173, and 5.3.16, p. 174, are easily generalized to multisummable series (cf. [MarR91, Prop. 8 p. 349]).

Without loss of generality we assume that the smallest level k_1 is greater than 1/2.

Theorem 7.7.5 *Let $\underline{k} = (k_1 \ldots, k_\nu)$ be a multi-level, $\underline{I}$ be a $\underline{k}$-multi-arc and $k_{\nu+1} = \infty$. Suppose k' satisfies $k_j \le k' < k_{j+1}$ for some $j \in \{1, 2, \ldots, \nu\}$.*

Suppose $\widetilde{f}(x)$ is a series both $\underline{k}$-summable on $\underline{I}$ and $1/k'$-Gevrey. Then, $\widetilde{f}(x)$ is also $(k', k_{j+1}, \ldots, k_\nu)$-summable on $(I_j, I_{j+1}, \ldots, I_\nu)$.

Proof. Denote by $\underline{k}'$ the multi-level $\underline{k}$ augmented by k' in case $k' \neq k_j$. The proof is performed in the space $(X^{\underline{k}'}, \mathscr{F}^{\underline{k}'})$. The series $\widetilde{f}(x)$ being $1/k'$-Gevrey it can be continued as a section of $\mathscr{F}^{\underline{k}'}$ to the closed disc $D_{k'}$ with radius $\{k', 0\}$. Being $\underline{k}$-summable on $\underline{I}$ it can be continued to the $\underline{k}$-sector $\boldsymbol{\Delta}_{\underline{k},\underline{I}}$ built on $\underline{I}$.

With the same arguments as in section 5.4.2, p. 181, one proves that the two continuations agree on $D_{k'} \cap \boldsymbol{\Delta}_{\underline{k},\underline{I}}$. Hence, $\widetilde{f}(x)$ can be continued to the $(k', k_{j+1}, \ldots, k_\nu)$-sector $\boldsymbol{\Delta}' = D_{k'} \cup \boldsymbol{\Delta}_{\underline{k},\underline{I}}$ and the result follows. $\square$

Given two multi-levels $\underline{k}' = (k'_1, k'_2, \ldots, k'_{\nu'})$ and $\underline{k}'' = (k''_1, k''_2, \ldots, k''_{\nu''})$ we define the multi-level $\underline{K} = (K_1, \ldots, K_\nu)$ as being the shuffle of $\underline{k}'$ and $\underline{k}''$ starting from $K_1 = \max(k'_1, k''_1)$.

Proposition 7.7.6 *A series $\widetilde{f}(x)$ which is both $\underline{k}'$- and $\underline{k}''$-summable in the θ direction is $\underline{K}$-summable in the θ direction and the sums agree.*

The proof is similar to the previous one.

Such a result is of poor interest since, in general, K has more levels than both $\underline{k}'$ and $\underline{k}''$ and being $\underline{K}$-summable in the θ direction is more complicated than being $\underline{k}'$- or $\underline{k}''$-summable in the θ direction. It would be more interesting to get $\underline{K}$-summable series for a $\underline{K}$ equal to the intersection of $\underline{k}'$ and $\underline{k}''$. This is impossible when one considers summable series in a given direction. However, in the case of series that are both $\underline{k}'$- and $\underline{k}''$-summable (in the global meaning of summability in almost all directions) one has the following generalization of the Tauberian theorem 5.3.16, p. 174.

Theorem 7.7.7 *With notations as before let $\underline{\kappa} = \underline{k}' \cap \underline{k}''$ be the multi-level defined on the common values of $\underline{k}'$ and $\underline{k}''$: $\underline{\kappa} = (\kappa_1, \ldots \kappa_{\nu_0})$ satisfies*

$$\{\kappa_1, \ldots, \kappa_{\nu_0}\} = \{k_1', k_2', \ldots, k_{\nu'}'\} \cap \{k_1'', k_2'', \ldots, k_{\nu''}''\}.$$

A series $\widetilde{f}(x)$ which is both $\underline{k}'$- and $\underline{k}''$-summable satisfies the following properties:

(i) *if $\underline{\kappa} = \emptyset$ then $\widetilde{f}(x)$ is convergent;*

(ii) *if $\underline{\kappa} \neq \emptyset$ then $\widetilde{f}(x)$ is $\underline{\kappa}$-summable.*

Proof. (i) Case when $\underline{\kappa}$ is empty. Suppose for instance that $\underline{k}'$ and $\underline{k}''$ satisfy

$$k_1' < \cdots < k_{j_1'-1}' < k_1'' < \cdots < k_{j_1''-1}'' < k_{j_1'}' < \cdots < k_{j_2'-1}' < k_{j_1''}'' < \cdots$$

It suffices to prove that the series has no singular arc in $(X^{\underline{k}' \cup \underline{k}''}, \mathscr{F}^{\underline{k}' \cup \underline{k}''})$.

Prove first that a $\underline{k}$-summable series ($\underline{k} = (k_1, k_2, \ldots, k_\nu)$) with no singular arc of level k_1 in $(X^{\underline{k}}, \mathscr{F}^{\underline{k}})$ is $(k_2, \ldots, k_\nu)$-summable. Indeed, in that case, one can choose a covering of $X^{\underline{k}}$ by $\underline{k}$-sectors $\boldsymbol{\Delta}^\ell$ based on $\underline{k}$-arcs $(I_1^\ell, I_2^\ell, \ldots, I_\nu^\ell)$ with the following properties: the sectors $\boldsymbol{\Delta}^\ell$ are sectors of $\underline{k}$-summation of the series; the I_j^ℓ's form a cyclic covering of S^1 and the consecutive intersections $I_1^\ell \cap I_1^{\ell+1}$ consist of arcs of length larger than π/k_1. From the Ramis-Sibuya corollary 5.2.2, p. 140, and the relative Watson lemma 7.2.1, p. 200, we conclude like in section 5.4.2, p. 181, that the corresponding sums glue together into a section of $\mathscr{F}^{\underline{k}}$ over the closed disc D_{k_2} with radius $\{k_2, 0\}$. The series is then $1/k_2$-Gevrey and, by theorem 7.7.5, p. 233, it is $(k_2, \ldots, k_\nu)$-summable.

The series $\widetilde{f}(x)$ being both $\underline{k}'$-summable and $1/k_1''$-Gevrey we know from theorem 7.7.5, p. 233 that it is at worst $(k_1'', k_{j_1'}', \ldots, k_{\nu'}')$-summable. As a $\underline{k}'$-summable series it has then no singular arc of level $< k_1''$ and then of level $< k_{j_1'}'$. This proves that $\widetilde{f}(x)$ is $(k_{j_1'}', k_{j_1'+1}', \ldots, k_{\nu'}')$-summable. Exchanging the role of the k''s and of the k'''s we show in the same way that the series is $(k_{j_1''}'', k_{j_1''+1}'', \ldots, k_{\nu''}'')$-summable; then $(k_{j_2'}', k_{j_2'+1}', \ldots, k_{\nu'}')$-summable and so on… until no singular arc is left.

(ii) Suppose for instance that $k_1'' = k_{j_1'}'$.

The previous reasoning remains valid on any arc I_1'' of summability. Instead of a continuation to the full infinitesimal neighborhood $(X^{\underline{k}' \cup \underline{k}''}, \mathscr{F}^{\underline{k}' \cup \underline{k}''})$ we obtain the continuation to any $\underline{\kappa}$-multi-sector but the finitely many ones that are based on arcs of levels in $\underline{k}' \cap \underline{k}''$ that are singular for the $\underline{k}'$- and the $\underline{k}''$-summation of $\widetilde{f}(x)$. This ends the proof of the theorem. □

With this result we see that any multisummable series is $\underline{k}$-summable for a unique $\underline{k}$ of smallest length. This is no more true for the directional summability. Recall, for intance, the case of the Leroy series which is both 1- and 2- summable in infinitely many directions (cf. Exa. 5.3.17, p. 174).

Let us end with the following note.

Remark 7.7.8 (Case of non-linear ordinary differential equations)
It was proved by B.L.J. Braaksma in [Bra92] that a series $\widetilde{f}(x)$ satisfying a non-linear (analytic) ordinary differential equation is always multisummable. The proof is given in terms of Écalle's accelerators and the levels of multisummability to be taken into account are the levels of the linearized equation along $\widetilde{f}(x)$.

Chapter 8
Exercises

Exercise 1 Study the asymptotic behavior at 0, the analytic continuation and the variation of the function

$$F(x) = \int_0^{+\infty} \frac{e^{-\xi/x}}{\xi^2+3\xi+2}\, d\xi.$$

Exercise 2 Prove the injectivity of the map μ in theorem 3.5.4, p. 94 without calling upon the exact sequence of cohomology.

Exercise 3 Consider a linear differential equation

$$Dy \equiv a_n(x)x^n \frac{\mathrm{d}^n y}{\mathrm{d}x^n} + a_{n-1}(x)x^{n-1}\frac{\mathrm{d}^{n-1}y}{\mathrm{d}x^{n-1}} + \cdots + a_0(x)y = g(x) \tag{1}$$

with formal series coefficients $a_j(x) \in \mathbb{C}[[x]]$ and $g(x) \in \mathbb{C}[[x]][1/x]$. We suppose that $g(x)$ is non-zero and we denote by $p \in \mathbb{Z}$ its valuation.
Consider the operator $D' = g(x)^2 \frac{\mathrm{d}}{\mathrm{d}x}\big(\frac{1}{g(x)}D\big)$ so that the equation

$$D'y \equiv b_{n+1}(x)x^{n+1}\frac{\mathrm{d}^{n+1}y}{\mathrm{d}x^{n+1}} + b_n(x)x^n\frac{\mathrm{d}^n y}{\mathrm{d}x^n} + \cdots + b_0(x)y = 0 \tag{2}$$

is the homogeneous form of equation (1).
Denote by $\mathscr{N}(D)$ and $\mathscr{N}(D')$ the Newton polygons at 0 of D and D' respectively, by ℓ and ℓ' the lengths of their horizontal side and by $\pi(\lambda)$ and $\pi'(\lambda)$ their indicial equations.

(a) Prove that $\ell' = \ell + 1$.

(b) Prove that $\pi'(\lambda) = C(\lambda - p)\,\pi(\lambda)$ for a convenient constant $C \neq 0$.

(c) When $\pi(\lambda)$ has no integer root conclude that equation (1) admits a solution in $\mathbb{C}[[x]][1/x]$. What happens when there exists $r \in \mathbb{Z}$ such that $\pi(r) = 0$?

M. Loday-Richaud, *Divergent Series, Summability and Resurgence II*,
Lecture Notes in Mathematics, DOI 10.1007/978-3-319-29075-1_8

Exercise 4 An example of factorization in the sheaf $\Lambda^{<0}(B_0)$ (cf. Lem. 3.5.13).

We consider a 3-dimensional system $(S): \ \mathrm{d}Y/\mathrm{d}x = B(x)Y$ with a formal fundamental solution $\widetilde{Y}(x) = \widetilde{F}(x)x^L e^{Q(1/x)}$ where $Q = \mathrm{diag}(q_1, q_2, q_3)$ consists of distinct unramified polynomials of the form $q_i(1/x) = -a_i/x^k + l.o.t.$. We suppose that the system has the unique level k, which means that the three a_i are distinct.

We denote by $(S_0): \ \mathrm{d}Y/\mathrm{d}x = B_0(x)Y$ the associated normal form with the formal fundamental solution $\widetilde{Y}_0(x) = x^L e^{Q(1/x)}$.

We choose a direction from 0, say $\theta = 0$, and a germ $\varphi(x) \in \Lambda_0^{<0}(B_0)$ of flat isotropy of (S_0) in the $\theta = 0$ direction. There are 3 possible Stokes arcs containing the direction 0. Suppose these Stokes arcs are distinct. They are then bisected by 3 distinct anti-Stokes directions $\alpha < \beta < \gamma$ in that order on S^1 when oriented clockwise.

Prove that $\varphi(x)$ can be factored in the form

$$\varphi(x) = \varphi_\alpha(x)\,\varphi_\beta(x)\,\varphi_\gamma(x).$$

where $\varphi_\alpha(x)$, $\varphi_\beta(x)$, $\varphi_\gamma(x)$ are germs of $\Lambda_\alpha^{<0}(B_0)$, $\Lambda_\beta^{<0}(B_0)$, $\Lambda_\gamma^{<0}(B_0)$ respectively.

Exercise 5

(1) Check that the function $F(x) = \int_0^{+\infty} \frac{e^{-\xi/x}}{\xi^2+3\xi+2}\, d\xi$ of exercise 1, p. 237 satisfies the linear differential equation

$$x^4 y''' + (2x^3 + 3x^2)\,y' + 2y = x \tag{3}$$

and explain the appearance of the exponential terms in the analytic continuation of $F(x)$ over the Riemann surface of the logarithm.

Put equation (3) in homogeneous form

$$D_1\, y = 0. \tag{4}$$

Draw its Newton polygon at 0 and write its characteristic and indicial equations.

Determine a fundamental set of formal solutions.

Write down the companion system of equation (4), a normal form and a formal fundamental solution.

Compute its Stokes matrix or matrices.

(2) Consider the linear differential equation

$$D_2\, y \equiv \Big(x^2(x+2)\frac{\mathrm{d}}{\mathrm{d}x} + 4(x+1)\Big)\Big(x^2\frac{\mathrm{d}^2}{\mathrm{d}x^2} + (x^2+x)\frac{\mathrm{d}}{\mathrm{d}x} - 1\Big)y = 0 \tag{5}$$

where the factor to the right is the homogeneous Euler operator $\mathscr{E}_0$ (Exa. 2.1.27, p. 44).

Show that $y = \mathrm{e}^{2/x}$ satisfies $D_2 y = 0$ and conclude that the equations $D_1 y = 0$ and $D_2 y = 0$ admit a same normal form (i.e., belong to the same formal class).

Compute the Stokes matrices of $D_2 y = 0$ and conclude that the equations $D_1 y = 0$ and $D_2 y = 0$ do not belong to the same meromorphic class.

Exercise 6 Given a system $(S): \; \mathrm{d}Y/\mathrm{d}x = B(x)Y$, let $(S_0): \; \mathrm{d}Y/\mathrm{d}x = B_0(x)Y$ be a normal form and $\widetilde{F}(x)$ be a gauge transformation from (S_0) to (S).

Let us recall briefly the construction of the cohomology class $c(\widetilde{F}) = \exp_\mu(\widetilde{F})$ associated with the formal gauge transformation $\widetilde{F}$ by the Malgrange-Sibuya classification theorem 3.5.6, p. 97. One considers a good covering $\mathscr{V} = \{V_j\}_{j\in\mathbb{Z}/J\mathbb{Z}}$ of S^1. Assuming that the V_j's are small enough the main asymptotic existence theorem provides analytic gauge functions F_j that change (S_0) into (S) and are asymptotic to $\widetilde{F}$ on V_j for all j. Then, setting $\varphi_j = F_j^{-1} F_{j+1}$ on $V_j \cap V_{j+1}$ for all j provides a 1-cocycle $\varphi = (\varphi_j)_{j\in\mathbb{Z}/J\mathbb{Z}}$ of $\mathscr{V}$ with values in the sheaf $\Lambda^{<0}(B_0)$ of flat isotropies of the normal form. The cohomology class $c(\widetilde{F})$ it induces in $H^1\big(S^1\,;\Lambda^{<0}(B_0)\big)$ does not depend on the choice of V_j and F_j for all j.

Suppose $\psi = (\psi_j)_{j\in\mathbb{Z}/J\mathbb{Z}}$ is another 1-cocycle of $\mathscr{V}$ with values in $\Lambda^{<0}(B_0)$ which is cohomologous to φ in $H^1\big(\mathscr{V}\,;\Lambda^{<0}(B_0)\big)$.

Prove that there exist gauge transformations G_j from (S_0) to (S) that are asymptotic to $\widetilde{F}$ and satisfy $\psi_j = G_j^{-1} G_{j+1}$ for all j.

Exercise 7 We consider a system $(S): \; \mathrm{d}Y/\mathrm{d}x = B(x)Y$ with a formal fundamental solution $\widetilde{Y}(x) = \widetilde{F}(x)x^L \mathrm{e}^Q$ and its normal form $(S_0): \; \mathrm{d}Y/\mathrm{d}x = B_0(x)Y$ with the formal fundamental solution $\widetilde{Y}_0(x) = x^L \mathrm{e}^Q$.

We denote by $\Lambda^{<0}(B_0)$ the sheaf over S^1 of flat isotropies of the normal form (cf. page 95).

We assume that these systems have a single integer level $k > 0$ and consider the covering $\mathscr{U} = (U_j)$ defined as follows. We order the anti-Stokes directions α_j for $j \in \mathbb{Z}/J\mathbb{Z}$ clockwise. This also orders the Stokes arcs $\mathfrak{S}_j$, meaning the maximal closed arcs where at least one exponential $\mathrm{e}^{q_m - q_\ell}$ is bounded. For all j, the Stokes arc $\mathfrak{S}_j$ has length π/k and the anti-Stokes direction α_j as middle point. The arc U_j is the interior of $\mathfrak{S}_{j-1} \cup \mathfrak{S}_j$ so that $\dot{U}_j = U_j \cap U_{j+1}$ is the interior of $\mathfrak{S}_j$ (where at least one exponential $\mathrm{e}^{q_m - q_\ell}$ is asymptotic to 0). Observe that $\mathscr{U}$ is not a good covering in general since there may exist non empty intersections of more than two U_j's.

(a) Prove that a 1-cocycle of $\mathscr{U}$ with values in a sheaf $\mathscr{F}$ is well defined by the data of arbitrary sections f_j of $\mathscr{F}$ over $\dot{U}_j$ for all $j \in \mathbb{Z}/J\mathbb{Z}$.

(b) Prove that the covering $\mathscr{U}$ is acyclic for $\Lambda^{<0}(B_0)$, that is, since the dimension of the base S^1 equals 1, that $H^1\big(U_j\,;\Lambda^{<0}(B_0)\big) = 0$ for all j. What does this imply for cohomology classes in $H^1\big(S^1\,;\Lambda^{<0}(B_0)\big)$?

(c) We recall that a Stokes cocycle is defined by the data of a Stokes germ in each anti-Stokes direction α_j. Prove that it can be identified canonically to a 1-cocycle of $H^1\big(\mathscr{U}\,;\Lambda^{<0}(B_0)\big)$.

(d) Observe that there exist no section of $\Lambda^{<0}(B_0)$ defined on U_j for all j. What does this imply for Stokes cocycles seen as 1-cocycles of $H^1\big(\mathscr{U}\,;\Lambda^{<0}(B_0)\big)$?

(e) Prove that $\mathscr{U}$ is the less fine covering allowing to represent all elements of $H^1\big(S^1\,;\Lambda^{<0}(B_0)\big)$ as 1-cocycles of $H^1\big(\mathscr{U}\,;\Lambda^{<0}(B_0)\big)$.

Exercise 8 Stokes cocycle and sums.

With notations as in exercise 7 we assume again that system (S) and, necessarily, system (S_0) have a single integer level k.
We denote by $c(\widetilde{F})$ the cohomology class associated with $\widetilde{F}(x)$ in $H^1\big(S^1\,;\Lambda^{<0}(B_0)\big)$ by the Malgrange-Sibuya classification theorem (Thm. 3.5.6, p. 97). We know from the Stokes cocycle theorem (Thm 3.5.11, p. 101) and the previous exercise that $c(\widetilde{F})$ admits a representation as a 1-cocycle $c = (c_{\alpha_j})_{j\in\mathbb{Z}/J\mathbb{Z}}$ in $H^1\big(\mathscr{U}\,;\Lambda^{<0}(B_0)\big)$.

(a) Prove that there exist gauge transformations F_j from (S_0) to (S) that are defined on U_j, are asymptotic to $\widetilde{F}(x)$ and satisfy $c_{\alpha_j} = F_j^{-1}F_{j+1}$ for all j.

(b) Conclude that the gauge transformation $\widetilde{F}(x)$ is k-summable with sum $F_j(x)$ in each direction θ satisfying $\alpha_{j-1} < \theta < \alpha_j$. In particular, given two non-anti-Stokes directions θ and θ' that are separated by no anti-Stokes direction, say

$$\alpha_{j-1} < \theta < \theta' < \alpha_j,$$

the two sums $S_\theta(\widetilde{F})$ and $S_{\theta'}(\widetilde{F})$ are analytic continuation of each other.

N.B. Note that, in this approach, the sums $F_j(x)$ have been obtained from the Stokes cocycle *without* the help of *any theory of summation*. This can be done in general, without the assumption of a single integer level.

Exercise 9 The Airy equation $(\mathscr{A}_\infty):\ \ y'' - zy = 0$ was already considered in volume I of this book [MS16]: first, in example 2.45 to illustrate Ramis's density theorem; second, to show the resurgent analysis at work on an example (Sect. 6.14). It is also briefly mentioned in volume III [Dela16, Exa. 2, Sect. 1.2.3.2]. The Airy equation is well known both from mathematicians and physicists. It was stated to model the position of the fringes in the theory of diffraction. It is the example on which Stokes discovered the phenomenon called now by his name. It is also well known for having solutions that are not "integrable by quadratures" (cf. [MS16, Exa. 2.45] where its differential Galois group is proved to be $SL(2,\mathbb{C})$ and remark 2.46 following example 2.45).

In the form above, the Airy equation has a unique singular point at infinity which is an irregular singular one. This singular point is moved to the origin by the change of variable $z = 1/x$. The Airy equation reads then

$$(\mathscr{A}_0): \ x^5 y'' + 2x^4 y' - y = 0.$$

A basis of formal solutions is given by the following two series (asymptotic expansions of the Airy functions Ai(x) and Bi(x), up to multiplicative constants that we neglect):

$$\begin{aligned}\widetilde{A}(x) &= \Big(\sum_{n\geq 0}(-1)^n a_n x^{3n/2}\Big) x^{1/4}\, \mathrm{e}^{-2/(3x^{3/2})}\\ \widetilde{B}(x) &= \Big(\sum_{n\geq 0} a_n x^{3n/2}\Big) x^{1/4}\, \mathrm{e}^{+2/(3x^{3/2})}\end{aligned}$$

where $a_0 = 1$ and for $n \geq 1$,

$$\begin{aligned}a_n &= \frac{(2n+1)(2n+3)\cdots(6n-1)}{144^n\, n!}\\ &= \frac{1}{2\pi}\Big(\frac{3}{4}\Big)^n \frac{1}{n!}\Gamma\Big(n+\frac{5}{6}\Big)\Gamma\Big(n+\frac{1}{6}\Big).\end{aligned}$$

The companion system (S) of equation $(\mathscr{A}_0)$ has the matrix $\widetilde{Y}(x) = \begin{bmatrix}\widetilde{A} & \widetilde{B}\\ \mathrm{d}\widetilde{A}/\mathrm{d}x & \mathrm{d}\widetilde{B}/\mathrm{d}x\end{bmatrix}$ as formal fundamental solution. The series in $\widetilde{A}(x)$ and $\widetilde{B}(x)$ involve powers of the fractional variable $x^{1/2}$. According to theorem 3.3.1, after symmetrization by means of the van der Monde matrix $U = \begin{bmatrix}1 & 1\\ 1 & -1\end{bmatrix}$, one obtains a formal fundamental solution of the form

$$\widetilde{Y}(x) = \widetilde{F}(x) x^J U \mathrm{e}^{Q(1/x)}$$

where the formal gauge matrix $\widetilde{F}(x)$ is a formal meromorphic matrix in integer powers of x, the matrix J is diagonal (in Jordan form) $J = \mathrm{diag}(1/4, 3/4)$, and the matrix $Q = \mathrm{diag}(q_1, q_2)$ has diagonal entries, $q_1 = -2/(3x^{3/2})$ and $q_2 = +2/(3x^{3/2})$.

(a) Check that, setting $a_{-1} = 0$,

$$\widetilde{F}(x) = \begin{bmatrix}\displaystyle\sum_{n\geq 0} a_{2n}x^{3n} & \displaystyle\sum_{n\geq 0} -a_{2n+1}x^{3n+1}\\ \displaystyle\sum_{n\geq 0}\Big(\big(3n-\frac{1}{4}\big)a_{2n} - a_{2n+1}\Big)x^{3n-1} & \displaystyle\sum_{n\geq 0}\Big(a_{2n} - \big(3n-\frac{5}{4}\big)a_{2n-1}\Big)x^{3(n-1)}\end{bmatrix}$$

(b) Draw the Newton polygons of $(\mathscr{A}_\infty)$ and $(\mathscr{A}_0)$ and comment the results. Explain how to compute the matrix Q.

(c) Compute the matrix of the formal monodromy $\widetilde{M}$ with respect to the formal fundamental solution $\widetilde{Y}(x)$. We recall that $\widetilde{M}$ is, by definition, the unique matrix such that $\widetilde{Y}(x\mathrm{e}^{2\pi \mathrm{i}}) = \widetilde{Y}(x)\widetilde{M}$ (cf. volume I, [MS16, Sect. 2.2.3.1]).

(d) Determine the anti-Stokes directions of the companion system (S).

Given a determination α' of an anti-Stokes direction α we denote by $I_2+C_{\alpha'}$ the Stokes matrix of the Stokes automorphism at α in the basis

$$\mathscr{Y}_{\alpha'}^{-}(x)=F_{\alpha}^{-}(x)\,\mathscr{Y}_{0,\alpha'}(x)$$

where $F_{\alpha}^{-}(x)$ denotes the sum of $\widetilde{F}(x)$ to the left (cf. N.B. below) and $\mathscr{Y}_{0,\alpha'}(x)$ denotes the analytic function defined for $\arg(x)\approx\alpha'$ as follows: the (formal) normal solution $\widetilde{Y}_0(x)=x^J\,U\,\mathrm{e}^{Q(1/x)}$ is changed into analytic by choosing the determination $\arg(x)\approx\alpha'$ of the argument in x^J and $Q(1/x)$, and replacing the formal exponential by the usual analytic function exp. Recall that $F_{\alpha}^{-}(x)$ depends on α only, not on the choice of its determination α'.

(e) Choose the principal determination $\arg(x)=0$ on $\mathbb{R}^+$ and determine the form of the Stokes matrix I_2+C_0.

(f) Choose now the determination $\arg(x)=2\pi$ on $\mathbb{R}^+$, determine the form of the corresponding Stokes matrix $I_2+C_{2\pi}$ and compare the Stokes multipliers in I_2+C_0 and in $I_2+C_{2\pi}$.

(g) Symmetries. Set $x=\mathrm{e}^{-2\pi\mathrm{i}/3}X$ and compute $\widetilde{Y}(X)$ in terms of $\widetilde{Y}(x)$.

Compare the Stokes matrices $I_2+C_{-2\pi/3}$, I_2+C_0 and $I_2+C_{2\pi/3}$.

(h) Explain the cyclic relation (cf. volume I [MS16, Exe. 2.38])

$$\widetilde{\mathrm{M}}\,(I_2+C_{2\pi/3})\,(I_2+C_0)\,(I_2+C_{-2\pi/3})=I_2 \tag{6}$$

and determine the value of the Stokes matrices $I_2+C_{-2\pi/3}$, I_2+C_0 and $I_2+C_{2\pi/3}$.

N. B. We recall that, for compatibility of the orientations at 0 and at infinity on the Riemann sphere, the anti-Stokes directions are numbered clockwise around 0 and counter-clockwise around infinity. The sum F_0^- is taken to the left of $\mathbb{R}^+$, that is, in a $\varepsilon>0$ direction whereas the sum F_0^+ is taken in a direction $-\varepsilon<0$.

Exercise 10 We consider a linear differential system (S) with an irregular singular point at 0 and a formal fundamental solution $\widetilde{Y}(x)=\widetilde{F}(x)\,x^L\,\mathrm{e}^{Q(1/x)}$ (cf. Thm. 3.3.1). We assume that the system belongs to the case without ramification (meaning with no root of x in Q).

We denote by $\mathfrak{A}=\{\alpha_1,\alpha_2,\ldots,\alpha_N\}$ the set of the anti-Stokes directions ordered clockwise and by $\varphi_1,\varphi_2,\ldots,\varphi_N$ the corresponding Stokes automorphisms of the system (cf. Def. 3.5.10 and Rem. 3.5.9).

The (topological) monodromy $\mathscr{M}$ acts on the analytic solutions of the differential system by analytic continuation along a loop Γ turning anti-clockwise once around 0.

The formal monodromy $\widetilde{\mathscr{M}}$ can be seen as the action of a formal turn about 0 (one substitutes $x\mathrm{e}^{2\pi\mathrm{i}}$ for x in the formal solutions). It can also be seen as the topological monodromy of the normal form of the system.

These monodromies are thoroughly defined in volume I [MS16].

The general cyclic relation (cf. [MS16, Exe. 2.38]) reads

$$\widetilde{\mathscr{M}} \circ \varphi_1 \circ \varphi_{\alpha_2} \circ \cdots \circ \varphi_N = \mathscr{M}. \tag{7}$$

Give an interpretation of this formula in the infinitesimal neighborhood of 0 with an appropriate picture.

Chapter 9
Solutions to the Exercises

Exercise 1, p. 237 The function $F(x)=\displaystyle\int_0^{+\infty}\frac{\mathrm{e}^{-\xi/x}}{\xi^2+3\xi+2}\,\mathrm{d}\xi$ is well defined and continuous on the closed half plane $\Re(x)\geq 0$.

▷ Let us study its asymptotics on the open half-plane $\Pi_0=\{x\in\mathbb{C}\,;\Re(x)>0\}$. The example is somewhat similar to the case of the Euler function (cf. Exa. 1.1.4) and could be treated directly in the same way. We prefer to expand the integrand in partial fractions:

$$\frac{\mathrm{e}^{-\xi/x}}{\xi^2+3\xi+2}=\frac{\mathrm{e}^{-\xi/x}}{\xi+1}-\frac{\mathrm{e}^{-\xi/x}}{\xi+2}$$

so that $F(x)=F_1(x)-F_2(x)$ where $F_\alpha(x)=\int_0^{+\infty}\mathrm{e}^{-\xi/x}/(\xi+\alpha)\mathrm{d}\xi,\ \alpha>0$. To study $F_\alpha(x)$ write, for $N\geq 1$,

$$\frac{1}{\xi+\alpha}=\frac{1}{\alpha}\sum_{n=0}^{N-2}(-1)^n\Big(\frac{\xi}{\alpha}\Big)^n+(-1)^{N-1}\frac{(\xi/\alpha)^{N-1}}{\xi+\alpha}.$$

Taking into account the fact that $\int_0^{+\infty}\xi^n\,\mathrm{e}^{-\xi/x}\,\mathrm{d}\xi=x^{n+1}\,\Gamma(n+1)$ we obtain

$$F_\alpha(x)=\sum_{n=0}^{N-2}(-1)^n\frac{\Gamma(n+1)}{\alpha^{n+1}}x^{n+1}+(-1)^{N-1}\frac{1}{\alpha^{N-1}}\int_0^{+\infty}\frac{\xi^{N-1}\,\mathrm{e}^{-\xi/x}}{\xi+\alpha}\mathrm{d}\xi\,.$$

For any δ such that $0<\delta<\pi/2$, consider the proper sub-sector Δ_δ of Π_0 defined by $\Delta_\delta=\{x\,;|\arg(x)|<\pi/2-\delta\}$. For $x\in\Delta_0$, we can write

$$\Big|\int_0^{+\infty}\frac{\xi^{N-1}\,\mathrm{e}^{-\xi/x}}{\xi+\alpha}\mathrm{d}\xi\Big|\leq\int_0^{+\infty}\frac{\xi^{N-1}}{\alpha}\exp\frac{-\xi\sin(\delta)}{|x|}\,\mathrm{d}\xi\leq\frac{1}{\alpha}\Gamma(N)\Big(\frac{|x|}{\sin(\delta)}\Big)^N \tag{8}$$

Hence, for all $x\in\Delta_\delta$, we obtain the estimate

$$\Big|F_\alpha(x)-\sum_{n=1}^{N-1}(-1)^{n-1}\frac{\Gamma(n)}{\alpha^n}x^n\Big|\leq\Gamma(N+1)\frac{1}{\big(\alpha\sin(\delta)\big)^N}|x|^N\,.$$

M. Loday-Richaud, *Divergent Series, Summability and Resurgence II*,
Lecture Notes in Mathematics, DOI 10.1007/978-3-319-29075-1_9

Since Δ_δ is any proper sub-sector of Π_0 this proves that $F_\alpha(x)$ is 1-Gevrey asymptotic to the series $\sum_{n\geq 1}(-1)^n\Gamma(n)(x/\alpha)^n$ on Π_0 (cf. Def. 1.2.7) and consequently, the function $F(x)$ is 1-Gevrey asymptotic on Π_0 to the (divergent) series

$$T_0F(x) = \sum_{n\geq 1}(-1)^{n-1}\Gamma(n)\big(1-1/2^n\big)x^n.$$

▷ Analytic continuation with asymptotic condition.
Choose the determination $-\pi/2 < \arg(x) < +\pi/2$ for the argument on Π_0 (any other determination would be OK). Given $\theta \in]-\pi,+\pi[$, $\theta \neq 0$, consider the function

$$F^\theta(x) = \int_0^{+\infty e^{i\theta}} \frac{e^{-\xi/x}}{\xi^2+3\xi+2}\,d\xi$$

and prove that it is 1-Gevrey asymptotic to the same series as $F(x)$ on the half-plane $\Pi_\theta = \{x\,;\Re(xe^{-}i\theta) > 0\}$ bisected by the θ direction.
Taking $\xi = e^{i\theta}\zeta$ and $x \in \Pi_0$ in the integral and denoting $\alpha_1 = e^{-i\theta}$ and $\alpha_2 = 2e^{-i\theta}$ we obtain $F^\theta(xe^{i\theta}) = F_{\alpha_1}(x) - F_{\alpha_2}(x)$. With $\alpha = e^{-i\theta}$ not real positive, estimate (8) has to be replaced by

$$\Big|\int_0^{+\infty} \frac{e^{-\zeta/x}}{\zeta+e^{i\theta}}d\zeta\Big| \leq \frac{1}{|\sin\theta|}\Gamma(N)\Big(\frac{|x|}{\sin(\delta)}\Big)^N$$

(replace $\alpha > 0$ by $|\Im(\alpha)| = |\sin(\theta)| > 0$) and similarly for $\alpha = 2e^{-i\theta}$. The rest of the calculation above remains valid for the values α_1 and α_2 of α and so, we can assert that $F^\theta(xe^{i\theta})$ is 1-Gevrey asymptotic to $\sum_{n\geq 1}(-1)^{n-1}\Gamma(n)(1-1/2^n)e^{ni\theta}x^n$ on Π_0. It follows that $F^\theta(x)$ is 1-Gevrey asymptotic on Π_θ to the same series $T_0F(x)$ as $F(x)$ on Π_0.

Prove now that any $F_\theta(x)$ is the analytic continuation of $F(x)$. Fix $\theta \neq 0, \pi$ and consider the path $\gamma = \gamma_1 \cup \gamma_2 \cup \gamma_3$ oriented positively where $\gamma_1 = [0,R]$, $\gamma_3 = [0,Re^{i\theta}]$ and γ_2 is the shortest arc of circle (centered at 0) from R to $Re^{i\theta}$. For x in $\Pi_0 \cap \Pi_\theta$ the expression $-\Re(\xi/x)$ is negative and the integral $\int_{\gamma_3} e^{-\xi/x}/(\xi^2+3\xi+2)d\xi$ tends to 0 as R tends to infinity. Applying Cauchy's theorem to $e^{-\xi/x}/(\xi^2+3\xi+2)$ on γ and letting R tend to infinity shows that $F(x) = F_\theta(x)$. Hence the functions $F_\theta(x)$ for all $\theta \in]-\pi,+\pi[$ glue together into an analytic function defined for all x satisfying $|\arg(x)| < 3\pi/2$. We denote again by $F(x)$ this analytic continuation.

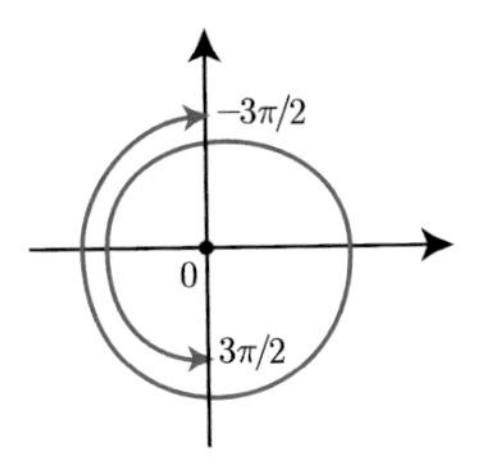

Fig. 7.7

▷ Show that the analytic continuation of $F(x)$ does not define an analytic function at 0. To this end it suffices to compute its variation. Let $\theta < \pi$ be close to π and choose x in $\Pi_{-\theta}$ so that $x\mathrm{e}^{2\pi\mathrm{i}}$ belongs to Π_θ. Consider as before a path $\gamma = \gamma_1 \cup \gamma_2 \cup \gamma_3$ oriented positively where $\gamma_1 = [0, R\mathrm{e}^{\mathrm{i}\theta}]$, $\gamma_3 = [0, R\mathrm{e}^{-\mathrm{i}\theta}]$ and γ_2 is the shortest arc of circle (centered at 0) from $R\mathrm{e}^{\mathrm{i}\theta}$ to $R\mathrm{e}^{-\mathrm{i}\theta}$. When $R > 2$ the path γ encloses the two poles -1 and -2 of $1/(\xi^2 + 3\xi + 2)$, the residues of $e^{-\xi/x}/(\xi^2 + 3\xi + 2)$ are $\mathrm{e}^{1/x}$ and $-\mathrm{e}^{2/x}$ and the integral on γ_2 tends to 0 as R tends to infinity. Applying Cauchy's residue theorem and letting R goes to infinity we obtain

$$\mathrm{var}(F)(x) \equiv F_{-\theta}(x) - F_\theta(x\mathrm{e}^{2\pi\mathrm{i}}) = 2\pi\mathrm{i}(\mathrm{e}^{2/x} - \mathrm{e}^{1/x}). \tag{9}$$

▷ We can now analytically continue $F(x)$ over the full Riemann surface of logarithm and determine the largest sector over which it has an asymptotic expansion. Denote by $F^+(x)$ the determination of $F(x)$ on $-\pi/2 \leq \arg(x) < +3\pi/2$ and by $F^-(x)$ its determination on $-3\pi/2 < \arg(x) \leq \pi/2$.

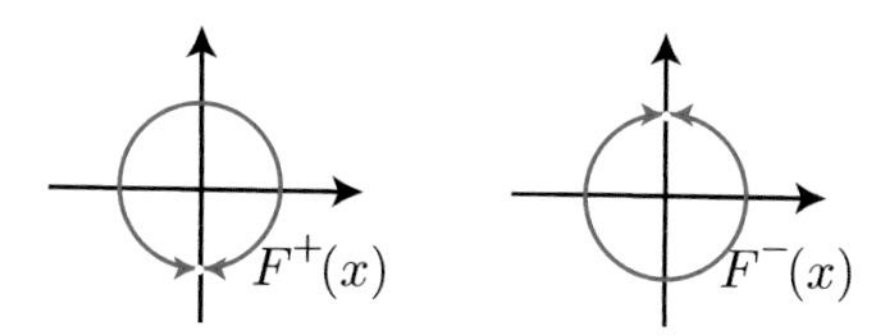

Fig. 7.8

Continuation of $F^+(x)$ to the next sheet $-\pi/2 + 2\pi \leq \arg(x) < 3\pi/2 + 2\pi$:
Taking into account Formula (9) above we see that the analytic continuation of $F^+(x)$ beyond $3\pi/2$ exists and is given by $F(x) = F^-(x\mathrm{e}^{-2\pi\mathrm{i}}) - 2\pi\mathrm{i}\big(\mathrm{e}^{2/x} - \mathrm{e}^{1/x}\big)$ and actually, since $F^-(x) = F^+(x)$ for $-\pi/2 < \arg(x) < +\pi/2$, it is also given, for $3\pi/2 \leq \arg(x) < 3\pi/2 + 2\pi$, by $F(x) = F^+(x\mathrm{e}^{-2\pi\mathrm{i}}) - 2\pi\mathrm{i}\big(\mathrm{e}^{2/x} - \mathrm{e}^{1/x}\big)$.

Continuation of $F^+(x)$ to higher sheets:
Iterating the process we see that, for $-\pi/2 + 2n\pi \leq \arg(x) < 3\pi/2 + 2n\pi$, the analytic continuation is

$$F(x) = F^+(x\mathrm{e}^{-2n\pi\mathrm{i}}) - 2n\pi\mathrm{i}\big(\mathrm{e}^{2/x} - \mathrm{e}^{1/x}\big).$$

Continuation of $F^+(x)$ to lower sheets is left to the reader.

Since the term $\big(\mathrm{e}^{2/x} - \mathrm{e}^{1/x}\big)$ is unbounded for $-\pi/2 < \arg(x) < \pi/2 \bmod 2\pi$ we can conclude that the largest sector where $F(x)$ has an asymptotic expansion is $-3\pi/2 < \arg(x) < +3\pi/2$ and the asymptotics is 1-Gevrey. □

Exercise 2, p. 237 The map μ being linear it suffices to prove that the condition $\mu(\widetilde{f}) = 0$ implies $\widetilde{f} = 0$ in $\mathbb{C}[[x]]/\mathbb{C}\{x\}$.

The cohomology class $\mu(\widetilde{f})$ is obtained by means of the Borel-Ritt theorem in the form of a 1-cochain $\varphi = (-f_j + f_{j+1}) \in \prod_j \Gamma(U_j \cap U_{j+1}\,;\mathscr{A}^{<0})$ over a good covering $\mathscr{U} = (U_j)_{j\in\mathbb{Z}/J\mathbb{Z}}$ of S^1. The functions f_j belong to $\Gamma(U_j\,;\mathscr{A})$ and admits $Tf_j = \widetilde{f}$ as Taylor series at 0.

Since the inclusion maps $\mathfrak{S}(\mathscr{V},\mathscr{U})$ are injective the hypothesis $\mu(\widetilde{f}) = 0$ means that there exists a covering $\mathscr{V} = (V_j)_{j\in\mathbb{Z}/J'\mathbb{Z}}$ finer than $\mathscr{U}$ (denoted by $\mathscr{V} \preceq \mathscr{U}$) such that the 1-cocycle $\psi = \mathfrak{S}(\mathscr{V},\mathscr{U})(\varphi)$ is cohomologous to 0 in $H^1(\mathscr{V}\,;\mathscr{A}^{<0})$. Again, the 1-cocycle ψ has the form $\psi_j = -f'_j + f'_{j+1}$ where the functions f'_j are obtained from the f_j's by restriction to the open arcs V_j of $\mathscr{V}$. The fact that it is cohomologous to 0 in $H^1(\mathscr{V}\,;\mathscr{A}^{<0})$ means that there exists a 0-cochain $(g_j) \in \prod_j \Gamma(V_j\,;\mathscr{A}^{<0})$ such that $-f'_j + f'_{j+1} = -g_j + g_{j+1}$ for all $j \in \mathbb{Z}/J'\mathbb{Z}$. Hence, the functions $f'_j - g_j$ glue together into a holomorphic function at 0. Moreover, they have an asymptotic expansion $T(f'_j - g_j) = Tf'_j - Tg_j = \widetilde{f} - 0 = \widetilde{f}$. Hence, the series $\widetilde{f}(x)$ is convergent and it induces 0 in $\mathbb{C}[[x]]/\mathbb{C}\{x\}$. □

Exercise 3. p. 237 (a) Without loss in generality we assume that the weight of D (cf. comment below Prop. 3.3.15) is 0 so that the assumption that the Newton polygon of D has horizontal length $\ell \geq 0$ is equivalent to the following conditions on the valuation of the a_j's:

$$\begin{cases} v(a_j) \geq 0 \ \text{ for all } j \\ v(a_\ell) = 0 \\ v(a_j) > 0 \text{ if } \ell+1 \leq j \leq n. \end{cases}$$

For simplicity, we set $g'(x) = \mathrm{d}g(x)/\mathrm{d}x$ and we still denote by p the valuation of $g(x)$ although it might have changed after rescaling the weight of D to 0.

Let us first determine the b_j's as functions of the a_j's.

Using $\frac{\mathrm{d}}{\mathrm{d}x}\Big(a_j(x)x^j\frac{\mathrm{d}^j}{\mathrm{d}x^j}\Big) = a_j(x)x^j\frac{\mathrm{d}^{j+1}}{\mathrm{d}x^{j+1}} + \Big(\frac{\mathrm{d}a_j(x)}{\mathrm{d}x}x^j + ja_j(x)x^{j-1}\Big)\frac{\mathrm{d}^j}{\mathrm{d}x^j}$ we obtain

$$b_j(x) = -g'(x)a_j(x) + g(x)\Big(\frac{a_{j-1}(x)}{x} + \frac{\mathrm{d}a_j(x)}{\mathrm{d}x} + j\frac{a_j(x)}{x}\Big). \tag{10}$$

The conditions on the valuations of the a_j's imply the conditions $v(b_j) \geq p-1$ for all j, $v(b_{\ell+1}) = v(ga_\ell/x) = p-1$ and $v(b_j) \geq p$ for all $j \geq \ell+2$. Hence, the Newton polygon of D' has length $\ell+1 > 0$.

(b) The indicial equations of D and D' read respectively

$$\begin{aligned} \pi(\lambda) &= \alpha_0 + \alpha_1\lambda + \alpha_2\lambda(\lambda-1) + \cdots + \alpha_\ell\lambda(\lambda-1)\ldots(\lambda-\ell+1), \\ \pi'(\lambda) &= \beta_0 + \beta_1\lambda + \beta_2\lambda(\lambda-1) + \cdots + \beta_{\ell+1}\lambda(\lambda-1)\ldots(\lambda-\ell) \end{aligned}$$

where $\alpha_j = a_j(0)$ for $j = 1,\ldots,\ell$ and the β_j's are defined as the lower coefficients in $b_j(x) = \beta_j x^{p-1} +$ h.o.t.. From formula (10) and setting $\alpha_{-1} = \alpha_{\ell+1} = 0$ we deduce the relation $\beta_j = \alpha_{j-1} - (p-j)\alpha_j$ for all $j = 1,\ldots,\ell+1$.

It results that

$$\begin{aligned}\pi'(\lambda) &= -p\alpha_0 \\ &\quad +(\alpha_0-(p-1)\alpha_1)\,\lambda \\ &\quad +(\alpha_1-(p-2)\alpha_2)\,\lambda(\lambda-1) \\ &\quad \dots \\ &\quad +(\alpha_{\ell-1}-(p-\ell)\alpha_\ell)\,\lambda(\lambda-1)\dots(\lambda-\ell+1) \\ &\quad +\alpha_\ell\,\lambda(\lambda-1)\dots(\lambda-\ell)\end{aligned}$$

and with the successive cancellations, we observe that $\pi'(p)=0$ and thus, $(\lambda-p)$ divides $\pi'(\lambda)$. On another hand, by construction, the vector space of formal solutions of the equation $D'y=0$ are generated by one non trivial solution of $Dy=g(x)$ together with a fundamental set of formal solutions of $Dy=0$. Hence, all roots of $\pi(\lambda)$ (with multiplicity) are roots of $\pi'(\lambda)$, that is, $\pi(\lambda)$ divides $\pi'(\lambda)$.

If p is not a root of $\pi(\lambda)$ we can conclude that $\pi'(\lambda)=C(\lambda-p)\pi(\lambda)$. Otherwise, we obtain the same conclusion by using a perturbation, say, by replacing α_0 by $\alpha_o+\varepsilon$ and letting ε tend to 0.

(c) To any root λ of the indicial equation corresponds a formal solution of the form $x^\lambda\times\widetilde{f}(x)$ where $\widetilde{f}(x)$ is a power series in x (in case of a multiple root modulo $\mathbb{Z}$ some solutions may include logarithms but at least one of them is of that form). To the root $\lambda=p$ of $\pi'(\lambda)$ corresponds then a formal meromorphic series (i.e., a Laurent series) solution of $D'y=0$.

If $\pi(\lambda)$ has no integer root then $\lambda=p$ is a simple root of $\pi'(\lambda)$ mod $\mathbb{Z}$ and corresponds to a power series solution of Eq. (1).

If $\pi(\lambda)$ has an integer root then the power series solution can be solution of $Dy=0$ instead of $Dy=g(x)$. Let us prove by identification that this might happen. Denote $g(x)=\sum_{j\geq p}g_jx^j$ where $g_p\neq 0$ and set $g_j=0$ for $j<p$. Look for a solution in the form $Y(x)=\sum_{j\geq k}u_jx^j$ with valuation k (hence $u_k\neq 0$). Substituting $Y(x)$ for y in equation (1) and identifying equal powers of x produces a system of equations of the following form :

$$\begin{cases}\pi(k)\,u_k &= g_k\\ \pi(k+1)\,u_{k+1}+h_{k+1}(u_k) &= g_{k+1}\\ \dots & \\ \pi(s)\,u_s+h_s(u_k,\dots,u_{s-1}) &= g_s\\ \dots &\end{cases} \tag{11}$$

where the functions h_s are linear functions of $(u_1,\dots,u_{s-1})$.

Consider the case when p is the unique integer root of $\pi(\lambda)$. The valuation of $DY(x)$ is at least equal to k and so, in view to satisfy $DY(x)=g(x)$, we must have $k\leq p$. If $k<p$, then the first equation of system (11) reads $\pi(k)\,u_k=0$ and since $\pi(k)\neq 0$ this implies $u_k=0$, a contradiction with the fact that k is the exact valuation of $Y(x)$. If $k=p$ then the first equation reduces to $0\,u_p=g_p$ and since $g_p\neq 0$ this is impossible. In that case, Eq. (1) has no meromorphic series solution.

One can check that a solution exists in the form $\widetilde{f}_1(x)+\widetilde{f}_2(x)\ln(x)$ where $\widetilde{f}_1(x)$ and $\widetilde{f}_2(x)$ are meromorphic series.

Without doing a full study look at what happens when instead of p, the indicial equation $\pi(\lambda)$ has $p+1$ as unique integer root. Again k must be equal to p. The first equation of system (11) provides the unique value $u_p = g_p/\pi(p)$ for u_p. The second equation becomes $h_{p+1}(u_p) = g_{p+1}$ since $\pi(p+1) = 0$. The already determined value of u_p might or might not satisfy this equality. If it does not satisfy it the problem has no solution. If it satisfies it then one can choose u_{p+1} arbitrarily and the next equations solve uniquely; there is a one parameter family of solutions.

Observe that when $\pi(\lambda)$ has no integer roots $\geq p$, in particular when it has no integer roots at all, system (11) written with $k = p$ has a unique solution providing thus a meromorphic series solution $Y(x)$ with valuation p. □

Exercise 4, p. 238 The germ $\varphi(x)$ reads in the form

$$\varphi(x) = x^L \mathrm{e}^Q (I_3 + C)\mathrm{e}^{-Q} x^{-L} = I_3 + x^L \left[c_{j,\ell}\,\mathrm{e}^{q_j - q_\ell}\right] x^{-L}$$

and its flatness implies that $c_{j,\ell} = 0$ when $\mathrm{e}^{q_j - q_\ell}$ is not flat in the $\theta = 0$ direction. It results that, permuting the columns of $\widetilde{Y}(x)$, one can order the polynomial q_j so that the matrix C is upper triangular. Since in particular, its diagonal entries are 0 the matrix is also nilpotent ($I_3 + C$ is unipotent). We denote

$$I_3 + C = \begin{bmatrix} 1 & c_{1,2} & c_{1,3} \\ 0 & 1 & c_{2,3} \\ 0 & 0 & 1 \end{bmatrix}$$

The anti-Stokes directions α, β and γ are the directions closest to 0 where the polynomials $q_1 - q_2$, $q_2 - q_3$ and $q_1 - q_3$ are real negative respectively. We observe that $(q_1 - q_2) + (q_2 - q_3) = (q_1 - q_3)$ and this implies that the anti-Stokes direction for $q_1 - q_3$ must be between the anti-Stokes direction for the other two polynomials $q_1 - q_2$ and $q_2 - q_3$. Hence, it is β. Suppose α is anti-Stokes for $q_1 - q_2$ and thus, γ anti-Stokes for $q_2 - q_3$.

To factor $I_3 + C$ we multiply it by an elementary matrix to the left to reduce $c_{1,2}$ to 0 and then by an elementary matrix to the right to reduce $c_{2,3}$ to 0. We obtain

$$\begin{bmatrix} 1 & -c_{1,2} & 0 \\ 0 & 1 & 0 \\ 0 & 0 & 1 \end{bmatrix} \begin{bmatrix} 1 & c_{1,2} & c_{1,3} \\ 0 & 1 & c_{2,3} \\ 0 & 0 & 1 \end{bmatrix} \begin{bmatrix} 1 & 0 & 0 \\ 0 & 1 & -c_{2,3} \\ 0 & 0 & 1 \end{bmatrix} = \begin{bmatrix} 1 & 0 & c_{1,3} - c_{1,2}c_{2,3} \\ 0 & 1 & 0 \\ 0 & 0 & 1 \end{bmatrix}$$

Therefore, we obtain

$$\begin{bmatrix} 1 & c_{1,2} & c_{1,3} \\ 0 & 1 & c_{2,3} \\ 0 & 0 & 1 \end{bmatrix} = \begin{bmatrix} 1 & c_{1,2} & 0 \\ 0 & 1 & 0 \\ 0 & 0 & 1 \end{bmatrix} \begin{bmatrix} 1 & 0 & c_{1,3} - c_{1,2}c_{2,3} \\ 0 & 1 & 0 \\ 0 & 0 & 1 \end{bmatrix} \begin{bmatrix} 1 & 0 & 0 \\ 0 & 1 & c_{2,3} \\ 0 & 0 & 1 \end{bmatrix}$$

Denote respectively by $I_3+C_\alpha, I_3+C_\beta$ and I_3+C_γ the matrices in the right-hand side (in that order). The result follows from the fact that the germs

$$\varphi_\alpha = x^L e^Q (I_3+C_\alpha) e^{-Q} x^{-L},\ \varphi_\beta = x^L e^Q (I_3+C_\beta) e^{-Q} x^{-L},\ \varphi_\gamma = x^L e^Q (I_3+C_\gamma) e^{-Q} x^{-L}$$

belong to $\Lambda_\alpha^{<0}(B_0), \Lambda_\beta^{<0}(B_0)$ and $\Lambda_\gamma^{<0}(B_0)$ respectively, and $\varphi = \varphi_\alpha \varphi_\beta \varphi_\gamma$ □

Exercise 5, p. 238 (1) We check that

$$x^4 F''(x) + (2x^3+3x^2) F'(x) + 2F(x) = \int_0^{+\infty} e^{-\xi/x} d\xi = x.$$

We observe that the two exponentials $e^{1/x}$ and $e^{2/x}$ are both solution of the homogeneous equation $x^4 y''' + (2x^3+3x^2) y' + 2y = 0$ and even form a basis of solutions. The difference of two determinations of $F(x)$ is a solution of this equation and then, is a linear combination of $e^{1/x}$ and $e^{2/x}$. The fact that there appear the coefficients $\pm 2\pi i$ is linked to the Stokes phenomenon and cannot be guessed in advance.

To get the homogeneous form of Eq. (3), divide by x and derivate. We obtain

$$D_1 y \equiv x^5 y''' + (5x^4+3x^3) y'' + (4x^3+3x^2+2x) y' - 2y = 0 \tag{12}$$

Denote by $\widetilde{F}(x) = \sum_{n\geq 1} (-1)^{n-1} \Gamma(n) (1-1/2^n) x^n$ the Taylor series of $F(x)$ at 0.

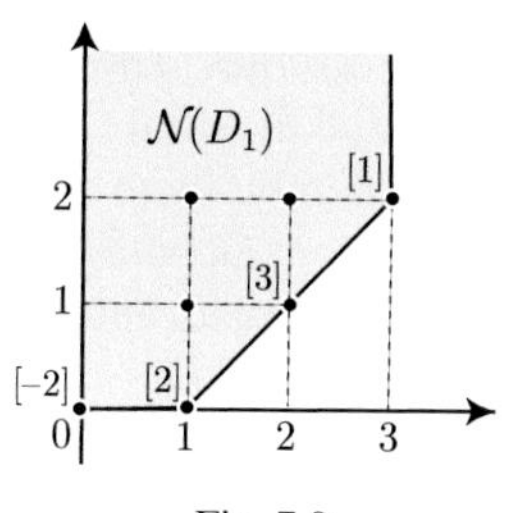

Fig. 7.9

The Newton polygon has a horizontal side of length 1. As indicated on figure 7.9 the indicial equation reads $-2+2\lambda = 0$ and it has a unique solution $\lambda = 1$. All this means that the equation has a unique series solution with valuation 1 (it is the series $\widetilde{F}(x)$).

The Newton polygon has another side with slope 1 and horizontal length 2. The associated characteristic equation reads $\lambda^2 + 3\lambda + 2 = 0$ and it has the two solutions $\lambda = -1$ and $\lambda = -2$. This means that the equation has a formal solution of the form $\widetilde{f}_1(x) x^{\alpha_1} e^{1/x}$ and another one of the form $\widetilde{f}_2(x) x^{\alpha_2} e^{2/x}$. The exponents α_1 and α_2 are the solution of the indicial equation associated with the equations satisfied by $z = y e^{-1/x}$ and $z = y e^{-2/x}$ respectively. We know that the exponentials $e^{1/x}$ and $e^{2/x}$ themselves are the followed solutions and thus we can avoid doing these calculations.

We can conclude that a fundamental set of formal solutions of the equation reads

$$\widetilde{F}(x), \quad e^{1/x}, \quad e^{2/x}$$

The companion system of Eq. (4) reads

$$x^5 \frac{dY}{dx} = \begin{bmatrix} 0 & x^5 & 0 \\ 0 & 0 & x^5 \\ 2 & -(4x^3+3x^2+2x) & -(5x^4+3x^3) \end{bmatrix} Y$$

It has a formal fundamental solution of the form $\widetilde{\mathscr{Y}_1}(x) = \widetilde{H}_1(x)\, e^{Q(1/x)}$ where

$$\begin{cases} Q(1/x) = \mathrm{diag}(0,\, 1/x,\, 2/x) \\ \widetilde{H}_1(x) \text{ is an invertible matrix with entries in } \mathbb{C}[[x]][1/x]. \end{cases}$$

To be more precise, the entries of the first column of $\widetilde{\mathscr{Y}_1}(x)$ are $(\widetilde{F}(x), \widetilde{F}'(x), \widetilde{F}''(x))$, those of the second column are $\left(e^{1/x}, (e^{1/x})', (e^{1/x})''\right)$ and the entries of the third column are $\left(e^{2/x}, (e^{2/x})', (e^{2/x})''\right)$. A normal solution reads $e^{Q(1/x)}$ and then, a normal form is

$$x^2 \frac{dY}{dx} = \begin{bmatrix} 0 & 0 & 0 \\ 0 & -1 & 0 \\ 0 & 0 & -2 \end{bmatrix} Y.$$

The first column of $\mathscr{Y}_1(x)$ admits the unique anti-Stokes direction $\theta = \pi$ which is also a singular direction since $\widetilde{F}(x)$ diverges. The second column has the two anti-Stokes directions $\theta = 0$ and $\theta = \pi$ and the third column $\theta = 0$ but these directions are not singular directions. Consequently, there is only a non trivial Stokes matrix, associated with the $\theta = \pi$ direction. Denote by $H_1^+(x)$ and $H_1^-(x)$ the "sums" of $H_1(x)$ on both sides of π. For the convergent entries the sums are the entries themselves ; for $\widetilde{F}(x)$ they are the functions $F^+(x) = F_{\pi-\varepsilon}(x)$ and $F^-(x) = F_{-\pi+\varepsilon}(x)$ defined in exercise (1).

Taking into account formula (9) the Stokes matrix in the $\theta = \pi$ direction defined by the relation $H_1^+(x) = H_1^-(x)\, e^{Q(x)}(I+C)\, e^{-Q(x)}$ reads

$$I + C_1 = \begin{bmatrix} 1 & 0 & 0 \\ 2\pi i & 1 & 0 \\ -2\pi i & 0 & 1 \end{bmatrix} \tag{13}$$

(2) One can check that $\mathscr{E}_0\, e^{2/x} = (1+2/x)\, e^{2/x}$ and that $x^2(x+2)d/dx + 4(x+1)$ applied to $(1+2/x)\, e^{2/x}$ yields 0.

As in example 1.1.4, denote by $\widetilde{\mathrm{E}}(x) = \sum_{n\geq 0}(-1)^n n!\, x^{n+1}$ the Euler series. A basis of formal solutions of Eq. (5) is given by

$$\widetilde{\mathrm{E}}(x), \quad e^{1/x}, \quad e^{2/x}$$

and a normal solution of its companion system is $e^{Q(1/x)}$ as for equation (4). Hence, equations (4) and (5) are formally equivalent.

Again, only the Stokes matrix in the $\theta = \pi$ direction is non-trivial. From the variation formula of the Euler function (1.3), p. 5, we deduce that this Stokes matrix is equal to

$$I + C_2 = \begin{bmatrix} 1 & 0 & 0 \\ 2\pi \mathrm{i} & 1 & 0 \\ 0 & 0 & 1 \end{bmatrix} \tag{14}$$

The fact that the entry at the third row and first column is non-zero in $I + C_1$ and zero is $I + C_2$ proves that the two equations are not meromorphically equivalent. □

Exercise 6, p. 239 We refer to notations of Sect. 3.5.2.

The fact that ϕ and ψ are cohomologous 1-cocycles in $H^1\big(\mathscr{V}\,;\Lambda^{<0}(B_0)\big)$ means that there exists a 0-cochain (c_j) of $\mathscr{V}$ with values in $\Lambda^{<0}(B_0)$ such that, for all j,

$$\psi_j = c_j^{-1}\,\varphi_j\,c_{j+1}.$$

The product $G_j(x) = F_j(x)\,c_j(x)$ consisting of an isotropy of the normal form composed with a gauge transformation is itself a gauge transformation of the normal form. It is asymptotic to $\widetilde{F}(x)I = \widetilde{F}(x)$ on V_j. Moreover, the 0-cochain $(G_j\,c_j)_{j\in\mathbb{Z}/J\mathbb{Z}}$ satisfies the relation $\psi_j = G_j^{-1}\,G_{j+1}$ for all j. Hence, the result. □

Exercise 7, p. 239

(a) A 1-cocycle of $H^1\big(\mathscr{U}\,;\mathscr{F}\big)$ is the data of sections $h_{j,\ell}$ of $\mathscr{F}$ on all non empty two-by-two intersections $U_j \cap U_\ell$. Given $\ell \neq j+1$ we must prove that $h_{j,\ell}$ is determined by the $h_{j,j+1} = f_j$'s. This results from the configuration of the arcs U_j. Indeed, as a first example, consider $\ell = j+2$ then the intersection $U_j \cap U_{j+2}$ satisfies the relation $U_j \cap U_{j+2} = U_j \cap U_{j+1} \cap U_{j+2}$. The cocycle condition which reads $h_{j,j+2} = h_{j,j+1}h_{j+1,j+2}$ implies that necessarily, one must take $h_{j,j+2} = f_j\,f_{j+1}$. If now, ℓ satisfies $j < \ell \le j + J/2$ (the Stokes arcs coming by pairs, J is even) and if $U_j \cap U_\ell \neq \emptyset$ then one has the relation $U_j \cap U_\ell = U_j \cap U_{j+1} \cap \cdots \cap U_\ell$ and one has to take $h_{j,\ell} = f_j f_{j+1} \dots f_{\ell-1}$. If $j + J/2 < \ell \le j + J$ we exchange the roles of j and ℓ. The result is indeed a 1-cocycle of $\mathscr{U}$.

(b) Recall the form of germs in $\Lambda^{<0}(B_0)$. By definition, an isotropy f of $\Lambda^{<0}(B_0)$ changes the fundamental solution $x^L \mathrm{e}^Q$ of the normal form into another fundamental solution. Hence, the existence of a constant invertible matrix $I + C$ such that

$$f(x)\,x^L\,\mathrm{e}^Q = x^L\,\mathrm{e}^Q\,(I + C).$$

The isotropy is flat if and only if $x^L\,\mathrm{e}^Q\,(I + C)\mathrm{e}^{-Q}x^{-L} = I + \big[c_{m,\ell}\,\mathrm{e}^{q_m - q_\ell}\big]$ is flat (in the multiplicative meaning of asymptotic to identity) and this means that $c_{m,\ell} = 0$ unless $\mathrm{e}^{q_m - q_\ell}$ is asymptotic to zero.

Fix j and consider first a covering $\mathscr{V}$ of U_j consisting of 2 open arcs, V_- to left and V_+ to the right. The arc U_j contains the two anti-Stokes directions α_{j-1} and α_j. It may contain other anti-Stokes directions but none lies from α_{j-1} to α_j. A 1-cocycle of $\mathscr{V}$ with values in $\Lambda^{<0}(B_0)$ consists of a section $f(x)$ of $\Lambda^{<0}(B_0)$ of the form above defined on $V_- \cap V_+$. If an entry $c_{m,\ell}\,\mathrm{e}^{q_m - q_\ell}$ of C is not 0 then $V_- \cap V_+$ is contained in the Stokes arc of $q_m - q_\ell$ centered either to the left of α_{j-1} (including α_{j-1}) or to the right of α_j (including α_j). From lemma 3.5.13, p. 104 (see also exercice 4) the section $f(x)$ can be factored in the form $f(x) = f_-(x)\,f_+(x)$ where $f_-(x)$ is an isotropy containing only exponential terms attached to anti-Stokes directions to the left of α_{j-1} and $f_+(x)$ only exponentials terms attached to anti-Stokes directions to the right of α_j. This implies that $f_-(x)$ can be extended to V_- and $f_+(x)$ to V_+ as elements of $\Lambda^{<0}(B_0)$. Hence, $f(x)$ is a coboundary for the covering $\mathscr{V}$, that is, $f(x)$ is cohomologous to the identity in $H^1\big(\mathscr{V}\,;\Lambda^{<0}(B_0)\big)$.

In the general case it suffices to consider good coverings $\mathscr{V}$ consisting of a finite number of arcs. We proceed iteratively: factor the first component to the right as previously; we can conjugate (by the cohomology relation) the cocycle so as to move the factor $f_-(x)$ to the next component, factor it and so on until the last component to the left. The result is again a 1-cocycle cohomologous to identity. We leave the details to the reader.

As a consequence of this result, by the analogue of Leray's theorem in non abelian cohomology, any cohomology class in $H^1\big(S^1\,;\Lambda^{<0}(B_0)\big)$ can be represented by a 1-cocycle in $H^1\big(\mathscr{U}\,;\Lambda^{<0}(B_0)\big)$.

(c) A Stokes germ attached to an anti-Stokes α_j lives and is flat inside the Stokes arc centered at α_j hence on $\dot{U}_j$. It can then canonically be identified with a component of a 1-cocycle of $H^1\big(\mathscr{U}\,;\Lambda^{<0}(B_0)\big)$.

(d) A section of $\Lambda^{<0}(B_0)$ lives on arcs of flatness of at least one of the exponentials $\mathrm{e}^{q_j - q_\ell}$; hence on arcs of length at most π/k. Since all arcs U_j are longer than π/k they admit no non trivial section of $\Lambda^{<0}(B_0)$.

This implies that there exists no 0-cochain on $\mathscr{U}$; hence, no cohomology relation for the 1-cocycles. In other words, two 1-cocycles of $H^1\big(\mathscr{U}\,;\Lambda^{<0}(B_0)\big)$ are cohomologous if and only if they are equal.

(e) Suppose the covering $\mathscr{U} = (U_j)_{j\in\mathbb{Z}/J\mathbb{Z}}$ is finer than $\mathscr{V} = (V_\lambda)_{\lambda\in L}$. And suppose that there exist a 1-cocycle $\varphi = (\varphi_j)$ on $\mathscr{U}$ and a 1-cocycle $\psi = (\psi_\lambda)$ on $\mathscr{V}$ that induce the same cohomology class in $H^1\big(S^1\,;\Lambda^{<0}(B_0)\big)$.

The covering $\mathscr{U}$ being finer than $\mathscr{V}$ (cf. Def. 2.2.3) there exists a simplicial map $\sigma : \mathbb{Z}/J\mathbb{Z} \to L$ such that for all $j \in \mathbb{Z}/J\mathbb{Z}$ one has the inclusion $U_j \subset V_{\sigma(j)}$. Suppose the component φ_j of φ is non trivial and at least one of the inclusions $U_j \subset V_{\sigma(j)}$ and $U_{j+1} \subset V_{\sigma(j+1)}$ is proper. It results that $U_j \cap U_{j+1}$ is properly included in $V_{\sigma(j)} \cap V_{\sigma(j+1)}$. Since $\dot{U}_j = U_j \cap U_{j+1}$ has the maximal size for the existence of non trivial sections of $\Lambda^{<0}(B_0)$ then, the cocycle ψ induces a trivial component (the identity) on $\dot{U}_j$. Since moreover, there exists no cohomology relations on $\mathscr{U}$ (cf. question (d)), the 1-cocycle φ and the 1-cocycle induced by ψ on $\mathscr{U}$ cannot be cohomologous. Hence, the contradiction which achieves the proof.

Exercise 8, p. 240

(a) We proved in the previous exercise that the Stokes cocycle $c = (c_{\alpha_j})_{j\in\mathbb{Z}/J\mathbb{Z}}$ can be seen as a 1-cocycle in $H^1\big(\mathscr{U}\,;\Lambda^{<0}(B_0)\big)$. Let us narrow the arcs U_j into smaller arcs V_j so that the covering $\mathscr{V} = (V_j)_{j\in\mathbb{Z}/J\mathbb{Z}}$ is a good covering of S^1 and that $\dot{V}_j = V_j \cap V_{j+1}$ contains the unique anti-Stokes direction α_j. The Stokes cocycle c induces on $\mathscr{V}$ the 1-cocycle (c'_j) where c'_j is the restriction of c_{α_j} to $\dot{V}_j$. The proof of the Malgrange-Sibuya isomorphism theorem provides a refinement of $\mathscr{V}$ (for simplicity, we denote again this finer covering by $\mathscr{V} = (V_j)_{j\in\mathbb{Z}/J\mathbb{Z}}$), and functions $f_j(x) \in \Gamma(V_j\,;\Lambda)$ that are asymptotic to $\widetilde{F}(x)$ and satisfy $c_j = f_j^{-1}\,f_{j+1}$ on $\dot{V}_j$ for all j.

Let us prove first that the f_j's can be continued as sections F_j over U_j with similar properties. This results as follows from the fact that the c'_j's can be continued to $\dot{U}_j$.
Denote $U_j =]A_j\,;B_j[$ for all j. Then, $\dot{U}_j =]A_{j+1}\,;B_j[$ and has the anti-Stokes direction α_j as middle point. Similarly, denote $V_j =]a_j\,;b_j[\subset U_j$. For all j, the arc V_j contains the anti-Stokes directions α_{j-1} and α_j and these only; the arc $\dot{V}_j =]a_{j+1}\,;b_j[$ contains the anti-Stokes direction α_j and no other one, and we could even assume that α_j is its middle point.
We have $A_j < a_j < a_{j+1} < b_j < b_{j+1}$. Since $f_j = f_{j+1}\,c_j^{-1}$ on $]a_{j+1}\,;b_j[$ and c_j lives on $\dot{U}_j =]A_{j+1}\,;B_j[$ we can continue f_j to the right from b_j to $b'_{j+1} = \min(b_{j+1}, B_j)$ by setting $f_j = f_{j+1}\,c_j^{-1}$ on $[b_j\,;b'_{j+1}[$. If $b_{j+1} < B_j$ we are finished. If not, we can continue f_{j+1} from b_{j+1} to $b'_{j+2} = \min(b_{j+2}\,, B_j)$ by setting $f_{j+1} = f_{j+2}c_{j+1}^{-1}$ on $[b_{j+1}\,;b'_{j+2}[$ and again $f_j = f_{j+1}\,c_j^{-1}$ on $[b_{j+1}\,;b'_{j+2}[$. We iterate the process until we reach B_j. We proceed similarly to continue f_j to the left from a_j to A_j. We denote by F_j the analytic continuation of f_j to U_j. By construction the F_j's satisfy the relations $F_j^{-1}\,F_{j+1} = c_j$ on $\dot{U}_j$ and are asymptotic to $\widetilde{F}(x)$ on U_j.

Let us now prove, like in the Malgrange-Sibuya classification theorem 3.5.6, that $F_j(x)$ is a gauge transformation of the normal form $\mathrm{d}Y/\mathrm{d}x = B_0(x)Y$. The matrix ${}^{F_{j+1}}B_0$ satisfies ${}^{F_{j+1}}B_0 = {}^{F_jc_j}B_0$ on $\dot{U}_j$ and the latter one is equal to ${}^{F_j}B_0$ since c_j is an isotropy of the normal form $\mathrm{d}Y/\mathrm{d}x = B_0(x)Y$. Hence, the matrices ${}^{F_j}B_0$ glue together into a meromorphic matrix $B(x)$ and it results that $\widetilde{F}(x)$ is a formal gauge transformation from the normal form to $\mathrm{d}Y/\mathrm{d}x = B(x)Y$. This ends the proof of this question.

(b) The 0-cochain $(F_j)_{j\in\mathbb{Z}/J\mathbb{Z}}$ satisfies the conditions of the Ramis-Sibuya theorem 5.2.1. Indeed, the 1-cocycle $c = (c_j)_{j\in\mathbb{Z}/J\mathbb{Z}}$ is asymptotic to the identity. We can then write $c_j = I + C_j$ where I is the identity matric C_j is asymptotic to 0 with exponential order k on $\dot{U}_j$ (all exponentials have degree k). From the relation $F_j^{-1}F_{j+1} = I + C_j$ we deduce that $F_{j+1} - F_j = F_jC_j$ is also asymptotic to 0 with exponential order k on $\dot{U}_j$. It results that F_j is s-Gevrey asymptotic ($s = 1/k$) to $\widetilde{F}$ on U_j for all j.

From the fact that U_j is more than π/k large and from Ramis's definition 5.1.6 of k-summability we can conclude that, for any θ direction located between the successive directions anti-Stokes $\alpha_{j-1} = A_j + \pi/(2k)$ and $\alpha_j = A_{j+1} + \pi/(2k)$, the matrix $\widetilde{F}(x)$ is k-summable with k-sum F_j. This means, in particular, that when two directions θ and θ' are separated by no anti-Stokes directions the k-sums of $\widetilde{F}(x)$ in the θ and θ' directions are analytic continuation from each other.

In this approach the "sums" F_j are given for free by the Stokes cocycle with, generically, the largest possible domain, and this is done independently of any theory of summation. By generically we mean that these domains might be larger in some particular cases; for instance when the Stokes matrix is trivial, meaning that $C_j = 0$, the two functions F_j and F_{j+1} glue together into an analytic gauge transformation living on a much larger domain.

The same result was already proved by another method relying on properties of k-sums (therefore, sums given by a theory of summation) in volume I [MS16, Lem. 2.37]. The reader is invited to look by himself or herself for a proof relying on k-Borel-Laplace definition of k-sums (using proposition 5.3.7) and also for a proof based on the interpretation of k-sums in the infinitesimal neighborhood X^k (see Sect. 5.4.2).

Exercise 9, p. 240

(a) is clear.

(b) The Newton polygons $\mathcal{N}_\infty$ of $(\mathcal{A}_\infty)$ and $\mathcal{N}_0$ of $(\mathcal{A}_0)$ are symmetric to each other with respect to a horizontal line. Recall that they are translated vertically after multiplication by a power of x. From now we refer to $\mathcal{N}_0$.

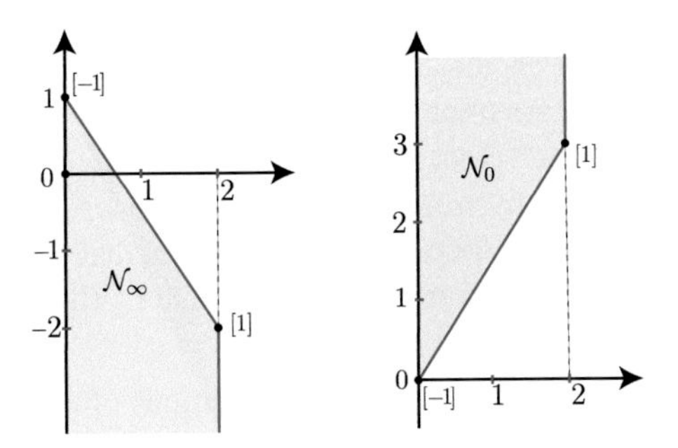

Fig. 7.10. Newton polygons of Airy equations $\mathcal{A}_\infty$ at infinity and $\mathcal{A}_0$ at 0

The degree of the determining polynomials q_1 and q_2 is equal to the (unique) slope of $\mathcal{N}_0$, that is $3/2$ and the characteristic equation $X^2 - 1 = 0$ has $+1$ and -1 as roots. This implies that the dominant coefficients in q_1 and q_2 are $-2/3$ and $+2/3$ (cf. Eq. (3.10), p. 85).

To prove that the polynomials q_1 and q_2 are monomials it suffices to prove that one of them is a monomial. Indeed, if $q_1(1/x)$ is a determining polynomial, so is $q_1\big(1/(x\mathrm{e}^{2\pi\mathrm{i}})\big)$. The two polynomials $q_1\big(1/(x\mathrm{e}^{2\pi\mathrm{i}})\big)$ and $q_1(1/x)$ are distinct since

$$q_1\big(1/(x\mathrm{e}^{2\pi\mathrm{i}})\big) = -2/3(x\mathrm{e}^{2\pi\mathrm{i}})^{-3/2} = +2/3x^{-3/2} \neq q_1(1/x).$$

Hence, $q_1\big(1/(x\mathrm{e}^{2\pi\mathrm{i}})\big)$ is the other determining polynomial $q_2(1/x)$.

The fact that q_1 and q_2 are monomials results from the fact that the change of variable $y \leftarrow \mathrm{e}^{-q_1(1/x)}y$ produces an equation with a Newton polygon that has no slope smaller than $3/2$ but the slope 0.

(c) The formal fundamental solution $\widetilde{Y}(x)$ reads $\widetilde{Y}(x) = \widetilde{F}(x)x^J U \mathrm{e}^{Q(1/x)}$. Since $\widetilde{F}(x)$ has no fractional powers of x it has no formal monodromy and it suffices to compute the formal monodromy of $\widetilde{F}_0(x) = x^J U \mathrm{e}^{Q(1/x)}$. To this end, we compute $\widetilde{F}_0(x\mathrm{e}^{2\pi\mathrm{i}}) = x^J \mathrm{diag}(\mathrm{i},-\mathrm{i})\,U P^{-1}\mathrm{e}^{Q(1/x)}P$ where P is the circulant matrix $P = \begin{bmatrix} 0 & 1 \\ 1 & 0 \end{bmatrix}$. It results that $\widetilde{F}_0(x\mathrm{e}^{-2\pi\mathrm{i}}) = \widetilde{F}_0(x)\begin{bmatrix} 0 & \mathrm{i} \\ \mathrm{i} & 0 \end{bmatrix}$ and $\widetilde{\mathrm{M}} = \begin{bmatrix} 0 & \mathrm{i} \\ \mathrm{i} & 0 \end{bmatrix}$.

(d) The anti-Stokes direction of $(\mathscr{A}_0)$ or (S) are the directions of maximal decay of the two polynomials $q_1 - q_2$ and $q_2 - q_1$. They are given by the conditions $\arg(x) = 0 \mod 4\pi/3$ and $\arg(x) = 2\pi/3 \mod 4\pi/3$. In the plane $\mathbb{C}$, this gives 3 anti-Stokes directions given by

$$\arg(x) = -2\pi/3, \quad \arg(x) = 0 \quad \text{and} \quad \arg(x) = 2\pi/3 \mod 2\pi.$$

(e) The degree of $q_1 - q_2$ being equal to 3/2 it results from corollary 5.2.7 that the series $\widetilde{F}(x)$ is 3/2-summable; more precisely, $3/2$-summable in any direction but the three directions $-2\pi/3$, 0 and $2\pi/3 \mod 2\pi$ since these are the anti-Stokes directions of the system. From theorem 3.5.14, p. 106, the Stokes matrix with respect to $Y_0^-(x)$ is defined by

$$F_0^+(x)\,\mathscr{Y}_{0,0}(x) = F_0^-(x)\,\mathscr{Y}_{0,0}(x)\,(I_2 + C_0)$$

where $F_0^+(x)$ denotes the sum of $\widetilde{F}(x)$ to the right of $\mathbb{R}^+$ and $\mathscr{Y}_{0,0}(x) = x^J U \mathrm{e}^Q$ computed with the principal determination of the argument in a neighborhood of $\arg(x) = 0$. The functions $F_0^-(x)$ and $F_0^+(x)$ are single-valued, not depending on the choice of the argument of x. Write the relation above in the form

$$F_0^-(x)^{-1}F_0^+(x) = \mathscr{Y}_{0,0}(x)\,(I_2 + C_0)\,\mathscr{Y}_{0,0}(x)^{-1}. \tag{15}$$

The left-hand side is flat (asymptotic to I_2), hence the right-hand side too. This means that only the flat terms in $\mathscr{Y}_{0,0}(x)\,C_0\,\mathscr{Y}_{0,0}(x)^{-1}$ might not be 0. This is the case of only one term, the term in position (1,2) in the right up corner. Indeed, for $\arg(x) \approx 0$ the real part of $q_1 - q_2$ is negative whereas the real part of $q_2 - q_1$ is positive. Hence, only the term which contains the exponential $\mathrm{e}^{q_1 - q_2}$ is flat and might be not 0. This implies that the Stokes matrix in the $\mathbb{R}^+$ direction with $\arg(x) \approx 0$ has the form

$$I_2 + C_0 = \begin{bmatrix} 1 & c_0 \\ 0 & 1 \end{bmatrix}.$$

(f) With the choice of $\arg(x) \approx 2\pi$ near $\mathbb{R}^+$ only the polynomial $q_2 - q_1$ has a negative real part and the Stokes matrix in the $\mathbb{R}^+$ direction with $\arg(x) \approx 2\pi$ reads in the form

$$I_2 + C_{2\pi} = \begin{bmatrix} 1 & 0 \\ c_{2\pi} & 1 \end{bmatrix}.$$

To find the relation which links c_0 and $c_{2\pi}$ recall the relation $\widetilde{Y}(x\mathrm{e}^{2\pi\mathrm{i}}) = \widetilde{Y}(x)\,\widetilde{\mathrm{M}}$. From relation (15) and the value of $\widetilde{\mathrm{M}}$ computed in question (c) it results that

$$I_2 + C_0 = \widetilde{M}\,(I_2 + C_{2\pi})\,\widetilde{M}^{-1} = \begin{bmatrix} 1 & c_{2\pi} \\ 0 & 0 \end{bmatrix}.$$

Hence, $c_0 = c_{2\pi}$.

(g) Set $x = \mathrm{e}^{-2\pi\mathrm{i}/3}X$.
From the expression of $\widetilde{F}(x)$ in question (1) we obtain $\widetilde{F}(X) = K^{-1}\,\widetilde{F}(x)\,K$ where the matrix K is the constant invertible diagonal matrix $K = \mathrm{diag}\big(\mathrm{e}^{-\pi\mathrm{i}/3}, \mathrm{e}^{\pi\mathrm{i}/3}\big)$. Moreover, we have $X^J\,U\,\mathrm{e}^{Q(1/X)} = x^J \begin{bmatrix} \mathrm{e}^{\pi\mathrm{i}/6} & 0 \\ 0 & \mathrm{e}^{-\pi\mathrm{i}/2} \end{bmatrix} \begin{bmatrix} 1 & 1 \\ -1 & 1 \end{bmatrix} \mathrm{e}^{Q(1/x)}P$ where P is the circulant matrix $\begin{bmatrix} 0 & 1 \\ 1 & 0 \end{bmatrix}$. It results that

$$\widetilde{F}(X)X^J\,U\,\mathrm{e}^{Q(1/X)} = \mathrm{diag}\big(\mathrm{e}^{\pi\mathrm{i}/6}, \mathrm{e}^{-\pi\mathrm{i}/2}\big)\widetilde{F}(x)x^J\,U\,\mathrm{e}^{Q(1/x)}\,P,$$

that is, that

$$\widetilde{Y}(X) = \mathrm{diag}\big(\mathrm{e}^{\pi\mathrm{i}/6}, \mathrm{e}^{-\pi\mathrm{i}/2}\big)\widetilde{Y}(x)\,P. \tag{16}$$

We observe that, because of the presence of the constant diagonal matrix in (16) the matrix $\widetilde{Y}(X)$ is not a fundamental solution of the same system as $\widetilde{Y}(x)$ (the companion system of the Airy equation) but of a meromorphically equivalent system. If $\arg(x)$ is close to 0 then $\arg(X)$ is close to $2\pi\mathrm{i}/3$. Denote by $F^{\pm}_{2\pi/3}$ the sums of $\widetilde{F}$ on each side of the anti-Stokes direction $\alpha = 2\pi/3 \mod 2\pi$. From the equality (16) we obtain the relation

$$\big(Y^-_{2\pi/3}\big)^{-1}Y^+_{2\pi/3} = P^{-1}\,\big(Y^-_0\big)^{-1}Y^+_0\,P, \text{ equivalent to } I_2 + C_{2\pi/3} = P^{-1}\,(I_2 + C_0)\,P.$$

The same calculation with $x = \mathrm{e}^{2\pi\mathrm{i}/3}X$ (change i into $-$i) leads to the relation

$$I_2 + C_{-2\pi/3} = P^{-1}\,(I_2 + C_0)\,P,$$

and therefore, to $I_2+C_{2\pi/3} = I_2+C_{-2\pi/3} = \begin{bmatrix} 1 & 0 \\ c_0 & 0 \end{bmatrix}$.

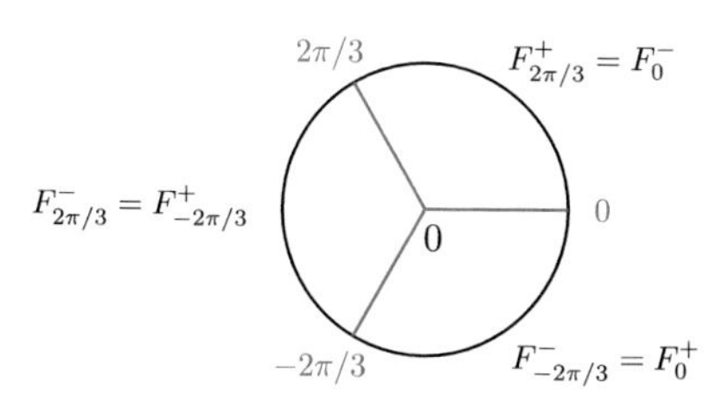

Fig. 7.11. Anti-Stokes directions and sums of $\widetilde{F}(x)$

(h) We saw in question (d) that the anti-Stokes directions of system (S) are the three directions $-2\pi/3, 0$ and $2\pi/3 \mod 2\pi$. From volume I [MS16, Lem. 2.37] and exercise 8 below, we know that the 3/2-sum $F_0^-(x)$ of $\widetilde{F}(x)$ to the left of $\alpha = 0 \mod 2\pi$ lives on the arc $J_1 =]-\pi/3\,; 2\pi/3+\pi/3 = \pi[$ and is equal to the sum $F^+_{2\pi/3}$ of $\widetilde{F}(x)$ to the right of the anti-Stokes direction $\alpha = 2\pi/3 \mod 2\pi$. Similarly, we have $F^-_{-2\pi/3} = F_0^+$ on $J_2 =]-\pi\,; \pi/3[$ and $F^+_{-2\pi/3} = F^-_{2\pi/3}$ on $J_3 =]\pi/3\,; 2\pi+2\pi/3[$. Recall that the sums do not depend on the choice of a determination of the argument and we should have denoted them by $F_{\alpha' \mathrm{mod}\, 2\pi}$. For simplicity and without any possible confusion, we denote by them by $F_{\alpha'}$.

The monodromy (cf. volume I [MS16, Sect. 1.1.2]) of the system (S) is trivial. Indeed, on the Riemann sphere, making a turn around 0 amounts to making a turn (in the opposite sense) around infinity. Infinity being an ordinary point of the system (S) all solutions are analytic on a neighborhood of infinity. Moreover, 0 being the unique singular point these analytic solutions have an analytic continuation on all of $\mathbb{C}^*$. This means that if we continue a local solution along a path around 0 the resulting solution glue together with the initial one. Let us make explicit such a continuation in terms of Stokes matrices.

From the definition of Stokes matrices the following relations hold:

$$\begin{cases} \mathscr{Y}^+_{2\pi/3} &= \mathscr{Y}^-_{2\pi/3}(I_2+C_{2\pi/3}) \\ \mathscr{Y}^+_0 &= \mathscr{Y}^-_0\,(I_2+C_0) \\ \mathscr{Y}^+_{-2\pi/3} &= \mathscr{Y}^-_{-2\pi/3}(I_2+C_{-2\pi/3}) \end{cases}$$

It results that the fundamental solution $\mathscr{Y}^+_{-2\pi/3}$ which lives on J_3 admits the fundamental solution $\mathscr{Y}^-_{-2\pi/3}(I_2+C_{-2\pi/3}) = \mathscr{Y}^+_0(I_2+C_{-2\pi/3})$ as its analytic continuation on J_1. It admits $\mathscr{Y}^-_0(I_2+C_0)(I_2+C_{-2\pi/3}) = \mathscr{Y}^+_{2\pi/3}(I_2+C_0)(I_2+C_{-2\pi/3})$ as its analytic continuation on J_2 and $\mathscr{Y}^-_{2\pi/3}(I_2+C_{2\pi/3})(I_2+C_0)(I_2+C_{-2\pi/3})$ as its analytic continuation on J_3.

Writing that this latter continuation is equal to $\mathscr{Y}^{+}_{-2\pi/3}$ on J_3 we obtain

$$F^{-}_{2\pi/3}(x)\mathscr{Y}_{0,2\pi/3}(x)(I_2+C_{2\pi/3})(I_2+C_0)(I_2+C_{-2\pi/3}) = F^{+}_{-2\pi/3}(x)\mathscr{Y}_{0,-2\pi/3}(x).$$

Now, consider this relation in restriction to $\mathbb{R}^-$. Then, $\mathscr{Y}_{0,2\pi/3}(x) = \mathscr{Y}_{0,+\pi}(x)$ whereas $\mathscr{Y}_{0,-2\pi/3}(x) = \mathscr{Y}_{0,-\pi}(x)$. Since $\pi = -\pi + 2\pi$, we have then, by definition of the formal monodromy, the relation $\mathscr{Y}_{0,2\pi/3}(x) = \mathscr{Y}_{0,-2\pi/3}(x)\,\widetilde{\mathrm{M}}$ with $\widetilde{\mathrm{M}}$ the formal monodromy matrix in the formal basis $\widetilde{Y}_0(x) = x^J\,U\,\mathrm{e}^{Q(1/x)}$ as well as in the analytic basis $\mathscr{Y}_{0,-2\pi/3}$ on J_3. On the other hand, as mentioned above, $F^{-}_{2\pi/3}(x) = F^{+}_{-2\pi/3}(x)$ on J_3, and in particular on $\mathbb{R}^-$.
Hence, we obtain the followed relation

$$\widetilde{\mathrm{M}}\,(I_2+C_{2\pi/3})\,(I_2+C_0)\,(I_2+C_{-2\pi/3}) = I_2.$$

The general relation with an arbitrary number of Stokes matrices and non trivial monodromy M in dimension n is proved in volume I [MS16, Exe. 2.38].

Let us now take into account the results of questions (c) and (g). We obtain

$$\begin{bmatrix} 0 & \mathrm{i} \\ \mathrm{i} & 0 \end{bmatrix} \begin{bmatrix} (1+c_0^2) & c_0 \\ (2c_0+c_0^3) & (1+c_0^2) \end{bmatrix} = I_2$$

which gives $c_0 = -\mathrm{i}$. Hence,

$$I_2 + C_0 = \begin{bmatrix} 1 & -\mathrm{i} \\ 0 & 1 \end{bmatrix} \quad \text{and} \quad I_2 + C_{-2\pi/3} = I_2 + C_{2\pi/3} = \begin{bmatrix} 1 & 0 \\ -\mathrm{i} & 1 \end{bmatrix}.$$

Remark 1 The calculation by this method was first performed by J. Martinet and J.-P. Ramis in [MarR89, p. 203]. There, Martinet and Ramis made the computation at infinity (at the singular point $z = \infty$ of the Airy equation $y'' - zy = 0$) with an orientation about infinity which, on the Riemann sphere, induces around 0 the inverse orientation of the one we chose at 0. It results that our Stokes matrices are inverse from each other, hence the change of i to $-$i.

The proof by D. Sauzin in volume I [MS16, Exe 6.100 and 6.101] is equivalent to this one: it makes explicit the $(3/2)$-sums of the power series in $\tilde{A}(x)$ and $\tilde{B}(x)$ (called $\tilde{\varphi}$ and $\tilde{\psi}$ respectively) and uses the fact that the sums are entire functions (hence, that the monodromy is trivial).

A variant of this consists in using the properties of Bessel functions and their link with the Airy equation. This approach can also be found in [MarR89].

Remark 2 The formulæ in [MS16, Thm. 6.102] provide another method to compute the Stokes coefficients above. In the basis $(\widetilde{A},\widetilde{B})$, corresponding to $(\widetilde{\varphi},\widetilde{\psi})$ in Sauzin's notation, the Stokes matrix reads $I_2 + C_0 = \begin{bmatrix} 1 & c_0 \\ 0 & 1 \end{bmatrix}$ with $c_0 = -\mathrm{i}$. From the resurgent point of view, to the θ_0 direction (choosing the principal determination) there is only one singular point ω in the Borel plane in that direction; this singular point is given by the Stokes value in $\mathrm{e}^{q_1-q_2} = \mathrm{e}^{-4/(3x^{3/2})}$, that is, 4/3 if we worked with the variable $x^{3/2}$. After the change of variable $z = \frac{2}{3}x^{-3/2}$ made by D. Sauzin this corresponds to the Stokes value $\omega = 2$. The Stokes phenomenon in the θ_0 direction is characterized by a Stokes automorphism with matrix $I_2 + C_0$. The alien derivative at $\omega = 2$ is then given by $\Delta_2 = \log(I_2 + C_0)$; hence $\Delta_2 = C_0 = \begin{bmatrix} 0 & -\mathrm{i} \\ 0 & 0 \end{bmatrix}$. The non-zero entry in Δ_2 corresponds to the $-\mathrm{i}$ in the formula $\Delta_2 \widetilde{\psi} = -\mathrm{i}\widetilde{\varphi}$ in [MS16, Eq. (6.122)]. For a general correspondence between the alien derivatives and the linear Stokes phenomenon we refer to [LR11] and its last two examples. □

Exercise 10, p. 242

Let us interpret the cyclic relation in the infinitesimal neighborhood $(X, \mathscr{F})$. An interpretation in $(X^{\underline{k}}, \mathscr{F}^{\underline{k}})$ is similar with a more complicated picture to draw with the $Y^{\underline{k}}$'s (cf. Fig. 3.6, p. 117).

To take into account the formal fundamental solution $\widetilde{Y}(x) = \widetilde{F}(x) x^L \mathrm{e}^{Q(1/x)}$ we extend the sheaf $\mathscr{F}$ of differential algebras into the sheaf $\mathscr{F}^*$ obtained from $\mathscr{F}$ by the adjunction of the entries of x^L and $\mathrm{e}^{Q(1/x)}$. Thus, to define $\mathscr{F}^*$ (see Sec 3.6.1), we set:

$$\overline{\mathscr{F}}^*\big(D(0,k)\big) = \mathbb{C}[[x]]_s[x^L, \mathrm{e}^Q] \quad (\text{with } s = 1/k) \tag{17}$$

where, for short, we denote by x^L and e^Q all entries of these matrices. Let us observe that, whereas the formal matrix x^L has a wild analytic continuation on all of X as a multivalued function, the exponentials of $\mathrm{e}^{Q(x)}$ cannot be continued on all of X but only on X deprived of some sectors with vertex 0 (cf. Exa. 3.6.2).

In X let us draw the singular points $a_j = (\alpha_j, k_{j_\ell})$ (with argument α_j on the circle of radius k_{j_ℓ} for all levels k_{j_ℓ} associated with α_j) and a circle C_r with radius $r < \min(k_{j_\ell})$ for all j_ℓ, oriented anti-clockwise. The path Γ can be taken as the boundary circle of X oriented anti-clockwise.

We denote as before by $F_\theta(x)$ the sum of $\widetilde{F}(x)$ in a non anti-Stokes direction θ and by F_α^- and F_α^+ the two lateral sums of $\widetilde{F}(x)$ respectively to the left and to the right of an anti-Stokes direction α. Given a non anti-Stokes direction θ, the sum F_θ of $\widetilde{F}(x)$ in the θ direction is the (wild) analytic continuation of $\widetilde{F}(x)$ along the ray d_θ from 0 to the boundary ∂X of X (where summable formal series become analytic functions in the classical meaning).

From theorem 3.5.14, p. 106, we know that $\varphi_{\alpha_j}(x) = \big(F_{\alpha_j}^-(x)\big)^{-1} F_{\alpha_j}^+(x)$. In other words, the Stokes automorphism in the α_j direction changes the basis of solutions $F_{\alpha_j}^-(x)$ into the basis $F_{\alpha_j}^+(x)$. With this formula the Stokes automorphism φ_{α_j} can be interpreted as the continuation of $\widetilde{F}(x)$ from 0 to ∂X along the ray $d_{\alpha_j}^+$ then along ∂X anti-clockwise to pass the direction α_j and eventually back to 0 along $d_{\alpha_j}^-$. It acts trivially on $x^L \mathrm{e}^{Q(1/x)}$. The loop thus followed can be changed homotopically into any loop turning anti-clockwise once around all a_j located on the anti-Stokes ray d_{α_j} and enclosing neither 0 nor other singular points. For all j, we choose a loop γ_j in X that starts at $r_j = (\alpha_j, r)$ on the circle C_r in the α_j direction and turns anti-clockwise around all singular points located on d_{α_j}.

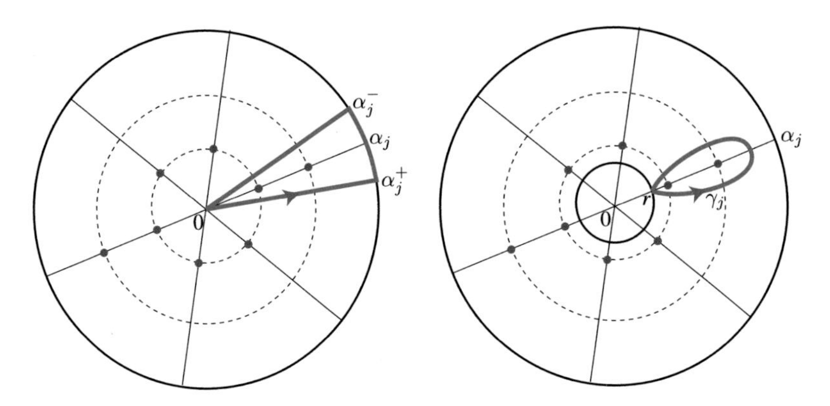

Fig. 7.12. Loop for the Stokes automorphism in the α_j direction in X (in red, the singular points)

The path Γ to compute the monodromy $\mathcal{M}$ can be deformed without passing any singular point, into a path γ as follows:

Let γ start, for instance, at r_N, follow γ_N, then, C_r from r_N to r_{N-1}, then γ_{N-1}, etc, until the loop γ_1 and finally reach r_N along C_r. Doing so, we have performed the Stokes automorphisms $\varphi_N, \varphi_{N-1}, \dots, \varphi_1$ in that order. Finally, ending with the path from r_1 to r_N on C_r completes a turn following C_r anti-clockwise; this results in the action of the formal monodromy (in the formal world cf. Eq. (17)).

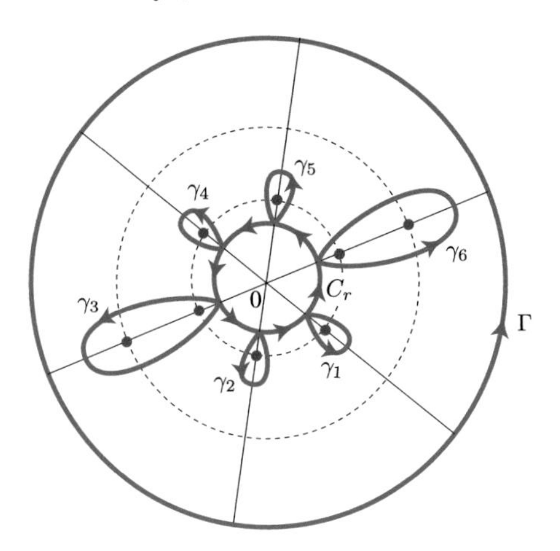

Fig. 7.13. The loops Γ and γ in the infinitesimal neighborhood X

We obtain so the followed relation

$$\widetilde{\mathscr{M}} \circ \varphi_1 \circ \varphi_{\alpha_2} \circ \cdots \circ \varphi_N = \mathscr{M}. \tag{7}$$

Observe that, in this formula, the Stokes automorphisms are ordered like the anti-Stokes directions. However, we can start with any of them. The formal monodromy comes last (when we close the loop C_r); hence, its position to the left.

References

[Bal92a] W. Balser. A different characterization of multi-summable power series. *Analysis*, 12(1-2):57–65, 1992.

[Bal92b] W. Balser. Summation of formal power series through iterated Laplace integrals. *Math. Scand.*, 70(2):161–171, 1992.

[Bal93] W. Balser. Addendum: "A different characterization of multi-summable power series" [Analysis **12** (1992), no. 1-2, 57–65; MR1159370 (93d:40009)]. *Analysis*, 13(3):317–319, 1993.

[Bal94] W. Balser. *From divergent power series to analytic functions*, volume 1582 of *Lecture Notes in Math.* Springer, 1994.

[Bal00] W. Balser. *Formal power series and linear systems of meromorphic ordinary differential equations*. Springer-Verlag, New York, 2000.

[BCL03] M. Barkatou, F. Chyzak, and M. Loday-Richaud. Remarques algorithmiques liées au rang d'un opérateur différentiel linéaire. In *From combinatorics to dynamical systems*, volume 3 of *IRMA Lect. Math. Theor. Phys.*, pages 87–129. de Gruyter, 2003.

[BH86] N. Bleistein and R. A. Handelsman. *Asymptotic expansions of integrals*. Dover Publications Inc., New York, 1986.

[Bir09] G. Birkhoff. Singular points of ordinary linear differential equations. *Trans. Amer. Math. Soc.*, 10-4:436–470, 1909.

[Bir39] G. Birkhoff. Déformations analytiques et fonctions auto-équivalentes. *Ann. Inst. H. Poincaré*, 9:51–122, 1939.

[BJL79a] W. Balser, W. Jurkat, and D. A. Lutz. A general theory of invariants for meromorphic differential equations. I. Formal invariants. *Funkcial. Ekvac.*, 22:197–221, 1979.

[BJL79b] W. Balser, W. B. Jurkat, and D. A. Lutz. A general theory of invariants for meromorphic differential equations. II. Proper invariants. *Funkcial. Ekvac.*, 22(3):257–283, 1979.

[Bra91] B. L. J. Braaksma. Multisummability and Stokes multipliers of linear meromorphic differential equations. *J. Differential Equations*, 92-1:45–75, 1991.

[Bra92] B. L. J. Braaksma. Multisummability of formal power series solutions of nonlinear meromorphic differential equations. *Ann. Inst. Fourier (Grenoble)*, 42(3):517–540, 1992.

[BV89] D. G. Babbitt and V. S. Varadarajan. Local moduli for meromorphic differential equations. *Astérisque*, (**169-170**):1–217, 1989.

[CL55] E. A. Coddington and N. Levinson. *Theory of ordinary differential equations*. McGraw-Hill Book Company, Inc., New York-Toronto-London, 1955.

[CNP93] B. Candelpergher, J.-C. Nosmas, and F. Pham. *Approche de la résurgence*. Actualités Mathématiques. Hermann, Paris, 1993.

[Cop36] F. Cope. Formal solutions of irregular linear differential equations. Part II. *Amer. J. Math.*, 58-1:130–140, 1936.

[Cos09] O. Costin. *Asymptotics and Borel summability*. CRC Monographs and Surveys in Pure and Applied Mathematics, CRC Press. Chapman & Hall, 2009.

[Dela16] É. Delabaere. *Divergent Series, Summability and Resurgence III, Resurgent Methods and the First Painlevé equation*, volume 2155 of *Lecture Notes in Mathematics*. Springer, Heidelberg, 2016.

[Deli70] P. Deligne. *Équations différentielles à points singuliers réguliers*, volume 163. Springer-Verlag, 1970.

[Die80] J. Dieudonné. *Calcul infinitésimal*. Collection Méthodes. Hermann Paris, 1980.

[DMR07] P. Deligne, B. Malgrange, and J.-P. Ramis. *Singularités irrégulières*. Documents Mathématiques (Paris). Société Mathématique de France, Paris, 2007. Correspondance et documents.

[Éca74] J. Écalle. *Théorie des invariants holomorphes*, volume 67, 74-04. Publ. Math. Orsay, 1974.

[Éca81] J. Écalle. *Les fonctions résurgentes, tome I : les algèbres de fonctions résurgentes*, volume 81-05. Publ. Math. Orsay, 1981.

M. Loday-Richaud, *Divergent Series, Summability and Resurgence II*,
Lecture Notes in Mathematics, DOI 10.1007/978-3-319-29075-1

[Éca85] J. Écalle. *Les fonctions résurgentes, tome III : l'équation du pont et la classification analytique des objets locaux*, volume 85-05. Publ. Math. Orsay, 1985.

[Éca93] J. Écalle. Cohesive functions and weak accelerations. *J. Anal. Math.*, 60:71–97, 1993.

[Fre57] J. Frenkel. Cohomologie non abélienne et espaces fibrés. *Bull. Soc. Math. France*, 85:135–220, 1957.

[FS13] Augustin Fruchard and Reinhard Schäfke. *Composite asymptotic expansions*, volume 2066 of *Lecture Notes in Mathematics*. Springer, Heidelberg, 2013.

[Gev18] M. Gevrey. Sur la nature analytique des solutions des équations aux dérivées partielles. *Ann. Sci. École Norm. Sup. (3)*, 25:129–190, 1918.

[God58] R. Godement. *Topologie algébrique et théorie des faisceaux*. Actualit'es Sci. Ind. No. 1252. Publ. Math. Univ. Strasbourg. No. 13. Hermann, Paris, 1958.

[HS99] P.-F. Hsieh and Y. Sibuya. *Basic Theory of Ordinary Differential Equations*. Springer-Verlag, 1999.

[Im96] G. K. Immink. On the summability of the formal solutions of a class of inhomogeneous linear difference equations. *Funkcial. Ekvac.*, 39:469–490, 1996.

[Inc44] E. L. Ince. *Ordinary Differential Equations*. Dover Publications, New York, 1944.

[Ive86] B. Iversen. *Cohomology of sheaves*. Universitext. Springer-Verlag, Berlin, 1986.

[Jac37] N. Jacobson. Pseudo-linear transformations. *Ann. of Math.*, 38-2:484–507, 1937.

[Kim71] T. Kimura. On the iteration of analytic functions. *Funkcial. Ekvac.*, 14:197–238, 1971.

[Lev75] A. H. M. Levelt. Jordan decomposition for a class of singular differential operators. *Ark. Mat.*, 13:1–27, 1975.

[Lod90] M. Loday-Richaud. Introduction à la multisommabilité. *Gazette des Mathématiciens, Soc. Math. France*, 44:41–63, 1990.

[Lod94] M. Loday-Richaud. Stokes phenomenon, multisummability and differential Galois groups. *Ann. Inst. Fourier (Grenoble)*, 44-3:849–906, 1994.

[Lod95] M. Loday-Richaud. Solutions formelles des systèmes différentiels linéaires méromorphes et sommation. *Expo. Math.*, 13:116–162, 1995.

[Lod01] M. Loday-Richaud. Rank reduction, normal forms and Stokes matrices. *Expo. Math.*, 19:229–250, 2001.

[Lod03] M. Loday-Richaud. Stokes cocycles and differential Galois groups. *Sovrem. Mat. Fundam. Napravl.*, 2:103–115, 2003. Translation in J. Math. Sci. (N.Y.), **124-5** (2004), p. 5262-5274.

[LP97] M. Loday-Richaud and G. Pourcin. On index theorems for linear ordinary differential operators. *Ann. Inst. Fourier (Grenoble)*, 47:1379–1424, 1997.

[LR11] M. Loday-Richaud and P. Remy. Resurgence, Stokes phenomenon and alien derivatives for level-one linear differential systems. *J. Differential equations*, 250-3:1591–1630, 2011.

[Mal74] B. Malgrange. Sur les points singuliers des équations différentielles. *L'Enseignement Mathématique*, XX, 1-2:147–176, 1974.

[Mal79] B. Malgrange. Remarques sur les équations différentielles à points singuliers irréguliers. In *Équations différentielles et systèmes de Pfaff*, volume 712 of *Lecture Notes in Math*, pages 77–86, 1979.

[Mal82] B. Malgrange. Travaux d'Écalle et de Martinet-Ramis sur les systèmes dynamiques. In *Bourbaki Seminar, Vol. 1981/1982*, volume 92 of *Astérisque*, pages 59–73. Soc. Math. France, 1982.

[Mal91a] B. Malgrange. *Équations différentielles à coefficients polynomiaux*, volume 96 of *Progress in Mathematics*. Birkhäuser, 1991.

[Mal91b] B. Malgrange. Fourier transform and differential equations. In *Recent developments in quantum mechanics (Poiana Braşov, 1989)*, volume 12 of *Math. Phys. Stud.*, pages 33–48. Kluwer Acad. Publ., 1991.

[Mal95] B. Malgrange. Sommation des séries divergentes. *Expo. Math.*, 13:163–222, 1995.

[MalR92] B. Malgrange and J.-P. Ramis. Fonctions multisommables. *Ann. Inst. Fourier (Grenoble)*, 42:353–368, 1992.

[MarR82] J. Martinet and J.-P. Ramis. Problèmes de modules pour des équations différentielles non linéaires du premier ordre. *Inst. Hautes Études Sci. Publ. Math.*, (55):63–164, 1982.

[MarR89] J. Martinet and J.-P. Ramis. Computer algebra and differential equations. In *Théorie de Galois différentielle et resommation*, pages 117–214. Academic Press, 1989.

[MarR91] J. Martinet and J.-P. Ramis. Elementary acceleration and multisummability. I. *Ann. Inst. H. Poincaré Phys. Théor.*, 54(4):331–401, 1991.

[Miy12] M. Miyake. Newton polygon and gevrey hierarchy in the index formulas for a singular system of ordinary differential equations. *Funkcial. Ekvac.*, 55(2):169–237, 2012.

[MS16] C. Mitschi and D. Sauzin. *Divergent Series, Summability and Resurgence I, Monodromy and Resurgence*, volume 2153 of *Lecture Notes in Mathematics*. Springer, Heidelberg, 2016.

[Nev19] F. Nevanlinna. Zur Theorie der Asymptotischen Potenzreihen. *Ann. Acad. Scient. Fennicæ, Ser. A*, XII:1–81, 1919.

[Ram80] J.-P. Ramis. Les séries k-sommables et leurs applications. In *Complex analysis, microlocal calculus and relativistic quantum theory (Proc. Internat. Colloq., Centre Phys., Les Houches, 1979)*, volume 126 of *Lecture Notes in Phys.*, pages 178–199. Springer, 1980.

[Ram84] J.-P. Ramis. Théorèmes d'indices Gevrey pour les équations différentielles ordinaires. *Mem. Amer. Math. Soc.*, 48(296):viii+95, 1984.

[Ram85] J.-P. Ramis. Phénomène de Stokes et resommation. *C. R. Acad. Sci. Paris Sér. I Math.*, 301(4):99–102, 1985.

[RS89] J.-P. Ramis and Y. Sibuya. Hukuhara's domains and fundamental existence and uniqueness theorems for asymptotic solutions of Gevrey type. *Asymptotic Analysis*, 2:39–94, 1989.

[Rud87] W. Rudin. *Real and complex analysis*. McGraw-Hill Book Co., New York, third edition, 1987.

[Sau05] D. Sauzin. Resurgent functions and splitting problems. *RIMS Kōkyūroku*, 1493:48–117, 2005.

[Sib77] Y. Sibuya. Stokes phenomena. *Bull. Amer. Math. Soc.*, 83-5:1075–1077, 1977.

[Sib90] Y. Sibuya. *Linear differential equations in the complex domain: problems of analytic continuation*, volume 82 of *Translations of Mathematical Monographs*. Amer. Math. Soc., 1990.

[Ten75] B. R. Tennison. *Sheaf Theory*, volume 20 of *London Mathematical Society Lecture Note Series*. Cambridge University Press, 1975.

[Val42] G. Valiron. *Théorie des Fonctions*. Masson et Cie., Paris, 1942.

[vdPS97] M. van der Put and M. Singer. Galois Theory of Difference Equations. In *Lecture Notes in Math.*, volume 1666. Springer, 1997.

[Was76] W. Wasow. *Asymptotic expansions for ordinary differential equations*. R.E. Krieger Publishing Co, 1976. First edited by InterScience, New York, 1965.

Glossary of Notations

$\widetilde{\cdot}$	notations with a tilde over refer to formal objects	
$\widehat{\cdot}$	notations with a hat over refer to Borel transformed objects	
$\Subset$	proper inclusion of sectors	2
$\mathfrak{A}$	set of anti-Stokes directions α of a linear differential equation or system	98
$\mathscr{A}$	sheaf over S^1 of germs of asymptotic functions	42
$\mathscr{A}_s$	subsheaf of $\mathscr{A}$ of s-Gevrey germs	42
$\mathscr{A}^{<0}$	subsheaf of $\mathscr{A}$ of flat germs, i.e., asymptotic to 0	42
$\mathscr{A}^{\leq -k}$	subsheaf of $\mathscr{A}^{<0}$ of germs exponentially flat of order k	42
$\mathscr{B}$	formal Borel transform	145
$\mathscr{B}_\theta$	Borel transform in the θ direction	145
$\mathscr{B}_{k,\theta}$	k-Borel transform in the θ direction	151
$\widetilde{\mathbb{C}}$	Riemann surface of logarithm	1
$\mathbb{C}[[x]]$	differential algebra of formal power series with complex coefficients and derivation d/dx	
$\mathbb{C}\{x\}$	differential subalgebra of $\mathbb{C}[[x]]$ restricted to convergent series	
$\mathbb{C}[[x]]_s$	differential subalgebra of $\mathbb{C}[[x]]$ restricted to Gevrey series of order s, i.e., of level $k = 1/s$	14
$\mathbb{C}\{x\}_{\{k,I\}}$	differential subalgebra of $\mathbb{C}[[x]]$ of k-summable series on I	138
$\mathbb{C}\{x\}_{\{\underline{k},\underline{I}\}}$	differential subalgebra of $\underline{k}$-multisummable series on $\underline{I}$	207
$\widetilde{\mathrm{E}}(x)$	Euler series	4
$\mathrm{E}(x)$	Euler function	4
$\mathscr{E}_0$	homogeneous Euler operator	44
$\mathrm{Ei}(x)$	exponential integral function	6
$\overline{\mathscr{G}}, \mathscr{G}$	presheaf and associated sheaf	38
$\mathscr{F}, \mathscr{F}^k$	sheaf of germs in the infinitesimal neighborhood	118
${}_3F_0$	example of a hypergeometric series	7
$\mathbb{G}$	$\mathbb{G} = \mathrm{GL}(n, \mathbb{C}\{x\}[1/x])$, group of convergent gauge transformations	72
$\widetilde{\mathbb{G}}(B)$	set of formal gauge transformations of $\mathrm{d}Y/\mathrm{d}x = B(x)Y$	72
$\mathbb{G}_0(B_0)$	group of formal isotropies of the normal form $\mathrm{d}Y/\mathrm{d}x = B_0(x)Y$	91
$\widetilde{g}(z)$	$\widetilde{g}(z) = z^{-4}{}_3F_0(\{3,4,5\}\|1/z)$	7

M. Loday-Richaud, *Divergent Series, Summability and Resurgence II*,
Lecture Notes in Mathematics, DOI 10.1007/978-3-319-29075-1

$\widetilde{h}(z)$	example of a solution of a mild difference equation	8
$\widetilde{\ell}(z)$	example of a solution of a wild difference equation	10
$k, k_1, \ldots$	levels (positive numbers)	
$\underline{k}, \underline{I}$	multi-level and multi-arc	205
Λ	sheaf $\Lambda = GL(n, \mathscr{A})$ over S^1 of germs of invertible matrices with entries in the sheaf of asymptotic functions at 0	95
$\Lambda^{<0}$	sub-sheaf of flat (asymptotic to identity) germs of Λ	95
$\Lambda(B_0)$	sub-sheaf of Λ consisting of the germs of isotropy of $\mathrm{d}Y/\mathrm{d}x = B_0(x)Y$	95
$\Lambda^{<0}(B_0)$	sub-sheaf of $\Lambda(B_0)$ of the germs of flat isotropies of $\mathrm{d}Y/\mathrm{d}x = B_0(x)Y$	95
$\mathscr{L}_\theta$	Laplace transform in the θ direction	145
$\mathscr{L}_{k,\theta}$	k-Laplace transform in the θ direction	151
$s, s_1, \ldots$	orders, i.e., inverses of levels $s = 1/k, s_1 = 1/k_1, \ldots$	
$\boldsymbol{\Delta}, \boldsymbol{\Delta}_1, \ldots$	open sectors in $\mathbb{C}^*$ or in its universal covering $\widetilde{\mathbb{C}}$ at 0	
$\overline{\boldsymbol{\Delta}}$	closure of the sector $\boldsymbol{\Delta}$ in $\mathbb{C}^*$ or $\widetilde{\mathbb{C}}$	
$\mathscr{S}$	open sector in $\mathbb{C}^*$ or in its universal covering at infinity	
$\boldsymbol{\Delta}_{\alpha,\beta}(R)$	open sector $\{x\,;\, \alpha < \arg(x) < \beta \text{ and } 0 < \lvert x\rvert < R\}$ in $\mathbb{C}^*$ or $\widetilde{\mathbb{C}}$	1
$\mathrm{Sto}_\alpha(B_0)$	Stokes group of $\mathrm{d}Y/\mathrm{d}x = B_0(x)Y$ in an anti-Stokes direction α	98
$\mathscr{S}_{\{k,I\}}$	k-summation operator on the arc I	137
$\mathscr{S}_{\{k,\theta\}}$	k-summation operator in the θ direction	137
$\mathscr{S}_{\{\underline{k},\underline{I}\}}$	$\underline{k}$-summation operator on the multi-arc $\underline{I}$	207
$T_{\boldsymbol{\Delta}}$	Poincaré asymptotic expansion at 0 on $\boldsymbol{\Delta}$ (usual Taylor map)	3
$T_{s,\boldsymbol{\Delta}}$	s-Gevrey asymptotic expansion at 0 on $\boldsymbol{\Delta}$ (Taylor map)	18
$\mathscr{V}, \mathscr{V}^{<0}$	sheaf of germs of solutions, resp. flat solutions of a linear diid. eq. or system	44
$\mathscr{V}^k$	subsheaf of $\mathscr{V}$ of germs with exponential growth k	
$\mathscr{V}^{\leq k}$	subsheaf of $\mathscr{V}$ of germs with exponential growth at most k	124
$\mathscr{V}^{\leq -k}$	subsheaf of $\mathscr{V}^{<0}$ of germs with exponential decay at least k	124
X	infinitesimal neighborhhood (exponential order)	114
X^k	infinitesimal neighborhhood (exponential order and type)	116
$x, z = 1/x$	coordinates on the Riemann sphere $x \approx 0$ and $z \approx \infty$ (Laplace planes)	145
ξ, ζ	corresponding coordinates after Borel transform (Borel planes)	145

Index

M. Loday-Richaud, *Divergent Series, Summability and Resurgence II*,
Lecture Notes in Mathematics, DOI 10.1007/978-3-319-29075-1

Printed in the United States
By Bookmasters